Nuclear Science and Technology

Volume I

Nuclear Science and Technology
Volume I

Edited by **Paul Patterson**

CLANRYE
INTERNATIONAL

New Jersey

Published by Clanrye International,
55 Van Reypen Street,
Jersey City, NJ 07306, USA
www.clanryeinternational.com

Nuclear Science and Technology: Volume I
Edited by Paul Patterson

International Standard Book Number: 978-1-63240-392-6 (Hardback)

Printed in the United States of America.

Contents

Preface

The world today is rapidly evolving and changing. This pace of evolution and development has pushed the boundaries of technology to a new level. Nuclear science is one such arena which has ushered in this era of development and change. What one often tends to sideline is the fact that nuclear energy or science requires specialized installations to be operated or researched upon. Thus there are specialized installations for the use of nuclear science that are called nuclear installations. To be more specific, nuclear installations refer to any nuclear fabrication plant, power reactor, waste disposal facility for radioactive waste, nuclear fuel reprocessing plant and any other facility handling the quantity of fissionable materials sufficient to form critical mass. They are quite distinct from thermal power plants and installations. This is a highly specialized industry and field of research that has quite a lot of ramifications in today's world. Their potential for being beneficial to us is quite high but there is also quite a potential for such installations being dangerous as well. Thus there are also various Nuclear Installation Acts and committees that regulate such installations. This also points to the demand for skilled workers and researchers.

This book attempts to compile and collate the research and data available in the field of nuclear installations. I am grateful to those who put in effort and hard work in this field. I am also thankful to those who supported me in this endeavour.

Editor

Computational Fluid Dynamics Modeling of Steam Condensation on Nuclear Containment Wall Surfaces Based on Semiempirical Generalized Correlations

Pavan K. Sharma, B. Gera, R. K. Singh, and K. K. Vaze

Reactor Safety Division, Bhabha Atomic Research Centre, Engg. Hall-7, Trombay, Mumbai 400085, India

Correspondence should be addressed to Pavan K. Sharma, pa1.sharma@gmail.com

Academic Editor: Leon Cizelj

In water-cooled nuclear power reactors, significant quantities of steam and hydrogen could be produced within the primary containment following the postulated design basis accidents (DBA) or beyond design basis accidents (BDBA). For accurate calculation of the temperature/pressure rise and hydrogen transport calculation in nuclear reactor containment due to such scenarios, wall condensation heat transfer coefficient (HTC) is used. In the present work, the adaptation of a commercial CFD code with the implementation of models for steam condensation on wall surfaces in presence of noncondensable gases is explained. Steam condensation has been modeled using the empirical average HTC, which was originally developed to be used for "lumped-parameter" (volume-averaged) modeling of steam condensation in the presence of noncondensable gases. The present paper suggests a generalized HTC based on curve fitting of most of the reported semiempirical condensation models, which are valid for specific wall conditions. The present methodology has been validated against limited reported experimental data from the COPAIN experimental facility. This is the first step towards the CFD-based generalized analysis procedure for condensation modeling applicable for containment wall surfaces that is being evolved further for specific wall surfaces within the multicompartment containment atmosphere.

1. Introduction

Steam condensation in the presence of noncondensable gases is a relevant phenomenon in many industrial applications, including nuclear reactors. Condensation on the containment structures during an accident and associated computations are important for the containment design of all the existing reactors for LOCA DBA, DBA, and BDBA hydrogen distribution and recombination and passive emergency systems in the nuclear reactors of new generation. Rate of steam condensation at containment walls affects the transient pressure in the containment after loss of coolant accident. Apart from this during a severe accident in a water-cooled power reactor nuclear power plant (NPP), large amounts of hydrogen would presumably be generated due to core degradation and released into the containment. The integrity of the containment could be threatened due to hydrogen combustion.

If composition of the hydrogen-steam-air mixture lies within a certain limits, the combustion will occur. The steam condensation phenomenon is important from hydrogen distribution point of view to locate the flammable region in the containment for adequate accident management procedures (Royl et al., 2000 [1]).

Currently use of CFD techniques to model such scenario is popular, since CFD codes provide more detailed information in such scenario. These commercially available CFD codes generally do not have built-in steam condensations models. Consequently, it is necessary to implement steam condensation via user-defined subroutines. Two main approaches have been proposed by various authors to model wall condensation in CFD codes. In the first approach (two-phase flow approach) separate momentum equations are solved for vapour and liquid phases. A fine mesh is required, and liquid film and the diffusion of steam towards wall through boundary layer formed by noncondensable gases are modeled. This approach is quite close to first-principle condensation modeling but requires very large computational time and will probably take some time to be used for any practical applications (full containment modeling) in future. In the second approach a single-fluid model is used where steam is modeled as a separate species via species conservation equation. For modeling steam condensation, mass sink and corresponding energy sink are modeled in the very first cell near to condensing wall. In this second approach there are two ways to calculate these sink terms. The first one is based on diffusion theory which requires a very fine mesh near the wall and computes steam condensation rate that diffuses towards wall through species boundary layer. Houkema et al., in 2008, [2] have used this approach without the use of engineering correlations for steam condensation on walls. Babić et al., 2008 [3], Kljenak et al., 2006 [4], and Kljenak et al., 2004 [5], have used individual HTC-based approach with CFD code CFX to simulate experiments on containment atmosphere behavior at accident conditions, which were performed in the TOSQAN and ThAI experimental facilities. Currently the most popular way is to include heat or mass transfer correlations that were originally developed for "lumped-parameter" (volume-averaged) calculations and apply them in the layer of cells contiguous to the condensation surface. These correlations use the temperature and steam mass fraction value from bulk flow and calculate the condensing heat transfer coefficient, since the corresponding sink term is applied for the very first fluid cell near the condensing wall, and the bulk flow parameters and thermophysical properties are evaluated at these cell centres. Thus in this approach rate of heat and mass transfer depends on the cell width near the condensing wall. CFD codes were developed to solve equations that are derived from the first principles, using local instantaneous description, so that the inclusions of correlations, which are based on averaged physical quantities and provide average condensation rates, are somehow contrary to the basic "philosophy" of CFD (Kljenak et al., 2006 [4]).

Wide ranges of HTCs have been reported which have come from condensation experimental setup largely different from actual containment accident situation during accident. Uses of the so-called conservative (safe values in context of peak pressure, temperature, or hydrogen concentration, etc.) individual HTC also have conflicting position in respect of containment pressure transient calculation and hydrogen distribution/management calculation. An HTC value which results in lower condensation of steam gives conservative results for containment pressure calculation, but gives nonconservative results for hydrogen management calculation due to artificial inerting atmosphere because of less condensation of steam. Authors try to overcome this weakness by trying to use many empirical correlations and theoretical calculations by making a generalized HTC from most of the generally reported approaches. The present combined approach allows relatively fast calculations and should be adequate for large industrial applications.

Steam condensation was modeled as a sink of mass and enthalpy by applying the correlation by Uchida et al., 1965 [6]. Valdepenas et al., 2007 [7], have also used the second approach for modeling steam condensation on walls where condensation rate was based on correlation developed by Terasaka and Makita, 1997 [8]. Mimouni et al., 2010 [9], have modeled steam condensation by a two-phase flow approach that takes into account the physical phenomena of importance like effect of liquid film thickness. In the CFD code TONUS developed for hydrogen risk analysis (Kudriakov et al., 2008 [10]), steam condensation was implemented as steam diffusion through a mass boundary layer based on heat and mass transfer analogy (Chilton-Colburn type).

The present paper is about development and comparison of the wall condensation model based on semiempirical correlation (existing and proposed based on curve fit of various existing correlations) in CFD-ACE+ [11]. The implemented condensation model will be described, and results of validation calculations are compared with experimental data. A brief overview will be given on the comparison of results obtained with CFD-ACE+ versus experimental results from COPAIN test facility. The simulation was performed with the knowledge of experimental results.

2. Condensation for Containment Thermal Hydraulics for Practical Applications

Review papers on condensation on the containment structures (Green and Almenas, 1996 [12], Rosa et al., 2009 [13]) enlist the various parameter of interest, that is, subcooling, interfacial shear, superheating, difference of temperatures, noncondensables gases, lighter gases, effect of pressure, effect of velocity, density difference between bulk and interface, leaning of the surface, length of the surface, and so forth, mostly for average heat transfer coefficient. Some other parameters which affect the containment condensation are the material of wall, structure porosity, wettability, use of epoxy paints, effect of aerosol, and so forth. In the context of average heat transfer coefficient, the dimension of the test setup at which the coefficient is experimentally derived is also an important factor which can be in the form of volume scaling in context of energy addition from hot steam and height scaling in context of buoyancy (both due to temperature and concentration density gradient). The authors have plotted most of the reported empirical correlations from the literature. Several of such HTCs have been used in lumped-parameter-and CFD-based analysis. Table 1 compiles few other reported empirical HTC.

The major limitations for individual HTC use in CFD codes which have been mentioned earlier come from the fact that, the use of global correlation as the local one, different length and energy addition/removal time scales, modeling of surface condensation phenomenon as volume phenomena, different reported range of experimental parameters like size, flow regimes of convection, type and quantity of combustible, type of walls, configuration of wall (mostly pipes), and so forth. Some of the HTCs are plotted in Figure 1 by taking parameters from the Copain experiment (described later in the paper) data especially the temperature difference between wall and the bulk mixture along with pressure values.

Computational Fluid Dynamics Modeling of Steam Condensation on Nuclear Containment Wall Surfaces Based on
Semiempirical Generalized Correlations

3

TABLE 1: Review of containment-specific steam condensation HTC.

S. no.	Name	Equation
1	Uchida (Rosa et al., 2009 [13])	$h = 380\eta^{-0.7}$, $h\,(\text{W/m}^2)$
2	Tagami steady state (Rosa et al., 2009 [13])	$h = 11.4 + 284(\eta/(1-\eta))$, $h\,(\text{W/m}^2)$
3	Debhi (Rosa et al., 2009 [13])	$h = L^{0.05}[(3.7 + 28.0P) - (243.8 + 458.3P)\log\eta]/((T_b - T_w)^{0.25})$, $h\,(\text{W/m}^2)$, $P\,(\text{Atm})$, $L\,(\text{m})$, $T\,(\text{K})$
4	Kataoka (Rosa et al., 2009 [13])	$h = 430\eta^{-0.8}$, $h\,(\text{W/m}^2)$
5	Murase (Rosa et al., 2009 [13])	$h = 0.47\eta^{-1.0}$, $h\,(\text{W/m}^2)$
6	Liu (Lee and Kim, 2008 [14])	$h = 55.635(1-\eta)^{2.344}P^{0.252}((T_b - T_w)^{0.307})$, $h\,(\text{W/m}^2)$, $P\,(\text{Pa})$, $T\,(\text{K})$
7	Green (Green and Almenas, 1996 [12])	$h = 316\eta^{-0.86}((T_b - T_w)^{-0.15})$, $h\,(\text{W/m}^2)$, $T\,(\text{K})$
8	Kawakubo (Kawakubo et al., 2009 [15])	$h = \min(0.33\eta^{-0.8}\times(T_b - T_w)^{0.25}, \eta^{-1.0}\times(T_b - T_w)^{-0.22\eta^{-0.25}})\times(P + 0.5)$, $h\,(\text{Kw/m}^2)$, $P\,(\text{MPa})$
9	Nusselt UCB Multiplier (Park et al., 1999 [16])	$h = f_1 \times f_2 \times h_{\text{nusselt}}$ where $f_2 = (1 + 2.88E^{-05}\text{Re}_{\text{mix}}^{1.18})$ and a maximum fixed value of 2.0 is used in present paper. $h\,(\text{W/m}^2)$ $h_{\text{nusselt}} = k_f/\delta$ and $f_2 = (1 - 10.0\eta)$ for $\eta < 0.063$ $f_2 = (1 - 0.938\eta^{0.13})$ for $0.063 < \eta < 0.6$ $f_2 = (1 - \eta^{0.22})$ for $\eta > 0.6$
10	Nusselt LEE Multiplier (Lee and Kim, 2008 [14])	$h = f_{\text{lee}} \times h_{\text{nusselt}}$ where $h_{\text{nusselt}} = k_f/\delta$ and $f_{\text{lee}} = \tau_{\text{mix}}^{0.3124}(1 - 0.964\eta^{0.402})$, $h\,(\text{W/m}^2)$
11	Proposed fit from all the curves	$h = 3522.70 - 14324.78\eta + 22090.84\eta^2 + 11393.13\eta^3$, $h\,(\text{W/m}^2)$

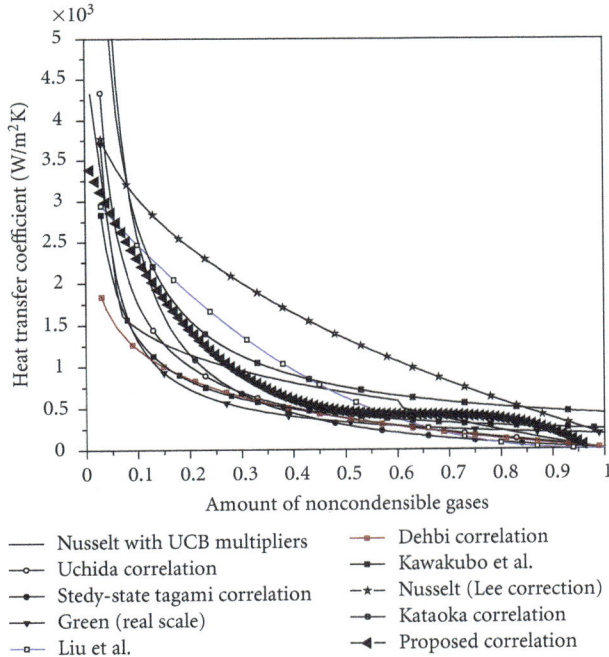

FIGURE 1: Various HTCs from the literature and proposed HTC based on curve fitting.

The classical Nusselt theory gives the HTC as the conductivity of the liquid film divided by the film thickness, so the classical HTC will be a fixed value. Subsequently Nusselt theory gets modified with the use of two multipliers (UCB multipliers): first one takes care of enhancement of the HTC due to liquid film shearing and the associated effects, and the second multiplier takes care of the degradation of the HTC due to presence of non condensable gases. Authors have also suggested and used the multiplier developed by Lee and Kim, 2008 [14], to modify the Nusselt HTC. The compilation of all such correlations, from various experimental correlations along with the classical Nusselt theory, Nusselt USB correlation and the almost real scale HDR test setup based on Green's HTC is depicted in Figure 1. In real situations for HDR, HTC is normally less as compared to the experimental scale and conditions.

Another issue is about the conservatism of the HTC in nuclear regulatory organization suggesting HTC and practical HTC due to large variations in reported values. We use a correlation with small HTC value (being conservative) for containment pressure and temperature calculation which will result in higher containment peak pressure values. But the same conservatism aspect may not be valid for hydrogen distribution in the reactor containment atmosphere as higher pressure differences in multiple-compartment configurations will result in increased intercompartmental mixing and reduction in hydrogen concentration. The lower HTC may also result in relatively uniform temperature field which will decrease the amount of hydrogen stratification due to buoyancy and also results in lower hydrogen concentration values. A lower HTC gives non-conservative results for hydrogen management calculation due to artificial inerting atmosphere due to low condensation of steam. Another issue of bias for use in the lower HTC values was with regard to the earlier correlations developed for low velocity of steam and calm and undisturbed atmosphere. The actual accident scenario may have more turbulent flow dynamics. In real accident situation the accident conditions will be very complex and uncertain. The most logical approach could be to fit a curve from all the possible HTC correlations for the most dominating parameter of interest, that is, amount of non condensable gases. A finite film of preidentified thickness is always assumed to be sticking to the wall.

3. Mathematical Model

The general-purpose CFD code CFD-ACE+ solves the local instantaneous transport equations for mass, momentum, energy, species, and turbulence parameters. The general transport equation for variable ϕ is

$$\frac{\partial(\rho\phi)}{\partial t} + \frac{\partial(\rho u_i \phi)}{\partial x_i} = \frac{\partial}{\partial x_i}\left(\Gamma_\phi \frac{\partial \phi}{\partial x_i}\right) + S_\phi, \qquad (1)$$

where ϕ is appropriate variable (mass, momentum, energy, species, and turbulence parameters), Γ_ϕ is diffusion coefficient and S_ϕ is the sink or source term. To model the mixture behaviour of air and steam, an approach where only one velocity field was defined using the average density of gas mixture was used. The independent behaviour of steam was considered using species transport equation. The discretisation of the equations in the CFD-ACE+ code is based on a conservative finite-volume method. A nonstaggered grid arrangement is employed, where all the variables (velocity components and scalars) are stored in the geometrical centers of control volumes (cells) that fill up the flow domain.

In the present work steam condensation was modeled as a sink for mass and enthalpy by applying the empirical correlation based on experiments on forced convection and implemented in a subroutine. The condensate film on structures was considered to have a fixed thickness. Basically, the steam condensation rate was obtained from the expression

$$m^0 = \text{HTC} \frac{A(T - T_{\text{wall}})}{h_{fg}}. \qquad (2)$$

In this correlation, values of all the physical variables were required from bulk flow except T_{wall} which was the condensing wall temperature. This correlation will always predict the condensation as long as bulk temperature is more than wall temperatures even if the steam partial pressure is lower than the saturation pressure at the condensing wall temperature. The steam partial pressure must be higher than the saturation pressure at the temperature of the wall for condensation to occur; this was ensured by applying the appropriate condition in user-defined subroutine. Thus, during the simulation, the amount of steam condensation was calculated only if the steam partial pressure was above the saturation pressure. The corresponding enthalpy sink H^0 (sink term in energy flow equation) due to condensation was calculated as follows:

$$H^0 = m^0 \left(C_{p,\text{steam}} T_{\text{cell}} - C_{p,\text{air}} T_{\text{ref}}\right). \qquad (3)$$

The specific heat of the air and reference temperature T_{ref} was used for calculation of reference enthalpy in the CFD-ACE+ code. A user-defined subroutine was added for modeling

steam condensation as sinks of mass and enthalpy were identified in cells contiguous to the condensing wall in the CFD-ACE+ computational tool. The mass sink was calculated from (2) for each cell where the "bulk flow" physical quantities (temperature, steam density, and noncondensible gas density) were evaluated at the cell centre. As the temperature of the gaseous mixture corresponding to the cell centre appears in (2) and (3), the calculated condensation rate and enthalpy sink necessarily depend on the width of cells contiguous to the condensation surface (Kljenak et al., 2006 [4]). The saturation pressure at condensing wall temperature was calculated, and it was ensured that steam partial pressure must be higher than the saturation pressure at the temperature of the wall for condensation to occur. The Antoine equation is used to describe the vapour pressure as a function of the surface temperature:

$$\ln\left(\frac{P}{1\,\text{Pa}}\right) = B + \frac{C}{T + D}. \qquad (4)$$

The coefficients B, C, and D were fitted based on data from steam tables, and the values used were $B = +23.1512$, $C = -3788.02\,\text{K}$, and $D = -47.3018\,\text{K}$. It was assumed that the condensing wall immediately removes the condensation heat. The air steam mixture was considered compressible, and density was modeled by ideal gas law. Standard k-ε model with standard wall function was used for modeling turbulence in all the simulation due to its wide use and acceptance in industrial applications. The use of standard wall functions is justified as the present condensation modeling is not from the first principle and not sensitive to the wall function.

4. COPAIN Experimental Facility

The COPAIN experimental facility (CEA, Grenoble) Cheng et al., 2001 [17] was a simple facility designed to study the phenomenon of wall condensation in the presence of non-condensible gases. The facility consists of a vertically placed rectangular channel of cross-section 0.6 m × 0.5 m. The vertical height is around 2.5 m where one vertical side served as a condensation plate up to 2 m height. The experiments were reported to be performed with pressure between 1 and 7 bar (100–700 kPa), temperature under 165°C, heat flux exchange with the plate between 1 and 30 kW/m², and maximum inwards velocity of 3 m/s, with different fraction of air, steam, and, helium. Inside the channel forced or natural convection was observed depending on prevailing conditions. The database selected to validate the wall condensation model is shown in Table 2 for various experiments. The full computational domain was simulated. A fine mesh was made with 93750 cells so that average cell size near the wall is 0.02 m as suggested by different authors (Babić et al., 2008 [3], Kljenak et al., 2006 [4]). The grids were uniform, including the nearest cells to the wall. In this simulation, the condensation acts as a sink of mass and energy. Thus, the liquid film and the influence of the non-condensible gas layer were reduced to a simple sink term.

Computational Fluid Dynamics Modeling of Steam Condensation on Nuclear Containment Wall Surfaces Based on
Semiempirical Generalized Correlations

5

TABLE 2: Parameters of the COPAIN tests.

Test no.	Convective heat transfer	Air velocity at inlet (m/s)	Pressure (bar)	Air temperature at inlet (K)	Wall temperature (K)	Mass fraction of noncondensible gases
P0441	Forced	3	1.02	353.23	307.4	0.767
P0443	Free	1	1.02	352.33	300.06	0.772
P0444	Natural	0.5	1.02	351.53	299.7	0.773
P0344	Natural	0.3	1.21	344.03	322	0.864

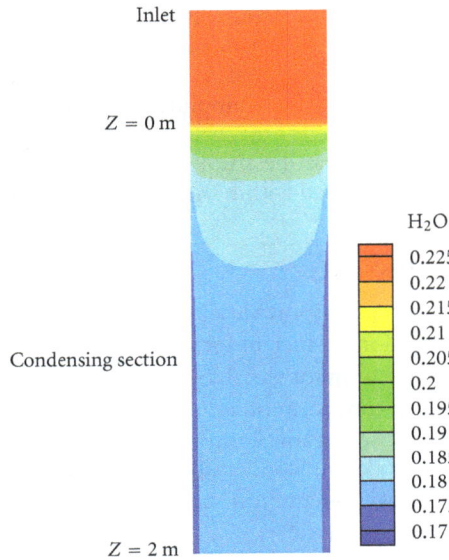

FIGURE 2: Steam mass fraction on the condensing wall for test P0444.

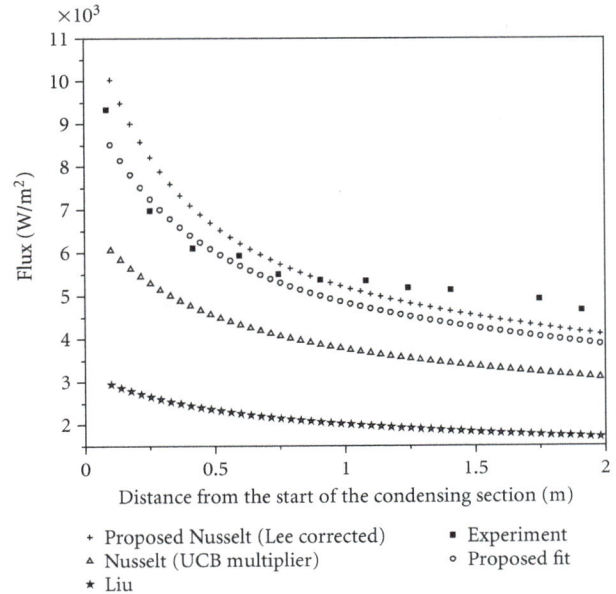

FIGURE 3: Variation of condensation flux with height, test P0441.

5. Results and Discussion

Figure 2 depicts the steam mass fraction on the layer of cell adjacent to the condensation wall for the test P0441. Figures 3, 4, 5, and 6 show the condensation flux at wall along the height in the condensing section for all the steady states. The amount of steam condensed (numerically estimated) is slightly overestimated in the case of free convective heat transfer (P0443) experimental data by the proposed fit and proposed Nusselt (Lee corrected); however, it was in good agreement in case of natural convection with lower velocity (P0444, P0344) and with forced convection (P0441) against experimental data. Use of individual empirical correlation predicted numerical results are expectedly different as compared to the experimental data. The proposed fit is derived from various experimental conditions and the parameters which reduce the uncertainty and ensure the consistent applicability of the correlation irrespective of the parameter ranges. The present fit has been derived taking a fixed value of liquid film thickness and temperature gradient. However this can be improved based on specific requirement, and inputs can be generated for complex intercompartmental configurations within CFD framework. This result also advocates the replacement of simplified condensation model (lumped

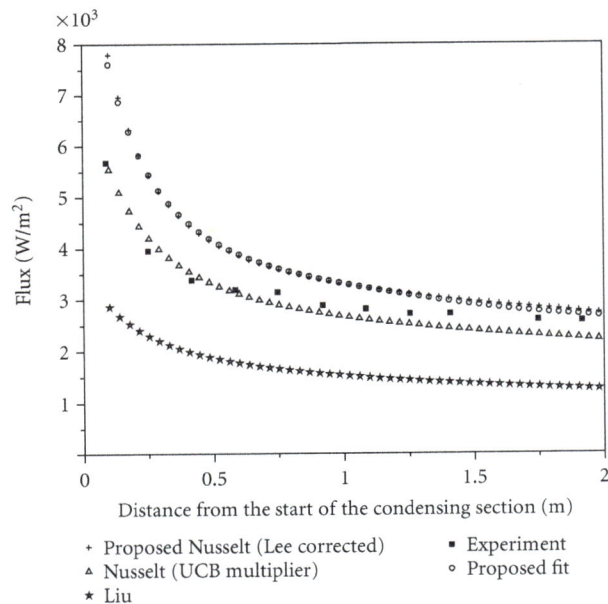

FIGURE 4: Variation of condensation flux with height, test P0443.

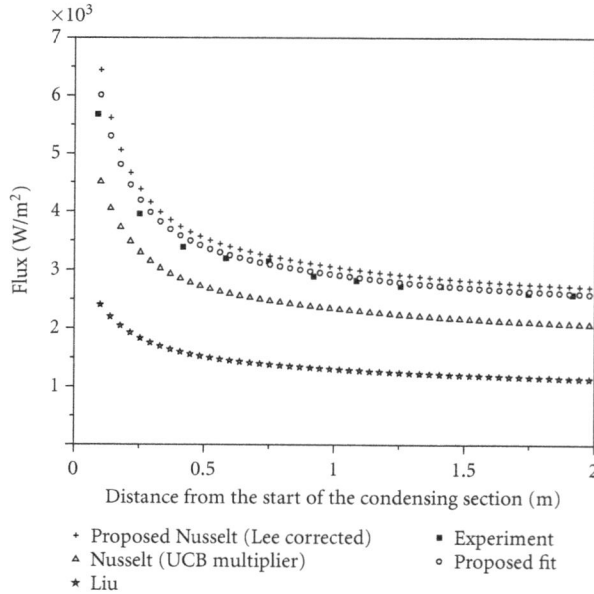

FIGURE 5: Variation of condensation flux with height, test P0444.

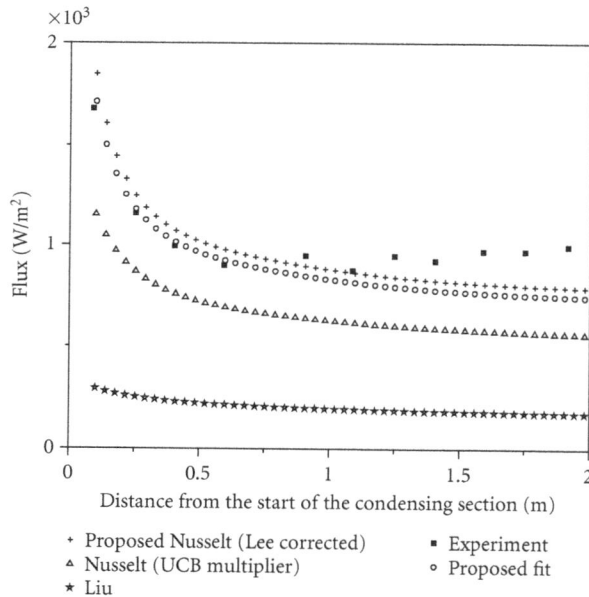

FIGURE 6: Variation of condensation flux with height, test P0344.

parameter code-like approach) with a new improved simple HTC in CFD-based approach which can be used to model real large containment geometry for safety evaluation in a more consistent manner.

6. Conclusions

A model for steam condensation was implemented and tested based on various correlations developed for volume-averaged approach in the computational fluid dynamics code CFD-ACE+. The CFD code CFD-ACE+ was used to simulate condensation experiments that were performed in the COPAIN facility. There was a reasonable agreement between simulated and experimental results for proposed approach derived from fitting of various experimental data. However further validations are needed against other experimental data. This approach does not solve the phenomenon from the first principle, but this approach can be considered as effective for industrial problem where solution for large computational domain is required. The present work suggests that the approach could be used for integrated hydrogen distribution/management calculation with steam condensation in nuclear reactor containment required for safety analysis. Another useful application of the present approach could be to obtain a first-order simplified generalized analysis result for a containment wall surface, and subsequently different wall surfaces can be studied in detail where specific correlations can be used within the CFD framework.

Nomenclature

ϕ: General transport variable
Γ_ϕ: Diffusion coefficient for variable ϕ
S_ϕ: Source term for variable ϕ
m^0:   Steam condensation rate (kg/s)
ρ_{steam}:   Density of steam (kg/m^3)
ρ_{nc}:   Density of noncondensible gases (kg/m^3)
A: Area of condensation surface (m^2)
T: Temperature (K)
T_{wall}:   Average temperature of condensation surface (K)
h_{fg}:   Latent heat of steam (J/kg)
H^0:   Enthalpy sink due to condensation (W)
$C_{p,\text{steam}}$: Specific heat of steam at constant pressure (J/kgK)
$C_{p,\text{air}}$:   Specific heat of air at constant pressure (J/kgK)
T_{ref}:   Reference temperature (K)
P: Pressure (Pa)
B, C, D: Constants.

References

[1] P. Royl, H. Rochholz, W. Breitung, J. R. Travis, and G. Necker, "Analysis of steam and hydrogen distributions with PAR mitigation in NPP containments," *Nuclear Engineering and Design*, vol. 202, no. 2-3, pp. 231–248, 2000.

[2] M. Houkema, N. B. Siccama, J. A. L. Nijeholt, and E. M. J. Komen, "Validation of the CFX4 CFD code for containment thermal-hydraulics," *Nuclear Engineering and Design*, vol. 238, no. 3, pp. 590–599, 2008.

[3] M. Babić, I. Kljenak, and B. Mavko, "Prediction of light gas distribution in experimental containment facilities using the CFX4 code," *Nuclear Engineering and Design*, vol. 238, no. 3, pp. 538–550, 2008.

[4] I. Kljenak, M. Babić, B. Mavko, and I. Bajsić, "Modeling of containment atmosphere mixing and stratification experiment using a CFD approach," *Nuclear Engineering and Design*, vol. 236, no. 14–16, pp. 1682–1692, 2006.

[5] I. Kljenak, I. Bajsić, and M. Babić, "Modelling of steam condensation on the walls of a large enclosure using a

Computational Fluid Dynamics Modeling of Steam Condensation on Nuclear Containment Wall Surfaces Based on Semiempirical Generalized Correlations

7

Computational Fluid Dynamics code," in *Proceedings of the ASME-ZSIS International Thermal Science Seminar II*, Slovenia, June 2004.

[6] H. Uchida, A. Oyama, and Y. Togo, "Evaluation of post-incident cooling systems of LWRs," in *Proceedings of the 13th International Conference on Peaceful Uses of Atomic Energy*, pp. 93–102, International Atomic Energy Agency, Vienna, Austria, 1965.

[7] J. M. M. Valdepenas, M. A. Jimenez, F. M. Fuertes, and J. A. Fernandez, "Improvements in a CFD code for analysis of hydrogen behaviour within containments," *Nuclear Engineering and Design*, vol. 237, pp. 627–647, 2007.

[8] H. Terasaka and A. Makita, "Numerical analysis of the PHEBUS containment thermal hydraulics," *Journal of Nuclear Science and Technology*, vol. 34, no. 7, pp. 666–678, 1997.

[9] S. Mimouni, A. Foissac, and J. Lavieville, "CFD modelling of wall steam condensation by a two-phase flow approach," *Nuclear Engineering and Design*, vol. 241, no. 11, pp. 4445–4455, 2011.

[10] S. Kudriakov, F. Dabbene, E. Studer et al., "The TONUS CFD code for hydrogen risk analysis: physical models, numerical schemes and validation matrix," *Nuclear Engineering and Design*, vol. 238, no. 3, pp. 551–565, 2008.

[11] *CFD-ACE+ V2009.2, User Manual*, ESI CFD Inc., Huntsville, Ala, USA, 2009.

[12] J. Green and K. Almenas, "An overview of the primary parameters and methods for determining condensation heat transfer to containment structures," *Nuclear Safety*, vol. 37, no. 1, pp. 26–48, 1996.

[13] J. C. de la Rosa, A. Escriva, L. E. Herranz, T. Cicero, and J. L. Muñoz-Cobo, "Review on condensation on the containment structures," *Progress in Nuclear Energy*, vol. 51, no. 1, pp. 32–66, 2009.

[14] K. Y. Lee and M. H. Kim, "Experimental and empirical study of steam condensation heat transfer with a noncondensable gas in a small-diameter vertical tube," *Nuclear Engineering and Design*, vol. 238, no. 1, pp. 207–216, 2008.

[15] M. Kawakubo, M. Aritomi, H. Kikura, and T. Komeno, "An Experimental Study on the Cooling Characteristics of passive containment cooling systems," *Journal of Nuclear Science and Technology*, vol. 46, no. 4, pp. 339–345, 2009.

[16] H. S. Park, H. C. No, and Y. S. Bang, "Assessment of two wall film condensation models of Relap/Mod3.2 in the presence of noncondensable gas in a vertical tube," *Journal of the Korean Nuclear Society*, vol. 31, pp. 465–475, 1999.

[17] X. Cheng, P. Bazin, P. Cornet et al., "Experimental data base for containment thermalhydraulic analysis," *Nuclear Engineering and Design*, vol. 204, no. 1–3, pp. 267–284, 2001.

2

Computational Method for Global Sensitivity Analysis of Reactor Neutronic Parameters

Bolade A. Adetula and Pavel M. Bokov

Research and Development Division, The South African Nuclear Energy Corporation (Necsa), Building 1900, P.O. Box 582, Pretoria 0001, South Africa

Correspondence should be addressed to Bolade A. Adetula, adetulap@yahoo.com

Academic Editor: Kostadin Ivanov

The variance-based global sensitivity analysis technique is robust, has a wide range of applicability, and provides accurate sensitivity information for most models. However, it requires input variables to be statistically independent. A modification to this technique that allows one to deal with input variables that are blockwise correlated and normally distributed is presented. The focus of this study is the application of the modified global sensitivity analysis technique to calculations of reactor parameters that are dependent on groupwise neutron cross-sections. The main effort in this work is in establishing a method for a practical numerical calculation of the global sensitivity indices. The implementation of the method involves the calculation of multidimensional integrals, which can be prohibitively expensive to compute. Numerical techniques specifically suited to the evaluation of multidimensional integrals, namely, Monte Carlo and sparse grids methods, are used, and their efficiency is compared. The method is illustrated and tested on a two-group cross-section dependent problem. In all the cases considered, the results obtained with sparse grids achieved much better accuracy while using a significantly smaller number of samples. This aspect is addressed in a ministudy, and a preliminary explanation of the results obtained is given.

1. Introduction

The apportioning of uncertainty in the output of a model (numerical or otherwise) to different sources of uncertainty in the model input is known as *sensitivity analysis* [1], and the associated quantitative values are known as *sensitivity indices*. The sensitivity indices can be used to rank the input variables of the model, based on the influence they have on the output. It thus becomes possible to recognize the probabilistically insignificant/unessential input variables that exert little influence on the output. This allows for the reduction of the dimensionality of the problem by fixing the unessential input variables, whilst more experiments, computations, research, and so forth can be done to determine the essential input variables with a higher degree of accuracy.

The focus of this study will be on *global sensitivity analysis* (GSA), which explores the full phase space of input parameters, as opposed to *local sensitivity analysis* (LSA) methods that are usually based on derivatives and analyse the behaviour of the model output around a chosen point. The implementation of GSA can be achieved by using either variance-[1–3] or entropy-[4, 5] based methods. In our study, we will use the Sobol's variance-based method [3]. This method is referred to as "variance-based" because within the framework of this approach, the uncertainty of the output is characterized by its (output) variance. The Sobol's method is robust, has a wide range of applicability, and, as stated in [6, 7], provides accurate sensitivity information for most models. However, the Sobol's method is defined for mutually independent input variables that are uniformly distributed. A modification of the method which allows one to deal with input variables that are blockwise correlated and normally distributed is presented in this work.

The modified method can then be applied to nuclear reactor calculations. Many reactor parameters of interest (such as the neutron multiplication factor, decay heat, reaction rates, etc.) are dependent on neutron cross-sections. These cross-sections are often described by only their first two statistical moments and are assumed to be normally distributed [8]. The uncertainties associated with the cross-sections are propagated to the final result of the calculated reactor parameters, and the uncertainty in a calculated reactor parameter can be apportioned to the different sources of uncertainty in the neutron cross-sections.

In this paper, we will present the method of global sensitivity analysis that will address the previous limitations and take into account the previously mentioned assumptions with an emphasis on the numerical/calculational aspects in implementing the method. The rest of the paper is organised in the following way. Section 2 contains two major parts: in the beginning, we give theoretical background and some mathematical derivations for the method we present, and in the second part of the section, we discuss its practical numerical implementation. The theory description is supported by two appendices: Appendix A is used to summarize the definitions and properties of the functional ANOVA decomposition, and Appendix B provides explanations concerning the sparse grid integration method. In Section 3, we describe the particularities of our implementation of the proposed method and the problem we use to test and characterise the method, as well as the results obtained. Finally, Section 4 is used to present our conclusions.

2. Method

2.1. Definitions and Assumptions.

Consider a problem in which some important reactor parameters, such as the neutron multiplication factor and the decay heat, depend on multigroup or few-group neutron cross-sections. We will use Y to denote the reactor parameter of interest and X_i ($i = 1, 2, \ldots, d$) to denote the cross-sections. The dependence of the parameter of interest on cross-sections can be written as a model

$$Y = f(X_1, X_2, \ldots, X_d), \tag{1}$$

where X_i are called *inputs* and Y is called the *output* or *response*. Model (1) is generally nonlinear and often calculated numerically in practice.

The cross-sections can be gathered in a column vector $\mathbf{X} = (X_1, X_2, \ldots, X_d)^T$, where the symbol "$T$" denotes the operation of transposing a row to a column. If input $\mathbf{X}$ is a random vector with a joint probability density function $p(\mathbf{x}) = p(x_1, x_2 \ldots, x_d)$, then the response Y is a random variable with the expected value $\mathrm{E}[Y]$ and the variance $\mathrm{Var}[Y]$ defined as

$$\mathrm{E}[f(\mathbf{x})] = \int_{\mathbb{R}^d} f(\mathbf{x}) p(\mathbf{x}) \mathrm{d}\mathbf{x},$$
$$\mathrm{Var}[f(\mathbf{x})] = \int_{\mathbb{R}^d} (f(\mathbf{x}) - \mathrm{E}[f(\mathbf{x})])^2 p(\mathbf{x}) \mathrm{d}\mathbf{x}, \tag{2}$$

correspondingly. Note that we will use, as it is the rule in statistics, a capital letter to denote a random variable and a lowercase letter to denote its value (realizations).

In this work, we will assume that the cross-sections are random variables distributed according to the normal law

with known means and covariances. The multivariate normal distribution for the probability $\mathrm{Pr}[X_i < x_i : i = 1, \ldots, d]$ is characterized by the probability density function [9]

$$p(\mathbf{x}) = \frac{1}{(2\pi)^{d/2} \det(\boldsymbol{\Sigma})^{1/2}} \exp\left[-\frac{1}{2}(\mathbf{x} - \boldsymbol{\mu})^T \boldsymbol{\Sigma}^{-1}(\mathbf{x} - \boldsymbol{\mu})\right], \tag{3}$$

where $\mathbf{X}$ is the column vector of random variables, $\boldsymbol{\mu} = \mathrm{E}[\mathbf{X}]$ is the column vector of their expected values, and $\boldsymbol{\Sigma} = \mathrm{E}[(\mathbf{x} - \boldsymbol{\mu})(\mathbf{x} - \boldsymbol{\mu})^T]$ is the covariance matrix.

2.1.1. Block-Correlated Random Variables. Let us assume that the input vector $\mathbf{X}$ can be partitioned into Γ subsets of variables, that is, $\mathbf{X} = (\mathbf{X}_1, \mathbf{X}_2, \ldots, \mathbf{X}_\Gamma)$, and that random vectors $\mathbf{X}_\alpha$ and $\mathbf{X}_\beta$ from this partitioning are mutually independent for $\alpha, \beta = 1, 2, \ldots, \Gamma$.

Using the definition of a covariance matrix, one can show [9] that in this case $\boldsymbol{\Sigma}_{\alpha\beta} = \boldsymbol{\Sigma}_{\beta\alpha} = 0$ for $\alpha \neq \beta$. Hence, the covariance matrix becomes block diagonal, that is, $\boldsymbol{\Sigma} = \mathrm{diag}(\boldsymbol{\Sigma}_{11}, \boldsymbol{\Sigma}_{22}, \ldots, \boldsymbol{\Sigma}_{\Gamma\Gamma})$, where $\boldsymbol{\Sigma}_{\alpha\alpha}$ is the covariance matrix of $\mathbf{X}_\alpha$ ($\alpha = 1, 2, \ldots, \Gamma$). The inverse of a block diagonal matrix is another block diagonal matrix, composed of the inverse of each block, that is, $\boldsymbol{\Sigma}^{-1} = \mathrm{diag}(\boldsymbol{\Sigma}_{11}^{-1}, \boldsymbol{\Sigma}_{22}^{-1}, \ldots, \boldsymbol{\Sigma}_{\Gamma\Gamma}^{-1})$. Moreover, taking into account that for block matrices $\det(\boldsymbol{\Sigma}) = \prod_{\alpha=1}^{\Gamma} \det(\boldsymbol{\Sigma}_{\alpha\alpha})$, one can write the expression for the joint probability density function defined in (3) in a form that reflects the block independence of variables:

$$p(\mathbf{x}) = \prod_{\alpha=1}^{\Gamma} \frac{1}{(2\pi)^{d_\alpha/2} \det(\boldsymbol{\Sigma}_{\alpha\alpha})^{1/2}}$$
$$\times \exp\left[-\frac{1}{2}(\mathbf{x}_\alpha - \boldsymbol{\mu}_\alpha)^T \boldsymbol{\Sigma}_{\alpha\alpha}^{-1}(\mathbf{x}_\alpha - \boldsymbol{\mu}_\alpha)\right], \tag{4}$$

where $p(\mathbf{x}_\alpha)$ is the joint probability density function of a subset α and $d_\alpha = \dim(\mathbf{x}_\alpha)$ is the number of variables in $\mathbf{X}_\alpha$.

2.2. Global Sensitivity Analysis.

The *variance-based global sensitivity analysis* method aims to quantify the relative importance of each input parameter in the response variance. It involves the calculation of the *global sensitivity indices*, sometimes called *Sobol's sensitivity indices* [2, 10].

In order to describe the global sensitivity indices, let us introduce the following notations: let $\{1, 2, \ldots, d\}$ be the set of input variable indices and let u be its arbitrary subset. Hence, $\mathbf{X}_u$ is a subset of variables whose indices are in u, whereas $\mathbf{X}_{-u}$ are the complimentary variables, that is, variables with indices not in u. Notation $|u|$ will be used for the cardinality of the set u. Variables X_i from non-overlapping sets u and $-u$ constitute the input vector $\mathbf{X} = (\mathbf{X}_u, \mathbf{X}_{-u})^T$.

Let us consider a subset $\mathbf{X}_u$ of input variables. Two types of sensitivity indices of the model response to the input random variables $\mathbf{X}_u$ can be introduced:

(i) the *main effect sensitivity index* $S_{\mathbf{X}_u}$, which describes the fraction of variance of the output Y that is expected to be *removed* if the true values of variables $\mathbf{X}_u$ become known.

(ii) the *total sensitivity index* $S_{\mathbf{X}_u}^{\text{tot}}$, which can be interpreted as the fraction of variance of the output Y that is expected to *remain* if the true values of variables $\mathbf{X}_{-u}$ become known.

In other words, $S_{\mathbf{X}_u}$ represents the effect due to $\mathbf{X}_u$ only, and $S_{\mathbf{X}_u}^{\text{tot}}$ represents the contribution to the variance of $\mathbf{X}_u$ with all the interactions of this variable with other variables.

The definition of sensitivity indices and their theoretical justification comes from functional ANOVA (analysis of variance). In Appendix A, we summarize formulae of the functional ANOVA decomposition, assuming that inputs are independent random variables with arbitrary continuous distributions.

Sobol [2, 3] introduced an alternative way of calculating sensitivity indices by sampling directly from $f(\mathbf{x})$, that is, without passing through the ANOVA decomposition. Sobol's alternative formulae are valid for uniformly distributed, independent random variables. Generalizing this result for continuous independent random variables with an arbitrary probability density function $p(\mathbf{x}) = p(\mathbf{x}_1) \cdots p(\mathbf{x}_d)$, one can write:

$$f_\varnothing = \int_{\mathbb{R}^d} f(\mathbf{x}) p(\mathbf{x}) d\mathbf{x}, \qquad D = \int_{\mathbb{R}^d} f^2(\mathbf{x}) p(\mathbf{x}) d\mathbf{x} - f_\varnothing^2, \tag{5}$$

$$D_{\mathbf{X}_u} = \int_{\mathbb{R}^{2d-|u|}} f(\mathbf{x}) f(\mathbf{x}_u, \mathbf{x}'_{-u}) p(\mathbf{x}) p(\mathbf{x}'_{-u}) d\mathbf{x} d\mathbf{x}'_{-u} - f_\varnothing^2, \tag{6}$$

$$D_{\mathbf{X}_u}^{\text{tot}} = \frac{1}{2} \int_{\mathbb{R}^{d+|u|}} [f(\mathbf{x}) - f(\mathbf{x}'_u, \mathbf{x}_{-u})]^2 p(\mathbf{x}) p(\mathbf{x}') d\mathbf{x} d\mathbf{x}'. \tag{7}$$

Here, the prime symbol over a variable (e.g., as in $\mathbf{x}'_u$) means that this variable has to be sampled independently from the corresponding marginal distribution ($p(\mathbf{x}'_u)$ in this case) of its unprimed analogue. Using the results from (5)–(7), the global sensitivity indices can be calculated as ratios:

$$S_{\mathbf{X}_u}^{\text{tot}} = \frac{D_{\mathbf{X}_u}^{\text{tot}}}{D}, \qquad S_{\mathbf{X}_u} = \frac{D_{\mathbf{X}_u}}{D}. \tag{8}$$

Note that $f_\varnothing$ and D correspond to the output mean and the output variance introduced in (2).

The independence condition for input variables can be relaxed. As discussed in [11], it is not necessary that *all* variables are mutually independent—this result holds when assuming independent blocks of input variables $\mathbf{X}_\alpha$ instead of single independent input variables X_i. Thus, if subsets of variables from $\mathbf{X}_u$ and $\mathbf{X}_{-u}$ are mutually independent, that is, $p(\mathbf{x}) = p(\mathbf{x}_u) p(\mathbf{x}_{-u})$, the sensitivity analysis formulas (6) and (7) are still applicable. Moreover, as one can see from (5), the formula for the output variance does not explicitly involve any particular subset of input variables. As a result, the variance of the output (D) can be calculated with the method presented here even in the case when all input variables are correlated. Since the variance is used to characterise the uncertainty in the output due to the uncertainty of the input, the method from this paper can be used for uncertainty analysis disregarding whether normally distributed inputs are correlated or not.

As follows from the previous description, the evaluation of sensitivity indices requires the calculation of the integrals in (5)–(7), which can be written in the following general form:

$$I_{d_{\text{eff}}}[g] = \int_{\mathbb{R}^{d_{\text{eff}}}} g(\widetilde{\mathbf{x}}) p(\widetilde{\mathbf{x}}) d\widetilde{\mathbf{x}}, \tag{9}$$

where $I_{d_{\text{eff}}}[\cdot]$ is the integration operator, $g(\widetilde{\mathbf{x}})$ represents a function being integrated, $d_{\text{eff}} = \dim(\widetilde{\mathbf{x}})$ is the effective dimensionality of the integral, and $p(\widetilde{\mathbf{x}})$ is the joint probability density function of $\widetilde{\mathbf{x}}$. For instance, in integral (6), function $g(\widetilde{\mathbf{x}})$ represents $[f(\mathbf{x}) f(\mathbf{x}_u, \mathbf{x}'_{-u})]$, $\widetilde{\mathbf{x}} = (\mathbf{x}, \mathbf{x}'_{-u}) = (\mathbf{x}_u, \mathbf{x}_{-u}, \mathbf{x}'_{-u})$, $p(\widetilde{\mathbf{x}}) = p(\mathbf{x}) p(\mathbf{x}'_{-u})$, and the effective dimensionality is $d_{\text{eff}} = 2d - |u|$.

2.3. Standard Normal Law Representation. Though the blockwise representation (4) of the joint probability density function (3) allows the exploiting of the independence of different subsets of variables, it gives no information about the practical way of a sensitivity index calculation. It is convenient to rewrite the expression in the so-called *standard form* in order to simplify future numerical evaluations of the global sensitivity indices.

Since covariance matrices are both symmetric and positive definite, for each $\Sigma_{\alpha\alpha}$ there is a nonsingular matrix $\mathbf{P}_{\alpha\alpha}$ such that $\Sigma_{\alpha\alpha} = \mathbf{P}_{\alpha\alpha}\mathbf{P}_{\alpha\alpha}^T$ (Cholesky factorization). Consider the linear transformation $\mathbf{z}_\alpha = \mathbf{P}_{\alpha\alpha}^{-1}(\mathbf{x}_\alpha - \boldsymbol{\mu}_\alpha)$. For any α, it leads to

$$\left(\mathbf{x}_\alpha - \boldsymbol{\mu}_\alpha\right)^T \Sigma_{\alpha\alpha}^{-1} \left(\mathbf{x}_\alpha - \boldsymbol{\mu}_\alpha\right) = \mathbf{z}_\alpha^T \mathbf{z}_\alpha, \tag{10}$$

and one can show that $\mathrm{E}[\mathbf{z}_\alpha] = 0$, $\mathrm{Cov}[\mathbf{z}_\alpha] = \mathbf{I}_\alpha$, where $\mathbf{I}_\alpha = \mathrm{diag}(1, 1, \ldots, 1)$ is the $d_\alpha \times d_\alpha$ identity matrix. Since $\sum_{\alpha=1}^{\Gamma} \mathbf{z}_\alpha^T \mathbf{z}_\alpha = \mathbf{z}^T \mathbf{z}$, the joint probability density function can be written in the standard form:

$$p(\widetilde{\mathbf{z}}) = \frac{1}{(2\pi)^{d_{\text{eff}}/2}} \exp\left(-\frac{1}{2}\widetilde{\mathbf{z}}^T\widetilde{\mathbf{z}}\right) = \frac{1}{(2\pi)^{d_{\text{eff}}/2}} \exp\left(-\frac{1}{2}\sum_{i=1}^{d_{\text{eff}}} z_i^2\right), \tag{11}$$

where $p(\widetilde{\mathbf{x}}) d\widetilde{\mathbf{x}} = p(\widetilde{\mathbf{z}}) d\widetilde{\mathbf{z}}$. New standard random variables Z_i ($i = 1, 2, \ldots, d$) have zero mean, standard deviations equal to one, and are not correlated, that is, $Z_i \sim N(0, 1)$.

Representation (11) can now be used for the calculation of sensitivity indices: variables Z_i can be sampled individually from $N(0, 1)$ and the corresponding $\widetilde{\mathbf{x}}$-points can be calculated as

$$\mathbf{x}_\alpha(\widetilde{\mathbf{z}}) = \boldsymbol{\mu}_\alpha + \mathbf{P}_{\alpha\alpha}\mathbf{z}_\alpha, \tag{12}$$

where α goes over all subsets of $\widetilde{\mathbf{X}}$. Nevertheless, in order to simplify the sampling procedure, to allow the use of a single calculational path and make a wider range of numerical integration techniques suitable for solving the problem, we do one more transformation from the normally distributed variables to the uniformly distributed ones.

Consider the following coordinate-wise change of variable from $z_i \in \mathbb{R}$ to $s_i \in (0, 1)$:

$$s_i(z_i) = \Phi(z_i), \tag{13}$$

where $\Phi(\cdot)$ is the cumulative distribution function for the normal distribution. From the properties of $\Phi(\cdot)$ follows $\lim_{z_i \to -\infty} s_i(z_i) = 0$, $\lim_{z_i \to +\infty} s_i(z_i) = 1$,

$$ds_i = \frac{1}{\sqrt{2\pi}} \exp\left(-\frac{1}{2}z_i^2\right) dz_i. \tag{14}$$

Applying this transformation coordinate-wise (i.e., for $i = 1, 2, \ldots, d_{\text{eff}}$) and introducing $h(\tilde{s}) = g(\tilde{x}[\tilde{z}(\tilde{s})])$ give the representation of the integral (9) in the form

$$I_{d_{\text{eff}}}[g] = \int_{[0,1]^{d_{\text{eff}}}} h(\tilde{s}) d\tilde{s}. \tag{15}$$

Here, $z_i(s_i) = \Phi^{-1}(s_i)$ for $i = 1, 2, \ldots, d_{\text{eff}}$, where $\Phi^{-1}(\cdot)$ is the inverse cumulative distribution function for the normal distribution, called the probit function, and $\tilde{x}(\tilde{z})$ is defined by (12).

2.4. Numerical Calculation of Sensitivity Indices

2.4.1. Numerical Quadratures. The integral in (15) can be approximated with a *quadrature* (sometimes called *cubature* in the literature), that can be written in the following general form:

$$I_{d_{\text{eff}}}[h(\tilde{s})] \approx Q_{d_{\text{eff}}}^N[h(\tilde{s})] = \sum_{n=1}^{N} w_n h(\tilde{s}_n), \tag{16}$$

where w_n are method-dependent quadrature weights, $h(\tilde{s}_n)$ are samples of the integrand at method-dependent nodes $\tilde{s}_n \in [0,1]^{d_{\text{eff}}}$, and N is the number of samples.

The integral in (15) is multidimensional, and, therefore, special numerical techniques, that can cope with the curse of dimension, are required to calculate it. Monte Carlo (including quasi-Monte Carlo) and sparse grid integration methods are suitable for this task and will be considered in our paper. Later, we will briefly introduce these methods and discuss their implementation in our work.

2.4.2. Monte Carlo and Quasi-Monte Carlo Quadratures. In the case of the traditional Monte Carlo method, the integral is sampled on a set of d_{eff}-dimensional *pseudo-random* points $\tilde{s}_n$, uniformly distributed in the unit hypercube $[0,1]^{d_{\text{eff}}}$. In the case of quasi-Monte Carlo, so-called *low discrepancy sequences* of *quasirandom* points (also uniformly distributed in $[0,1]^{d_{\text{eff}}}$), are used for integration. For both traditional Monte Carlo and quasi-Monte Carlo, the weights w_n are point-independent and equal, that is, $w_n = 1/N$. The quasi-Monte Carlo quadratures have a higher asymptotic convergence rate and often outperform the traditional Monte Carlo quadrature in practical applications [12].

There is a strong similarity between traditional Monte Carlo and quasi-Monte Carlo quadratures except for the type of sampling points (pseudo-random or quasi-random) and the way of error estimation. The error estimation will be done in the same way for both quadratures (see the discussion later). Hence, in this paper, both the traditional Monte Carlo and the quasi-Monte Carlo quadratures will be referred to as Monte Carlo quadratures.

In this work, we follow Sobol's recommendations [3] on the implementation of the Monte Carlo quadratures for the calculation of sensitivity indices. In particular, sampling is done from hypercube $[0,1]^{2d}$ instead of $[0,1]^{d_{\text{eff}}}$ and, in order to improve the accuracy of the estimation in (15), the function $f(\mathbf{x}) - c_0$ is evaluated instead of $f(\mathbf{x})$ in (5)–(7), where $c_0 \approx f_\varnothing$.

Our estimation of the accuracy of the Monte Carlo quadratures is based on a so-called randomization procedure [13]. This procedure consists of calculating R independent estimates, $\hat{I}_r^N$, of integral (15). The approximation to integral (15) is then calculated as an average of independent estimates, that is:

$$\hat{I}^N = \frac{1}{R} \sum_{r=1}^{R} \hat{I}_r^N, \tag{17}$$

and the error of such an approximation is characterized by the sample standard deviation, defined as

$$\hat{\epsilon}_{RN} = \sqrt{\frac{1}{R(R-1)} \sum_{r=1}^{R} \left(\hat{I}_r^N - \hat{I}^N\right)^2}. \tag{18}$$

Each estimate $\hat{I}_r^N$ is based on an independent sequence of N quasi- or pseudo-random points, where each new sequence of points is obtained from the initial one by a random modulo 1 shift [13].

2.4.3. Sparse Grid Quadratures. A sparse grid $\mathcal{H}_{\ell,d_{\text{eff}}}$ is a set of d_{eff}-dimensional points, which is generated using *Smolyak construction* [14] and is based on a chosen sequence of the univariate quadrature formulas Q_l, where $l \geq 0$ is the *accuracy level* of Q_l (see Appendix B for details). When applied to the integration of multivariate functions, the Smolyak construction is a multidimensional quadrature $Q_{\ell,d_{\text{eff}}}$ based on a tensor product of one-dimensional quadratures Q_l, which are combined in a special way in order to optimize the quadrature convergence rate [15, 16]. The sequence of univariate quadrature formulae Q_l leads to a sequence of sparse grid quadratures with an increasing *sparse grid accuracy level* $\ell \geq 0$.

$Q_{\ell,d_{\text{eff}}}[h(\tilde{s})]$ is a linear functional that depends on $h(\tilde{s})$ through function values at the set $\mathcal{H}_{\ell,d_{\text{eff}}}$, and the number of terms N in (16) is defined by its cardinality. The sparse grid points $\tilde{s}_n \in \mathcal{H}_{\ell,d_{\text{eff}}}$ and the quadrature weights w_n can be calculated using the procedure described in Appendix B.

If $\mathcal{H}_{\ell,d_{\text{eff}}} \subset \mathcal{H}_{\ell+1,d_{\text{eff}}}$, the quadrature is called *nested*. Nested quadratures permit the use of function values from previous levels, thus making integration less computationally expensive. Quadrature rules are said to be *open* when they do not include points on the boundary and *closed* otherwise. Points on the boundary (i.e., $s_i = 0$ or $s_i = 1$ for $i = 1, 2, \ldots, d_{\text{eff}}$) represent a problem for the numerical integration in (9), as a transformation $s_i \to z_i$ will lead to infinities in these points. Hence, only nested and strictly open sparse grid quadratures will be used in this work.

The sequence of sparse grid quadratures naturally leads to a formula for a practical estimation of the integration error

$$\hat{\epsilon}_\ell = \left| Q_{\ell,d_{eff}}[h(\tilde{\mathbf{s}})] - Q_{\ell-1,d_{eff}}[h(\tilde{\mathbf{s}})] \right|, \quad (19)$$

although this estimation is usually quite conservative.

Note that sparse grids are often defined on the hypercube $s^* \in [-1, 1]^{d_{eff}}$. In this case, they can easily be mapped to the unit hypercube $[0, 1]^{d_{eff}}$ using the transformation of variables $s_i = (s_i^* + 1)/2$. When this mapping is performed, all sparse grid quadrature weights w_n have to be adjusted by a factor of $2^{d_{eff}}$.

2.4.4. Inversion of the Standard Normal Cumulative Density Function. According to the methodology discussed in the previous section, each sample vector $\tilde{\mathbf{s}}_n$, generated with either Monte Carlo or sparse grid techniques, requires transformation to the corresponding $\tilde{\mathbf{z}}_n$ vector.

The traditional way to generate normally distributed points in conventional Monte Carlo is to sample from the uniform distribution and then to use the so-called Box-Muller transformation [17]. Unfortunately, it is not recommended [18] for use with quasi-Monte Carlo and is not suitable for use with sparse grids. An alternative way is to sample from the uniform distribution and then to use the inverse of the standard normal cumulative density function. It is recommended to use Moro's inversion algorithm [19], which is reported to be faster than the Box-Muller approach and has good accuracy for both the central region and the tails of the normal distribution [18].

In our work, Moro's algorithm is used for variable transformation $\tilde{\mathbf{s}}_n \rightarrow \tilde{\mathbf{z}}_n$ ($n = 1, 2, \ldots, N$) coordinate-wise (i.e., for each $s_{i,n}$, where $i = 1, 2, \ldots, d_{eff}$) for both Monte Carlo and sparse grid samples.

2.4.5. Algorithms for Calculation of Global Sensitivity Indices. Algorithms 1 and 2 provide examples of how to calculate sensitivity indices based on a Monte Carlo quadrature and a sparse grid quadrature, respectively. Note that these algorithms are given for the sake of illustration and do not contain details about possible memory management or performance enhancements.

3. Results

3.1. Test Problem Description. The OECD LWR UAM (OECD: Organization for Economic Co-operation and Development; LWR: light water reactor; UAM: Uncertainty Analysis in Modelling) benchmark [8] seeks to determine the uncertainty in LWR system calculations at all stages of coupled reactor physics/thermal hydraulics calculations. The benchmark specification consists of three phases, where the first phase is the neutronic phase.

The neutronic phase involved obtaining multigroup microscopic cross-section libraries. These libraries would then be used to calculate few group macroscopic cross-sections, which are to be used in criticality (steady state) stand-alone calculations. One of the reactors that was chosen as a reference LWR for the benchmark was the Peach Bottom

TABLE 1: Assembly homogenized 2-group cross-sections [21].

Variable	Notation	Value, cm^{-1}
Fast capture	Σ_c^1	$5.336 \cdot 10^{-3}$
Thermal capture	Σ_c^2	$2.693 \cdot 10^{-2}$
Fast fission	Σ_f^1	$1.9124 \cdot 10^{-3}$
Thermal fission	Σ_f^2	$2.8438 \cdot 10^{-2}$
Fast neutron production	$\nu\Sigma_f^1$	$4.920 \cdot 10^{-3}$
Thermal neutron production	$\nu\Sigma_f^2$	$6.929 \cdot 10^{-2}$
Fast removal	$\Sigma_s^{1 \to 2}$	$2.063 \cdot 10^{-2}$

TABLE 2: Test covariance matrix [21]. Values in bold correspond to Case A, in bold and non-italic correspond to Case B, and the full covariance matrix correspond to Case C.

	Σ_c^1	Σ_c^2	Σ_f^1	Σ_f^2	$\nu\Sigma_f^1$	$\nu\Sigma_f^2$	$\Sigma_s^{1 \to 2}$
Σ_c^1	**1.21**	0.23	−0.63	−0.04	−0.57	−0.03	0.77
Σ_c^2	0.23	**0.54**	−0.09	−0.48	−0.07	−0.34	−0.01
Σ_f^1	−0.63	−0.09	**0.68**	0.11	0.87	0.08	−0.68
Σ_f^2	−0.04	−0.48	0.11	**0.32**	0.06	0.72	0.04
$\nu\Sigma_f^1$	−0.57	−0.07	0.87	0.06	**0.98**	0.12	−0.64
$\nu\Sigma_f^2$	−0.03	−0.34	0.08	0.72	0.12	**0.45**	0.04
$\Sigma_s^{1 \to 2}$	0.77	−0.01	−0.68	0.04	−0.64	0.04	**1.11**

reactor. By energy collapsing and spatial homogenization of microscopic cross-section and covariance data [20], Williams et al. [21] obtained the 2-group homogenized neutron cross-section, with an energy boundary of 0.625 eV, and the corresponding covariance matrix for the Peach Bottom reactor fuel assembly. The neutron cross-sections are given in Table 1, they are assumed to be independent and normally distributed, and their mean values (given in the third column of Table 1) correspond to the vector $\boldsymbol{\mu}$ used in our methodology. The covariance matrix is shown in Table 2, where the diagonal terms are the percentage relative standard deviation and the off-diagonal terms are the correlation coefficients.

The global sensitivity analysis methodology discussed in Section 2 was applied to nuclear reactor calculations. The reactor parameter of interest that was chosen for this study is the infinite neutron multiplication factor, k_∞, and it was modelled as [22]

$$k_\infty = \frac{\nu\Sigma_f^1}{\Sigma_c^1 + \Sigma_f^1 + \Sigma_s^{1 \to 2}} + \frac{\nu\Sigma_f^2 \Sigma_s^{1 \to 2}}{\left(\Sigma_c^2 + \Sigma_f^2\right)\left(\Sigma_c^1 + \Sigma_f^1 + \Sigma_s^{1 \to 2}\right)}, \quad (20)$$

where the traditional notation for macroscopic cross-sections is used (see Table 1). To illustrate our methodology, three cases were considered (all three cases are shown in Table 2): one with a diagonal covariance matrix, another one with a block-diagonal covariance matrix, and the last one with the full covariance matrix.

In the first case (hereafter referred to as Case A), it was assumed that the input parameters (cross-sections) are not correlated, and the covariance matrix consisted of only the diagonal entries, highlighted in bold, while all other elements of the matrix were set to zero.

In the second case (hereafter referred to as Case B), we assumed a test block-diagonal covariance matrix. The test

$$
\begin{aligned}
&\textbf{input: } \boldsymbol{\mu}_1, \ldots, \boldsymbol{\mu}_\Gamma, \boldsymbol{\Sigma}_{11}, \ldots, \boldsymbol{\Sigma}_{\Gamma\Gamma}, u, R, N \\
&\textbf{for } \alpha = 1 \textbf{ to } \Gamma \textbf{ do} \\
&\quad \mathbf{P}_{\alpha\alpha} \leftarrow \text{Cholesky decomposition of } \boldsymbol{\Sigma}_{\alpha\alpha} \\
&\quad \mathbf{P}'_\alpha, \boldsymbol{\mu}'_\alpha \leftarrow \mathbf{P}_\alpha, \boldsymbol{\mu}_\alpha \\
&\textbf{end for} \\
&\widetilde{\mathbf{P}} \leftarrow \text{diag}(\mathbf{P}_{11}, \ldots, \mathbf{P}_{\Gamma\Gamma}, \mathbf{P}'_{11}, \ldots, \mathbf{P}'_{\Gamma\Gamma}) \\
&\widetilde{\boldsymbol{\mu}} \leftarrow \text{vec}(\boldsymbol{\mu}_1, \ldots, \boldsymbol{\mu}_\Gamma, \boldsymbol{\mu}'_1, \ldots, \boldsymbol{\mu}'_\Gamma) \\
&d_{\text{eff}} \leftarrow \dim(\widetilde{\boldsymbol{\mu}}) \\
&\textbf{for } n = 1 \textbf{ to } N \textbf{ do} \\
&\quad \textbf{for } r = 1 \textbf{ to } R \textbf{ do} \\
&\qquad \widetilde{\mathbf{s}}_n \leftarrow d_{\text{eff}}\text{-dimensional quasi- or pseudo-random point} \\
&\qquad w_n \leftarrow 1/N \\
&\qquad \widetilde{\mathbf{z}}_n \leftarrow \Phi^{-1}(\widetilde{\mathbf{s}}_n) \\
&\qquad \widetilde{\mathbf{x}}_n \leftarrow \widetilde{\boldsymbol{\mu}} + \widetilde{\mathbf{P}}\widetilde{\mathbf{z}}_n \\
&\qquad \mathbf{g}_n \leftarrow \mathbf{g}(\widetilde{\mathbf{x}}_n) \\
&\quad \textbf{end for} \\
&\quad (f_\varnothing, D, D_{\mathbf{X}_u}, D_{\mathbf{X}_u}^{\text{tot}})_r \leftarrow \sum_{n=1}^{N} w_n \mathbf{g}_n \\
&\textbf{end for} \\
&\text{calculate } S_{\mathbf{X}_u}, S_{\mathbf{X}_u}^{\text{tot}}, \hat{\epsilon}_{RN} \\
&\textbf{return } S_{\mathbf{X}_u}, S_{\mathbf{X}_u}^{\text{tot}}, \hat{\epsilon}_{RN}
\end{aligned}
$$

ALGORITHM 1: The calculation of sensitivity indices using Monte Carlo quadrature.

$$
\begin{aligned}
&\textbf{input: } \boldsymbol{\mu}_1, \ldots, \boldsymbol{\mu}_\Gamma, \boldsymbol{\Sigma}_{11}, \ldots, \boldsymbol{\Sigma}_{\Gamma\Gamma}, u, \ell_{\max} \\
&\textbf{for } \alpha = 1 \textbf{ to } \Gamma \textbf{ do} \\
&\quad \mathbf{P}_{\alpha\alpha} \leftarrow \text{Cholesky decomposition of } \boldsymbol{\Sigma}_{\alpha\alpha} \\
&\quad \mathbf{P}'_\alpha, \boldsymbol{\mu}'_\alpha \leftarrow \mathbf{P}_\alpha, \boldsymbol{\mu}_\alpha \\
&\textbf{end for} \\
&\widetilde{\mathbf{P}} \leftarrow \text{diag}(\mathbf{P}_{11}, \ldots, \mathbf{P}_{\Gamma\Gamma}, \mathbf{P}'_{11}, \ldots, \mathbf{P}'_{\Gamma\Gamma}) \\
&\widetilde{\boldsymbol{\mu}} \leftarrow \text{vec}(\boldsymbol{\mu}_1, \ldots, \boldsymbol{\mu}_\Gamma, \boldsymbol{\mu}'_1, \ldots, \boldsymbol{\mu}'_\Gamma) \\
&d_{\text{eff}} \leftarrow \dim(\widetilde{\boldsymbol{\mu}}) \\
&\textbf{for } \ell = 1 \textbf{ to } \ell_{\max} \textbf{ do} \\
&\quad \text{generate } \mathcal{H}_{\ell, d_{\text{eff}}} \\
&\quad N \leftarrow \text{size of } \mathcal{H}_{\ell, d_{\text{eff}}} \\
&\quad \textbf{for } n = 1 \textbf{ to } N \textbf{ do} \\
&\qquad \widetilde{\mathbf{s}}_n \leftarrow \text{node from } \mathcal{H}_{\ell, d_{\text{eff}}} \\
&\qquad w_n \leftarrow \text{sparse grid weight} \\
&\qquad \widetilde{\mathbf{z}}_n \leftarrow \Phi^{-1}(\widetilde{\mathbf{s}}_n) \\
&\qquad \widetilde{\mathbf{x}}_n \leftarrow \widetilde{\boldsymbol{\mu}} + \widetilde{\mathbf{P}}\widetilde{\mathbf{z}}_n \\
&\qquad \mathbf{g}_n \leftarrow \mathbf{g}(\widetilde{\mathbf{x}}_n) \\
&\quad \textbf{end for} \\
&\quad (f_\varnothing, D, D_{\mathbf{X}_u}, D_{\mathbf{X}_u}^{\text{tot}})_\ell \leftarrow \sum_{n=1}^{N} w_n \mathbf{g}_n \\
&\quad \text{calculate } S_{\mathbf{X}_u}, S_{\mathbf{X}_u}^{\text{tot}}, \hat{\epsilon}_\ell \\
&\quad \textbf{return } S_{\mathbf{X}_u}, S_{\mathbf{X}_u}^{\text{tot}}, \hat{\epsilon}_\ell \\
&\textbf{end for}
\end{aligned}
$$

ALGORITHM 2: The calculation of sensitivity indices using sparse grid quadrature.

matrix was artificially constructed based on the 2-group covariance matrix from [21] in such a way that the input variables can be partitioned into three mutually independent subsets $\{\Sigma_c^1, \Sigma_c^2, \Sigma_f^1, \Sigma_f^2\}$, $\{\nu\Sigma_f^1, \nu\Sigma_f^2\}$, and $\{\Sigma_s^{1\rightarrow 2}\}$, such that elements in the off-diagonal blocks are set to zero, that is, terms highlighted in italic are set to zero. It should be noted that the elements of the first subset correspond to those terms that contribute to the absorption cross-section. The elements of the second subset correspond to those terms that contribute to the production of neutrons, and the last subset corresponds to the removal of neutrons from the fast group to the thermal group.

For the last case (hereafter referred to as Case C), it was assumed that all the input parameters (cross-sections) are correlated with one another, and the full covariance matrix was used, that is, all entries highlighted in bold, italic, and non-italic.

It should be emphasised that neither of the first two examples (Cases A and B) considered pretends to reflect physical reality, but both the cross-section values and the

TABLE 3: Estimated uncertainty of the infinite multiplication factor in terms of variance and standard deviation (given in parenthesis).

Case	Traditional Monte Carlo	Quasi-Monte Carlo	Sparse grid
A	$3.680 \cdot 10^{-5}$ (607 pcm)	$3.680 \cdot 10^{-5}$ (607 pcm)	$3.671 \cdot 10^{-5}$ (606 pcm)
B	$3.575 \cdot 10^{-5}$ (598 pcm)	$3.576 \cdot 10^{-5}$ (598 pcm)	$3.567 \cdot 10^{-5}$ (597 pcm)
C	$3.115 \cdot 10^{-5}$ (558 pcm)	$3.116 \cdot 10^{-5}$ (558 pcm)	$3.108 \cdot 10^{-5}$ (558 pcm)

elements of the test covariance matrix are of a plausible order of magnitude (close to the values given in [21]): hence, this example is representative and suitable for testing of our method. Therefore, the results and conclusions will be given in order to characterise the method presented and not the neutron multiplication properties of the Peach Bottom reactor.

3.2. Method Implementation. A Fortran 90 program was written to implement all the steps of the methodology outlined in Section 2. The program was subdivided into blocks of code, where each block had an input to be evaluated to give an expected output and corresponded to step(s) along the calculational path of the methodology. The testing, verification, and validation of the program were done for each block of code using test functions, for which the corresponding results could be evaluated analytically.

Pseudo-random points were generated using the Fortran intrinsic subroutine `random_number()`. In implementing quasi-Monte Carlo, a Sobol quasi-random number generator written by J. Burkardt [23] was used. Furthermore, a randomization procedure was used in estimating the integration error, $\hat{\epsilon}_{RN}$, for the Monte Carlo quadratures, by considering $R = 100$ independent sequences with $N = 10^6$ samples in each sequence.

The implementation of sparse grid quadratures was greatly facilitated by subroutines written by J. Burkardt [23]. Different open sparse grid quadrature rules such as Fejer, Gauss-Patterson, and Gauss-Legendre rules were applied (note that closed rules were also tested and, as expected, numerical problems for the boundary points were encountered). The Gauss-Legendre quadrature outperformed the other rules in terms of computational time needed to achieve a given accuracy for the cases considered, and its results will be reported up to a sparse grid level of $\ell = 4$. A conservative procedure defined by (19) was used in estimating the integration error $\hat{\epsilon}_\ell$ for the sparse grid quadratures and is reported in this paper.

Variations were introduced into the neutron cross-sections by using a standardizing transformation as explained in (12), that is, $\tilde{\mathbf{x}}(\tilde{\mathbf{z}}) = \tilde{\mu} + \tilde{\mathbf{P}}\tilde{\mathbf{z}}$, where $\tilde{\mathbf{P}}$ is the extended Cholesky decomposed neutron cross-section covariance matrix, and $\tilde{\mathbf{z}}$ is obtained by using Moro's inversion of samples required by each of the implemented quadratures. Finally, in order to improve the accuracy of the Monte Carlo estimation of integral (15), a variance reduction technique [3], which consists of sampling function $\Delta f(\mathbf{x}) = [f(\mathbf{x}) - c_0]$ instead of $f(\mathbf{x})$ in (5)–(7), where $c_0 \approx f_\varnothing$, was used.

3.3. Computed Uncertainty and Sensitivity. The uncertainty of multiplication factor k_∞, computed in terms of variance D, are given in Table 3 for Cases A, B, and C. Though the output variance is the natural result of variance-based sensitivity analysis, the standard deviation is preferred in the literature because it allows an intuitive interpretation as the error bar for the value of the analysed parameter. The standard deviations are calculated as square root of variance, $\sqrt{D}$, and reported in Table 3 in parentheses for all cases and each quadrature. The uncertainty of the multiplication factor (expressed in relative units as $100\% \times \delta k/k$), which we obtained in Case C, was 0.51% for each of the three quadratures, and this result is in good agreement with the value of 0.49% reported in [21].

The computed sensitivity indices for each of the variables (cross-sections) in Case A and for each subset in Case B are given in Table 4. No sensitivity analysis was performed for Case C, since any sensitivity indices computed with our method would be meaningless because all input parameters are correlated in this case.

Considering the results of Case A, the input variable with the greatest influence on the infinite neutron multiplication factor is the thermal neutron production, $\nu\Sigma_f^2$, and the input variable with the least influence is the fast neutron fission, Σ_f^1. This is similar to what we anticipated, given the fact that the infinite neutron multiplication factor is highly dependent on the number of neutrons produced in the system. Since the system being considered is thermal, the thermal neutron production should account for most of the neutrons produced, and the effect of fast neutron fission was not expected to be significant.

Considering the results of Case B, the subset $\{\nu\Sigma_f^1, \nu\Sigma_f^2\}$, which corresponds to the neutron production, had the greatest influence on the infinite neutron multiplication factor. The subset $\{\Sigma_s^{1\to2}\}$, which corresponds to the fast neutron removal, was the least influential. It should be noted that the value of the sensitivity index for $\{\Sigma_s^{1\to2}\}$ is different in Cases A and B. This is because the off-diagonal terms of the covariance matrix influenced the results for $\{\Sigma_s^{1\to2}\}$. In other words, due to the off-diagonal terms in the correlation matrix, Cases A and B define different problems.

3.4. Error Analysis. The results for Cases A and B, in Tables 3 and 4, were obtained by using a high number of samples with all three numerical quadratures ($N = 10^6, R = 10^2$ in the case of Monte Carlo quadratures and $\ell = 4$, $N = 56785$ in the case of sparse grid quadrature), and these results are taken as the reference. There seems to be very good agreement of the computed uncertainties and sensitivity indices between all three quadratures.

TABLE 4: Estimated sensitivities of the infinite multiplication factor to different cross-sections (Case A) or their subsets (Case B).

Case	Subset of cross-sections $\mathbf{X}_u$	Traditional Monte Carlo		Quasi-Monte Carlo		Sparse grid	
		$S_{\mathbf{X}_u}$	$S_{\mathbf{X}_u}^{\text{tot}}$	$S_{\mathbf{X}_u}$	$S_{\mathbf{X}_u}^{\text{tot}}$	$S_{\mathbf{X}_u}$	$S_{\mathbf{X}_u}^{\text{tot}}$
A	Σ_c^1	0.1766	0.1766	0.1766	0.1766	0.1766	0.1766
	Σ_c^2	0.1624	0.1626	0.1626	0.1626	0.1626	0.1626
	Σ_f^1	0.0071	0.0072	0.0072	0.0072	0.0072	0.0072
	Σ_f^2	0.0640	0.0641	0.0641	0.0641	0.0641	0.0641
	$\nu\Sigma_f^1$	0.0807	0.0808	0.0808	0.0808	0.0808	0.0808
	$\nu\Sigma_f^2$	0.4676	0.4678	0.4677	0.4677	0.4677	0.4677
	$\Sigma_s^{1 \to 2}$	0.0409	0.0410	0.0410	0.0410	0.0410	0.0410
B	$\Sigma_c^1, \Sigma_c^2, \Sigma_f^1, \Sigma_f^2$	0.3453	0.3453	0.3453	0.3453	0.3453	0.3453
	$\nu\Sigma_f^1, \nu\Sigma_f^2$	0.6124	0.6126	0.6125	0.6125	0.6125	0.6125
	$\Sigma_s^{1 \to 2}$	0.0420	0.0422	0.0422	0.0422	0.0421	0.0421

The accuracy obtained for the reference results is much better than the accuracy needed to draw practical conclusions concerning the contribution of uncertainties of different cross-sections. By this we mean that the accuracy of the sensitivity index estimation has to be, at least, sufficient to discriminate between the contribution of different inputs and should also be able to discriminate between $S_{\mathbf{X}_u}^{\text{tot}}$ and $S_{\mathbf{X}_u}$ for a given input $\mathbf{X}_u$.

Therefore, an error estimation study was done in order to determine the influence of the number of samples on the absolute and relative quadrature error of the computed sensitivity indices, where the relative quadrature error is given by

$$\hat{\delta} = \frac{\hat{\epsilon}\left(S_{\mathbf{X}_u}^{(\text{tot})}\right)}{S_{\mathbf{X}_u}^{(\text{tot})}} \times 100 \, [\%], \qquad (21)$$

where the absolute quadrature error $\hat{\epsilon}$ is given by either (18) or (19). This study would help in determining the number of samples that is needed to get a good estimation of the sensitivity indices with the different numerical methods. For Monte Carlo methods, three different sample sizes were considered, $N = 10^2$, $N = 10^4$, and $N = 10^6$. In all cases, the number of independent sequences R was taken as 10^2. For the sparse grid, levels $\ell = 1$ to $\ell = 4$ were considered.

It was observed in both cases that the results obtained for $S_{\mathbf{X}_u}^{\text{tot}}$ and $S_{\mathbf{X}_u}$ are statistically similar for all subsets of the input variables, for all the three numerical methods that were used. This implies that the interaction effects can be neglected. It was also observed that the integration error for $S_{\mathbf{X}_u}^{\text{tot}}$ was smaller than for $S_{\mathbf{X}_u}$ in all the cases; hence, from now on, we will only consider $S_{\mathbf{X}_u}^{\text{tot}}$.

When considering Case A, it was observed that increasing N by a factor of 100 resulted, as expected, in a reduction of the integration error by a factor of approximately 10 for all the computed total sensitivity indices when using traditional Monte Carlo. The results for quasi-Monte Carlo showed that increasing N from 10^2 to 10^4, and subsequently from 10^4 to 10^6, resulted in a decrease of the integration error by a factor of about 30 and 40, respectively, for all the computed total sensitivity indices. For the sparse grid, a level change from

$\ell = 2$ to $\ell = 3$ and from $\ell = 3$ to $\ell = 4$ both resulted in a decrease of the integration error by a factor of about 3.

The maximal absolute and relative errors for the total sensitivity indices computed with different number of samples are reported in Table 5 for Monte Carlo quadratures and Table 6 for sparse grid quadrature. These maximal absolute errors are obtained by taking the maximal absolute error of all the sensitivity indices for a given case, a given number of samples, and a given quadrature. The maximal relative error is obtained in the same way.

As one can see from Table 5, a relatively small number of samples (100×100) in the case of Monte Carlo gave fairly good accuracy (about 2%) in the estimation of the total sensitivity indices.

It should be noted that levels $\ell = 0$ and $\ell = 1$ for the sparse grid were not considered in the error estimation. This is because for level $\ell = 0$, the abscissa consists of only one point, and the variance is zero; hence, the total sensitivity index will be undefined. For the same reason, the application of (19) cannot give reasonable results for level $\ell = 1$. However, looking at Table 6, it can be seen that the maximal difference between the results obtained for levels $\ell = 1$ and $\ell = 2$ is smaller than $2 \cdot 10^{-5}$, and the maximum relative quadrature error obtained when moving from level $\ell = 1$ to $\ell = 2$ is smaller than $3.8 \cdot 10^{-2}\%$. Hence, this shows that for both cases, level $\ell = 1$, which contains only 29 points, is sufficient to estimate the total sensitivity indices with a very good accuracy.

The relatively small number of sparse grid points needed for an accurate estimation of the sensitivity indices as well as the absence of interactions between input variables (as discussed earlier) was unexpected. This result can potentially be explained in the following way: the uncertainty in cross-sections is so small that only the vicinity of the cross-section mean values contributes to the integrals used in the estimation of sensitivity indices. In this vicinity, the neutron multiplication factor, which is used as the example, can be approximated with a fairly linear function.

A small numerical experiment was done to clarify this aspect. The standard deviations given in Table 1 were initially multiplied by arbitrary factors between 1 and 10 and, in the second phase, by arbitrary factors between 1 and 20,

TABLE 5: Maximal error of Monte Carlo quadratures.

Case	Samples $N \times R$	Traditional Monte Carlo		Quasi-Monte Carlo	
		$\hat{\epsilon}_{RN}$	$\hat{\delta}$, %	$\hat{\epsilon}_{RN}$	$\hat{\delta}$, %
A	$10^2 \times 10^2$	$9.0 \cdot 10^{-3}$	2.2	$7.4 \cdot 10^{-3}$	1.6
	$10^4 \times 10^2$	$8.0 \cdot 10^{-4}$	$1.9 \cdot 10^{-1}$	$2.7 \cdot 10^{-4}$	$5.8 \cdot 10^{-2}$
	$10^6 \times 10^2$	$7.8 \cdot 10^{-5}$	$2.0 \cdot 10^{-2}$	$6.8 \cdot 10^{-6}$	$1.8 \cdot 10^{-3}$
B	$10^2 \times 10^2$	$1.2 \cdot 10^{-2}$	2.1	$9.8 \cdot 10^{-3}$	1.6
	$10^4 \times 10^2$	$1.1 \cdot 10^{-3}$	$1.9 \cdot 10^{-1}$	$3.5 \cdot 10^{-4}$	$5.8 \cdot 10^{-2}$
	$10^6 \times 10^2$	$9.1 \cdot 10^{-5}$	$2.0 \cdot 10^{-2}$	$1.2 \cdot 10^{-5}$	$1.9 \cdot 10^{-3}$

TABLE 6: Maximal error for the sparse grid quadrature.

ℓ	N	Case A		Case B	
		$\hat{\epsilon}_\ell$	$\hat{\delta}$, %	$\hat{\epsilon}_\ell$	$\hat{\delta}$, %
1	29	N/A	N/A	N/A	N/A
2	477	$2.0 \cdot 10^{-5}$	$3.8 \cdot 10^{-2}$	$1.3 \cdot 10^{-5}$	$3.0 \cdot 10^{-2}$
3	5769	$6.4 \cdot 10^{-6}$	$1.3 \cdot 10^{-2}$	$4.3 \cdot 10^{-6}$	$1.0 \cdot 10^{-2}$
4	56785	$2.0 \cdot 10^{-6}$	$4.1 \cdot 10^{-3}$	$1.4 \cdot 10^{-6}$	$3.6 \cdot 10^{-3}$

and the sensitivity indices were recalculated. These factors were chosen to make the effect of a wider distribution more prominent without introducing a significant nonphysical effect due to negative cross-section values at the left tail of the distributions. It was observed that as the values of the standard deviations increase, interaction effects can be observed, that is, $S_{\mathbf{X}_u}^{\text{tot}}$ becomes statistically different from $S_{\mathbf{X}_u}$. Furthermore, a larger number of points (higher levels) is needed to achieve the same accuracy as in the reference case. These results may be used to confirm our assumption on the nature of the good performance of the sparse grid quadrature. However, a proper study was done to confirm our conclusion, and the results are reported in [24].

4. Conclusions

In this paper, the global variance-based sensitivity and uncertainty analysis of reactor parameters dependent on few-group or multigroup neutron cross-sections was discussed. It was assumed that the cross-sections are normally distributed random variables, with known means and correlation matrices, which can be partitioned into statistically independent blocks of variables and that this partitioning allows one to formulate scientifically and practically sound sensitivity analysis problems. The theoretical and mathematical aspects of the calculation of the global sensitivity indices under the previous assumptions have been discussed. The problem of practical numerical calculations of the variance-based global sensitivity indices was addressed; namely, different options for numerical integration were considered. A consistent overall path for the calculation of sensitivity indices was proposed and described.

The method was successfully implemented in practice and was tested on a problem that involved two-group assembly homogenised cross-sections as input variables. The performance of different numerical integration techniques was tested on a reactor problem with arbitrary, but plausible, two-group cross-sections and covariance matrices. Different implementations gave consistent results for the test problem under consideration. The implementation based on sparse grid quadrature demonstrated the best accuracy with as low as a few dozen samples.

This good performance of sparse grid integration was not expected and a special mini-study was performed with the purpose of explaining its origin as well as the absence of interactions in the obtained sensitivity indices. The results of this study confirmed our hypothesis that the observed results can be explained by the very small cross-section error. Nevertheless, this conclusion still has to be supported by a theoretical explanation.

From the methodological point of view, the method presented in the paper is applicable to problems with an arbitrary number of input variables. Nevertheless, one has to be cautious when dealing with multivariate problems in order to escape the curse of dimension. In this work, the applicability of our method to a few-group problem was demonstrated, but its applicability to multigroup reactor problems will be the topic of future studies.

Appendices

A. Functional ANOVA Decomposition for Independent Random Variables

Let $p(x_1, x_2 \ldots, x_d)$ be a joint probability density function of d random variables X_i:

$$P[X_1 \leq x_1, \ldots, X_d \leq x_d]$$
$$= \int_{-\infty}^{x_d} \cdots \int_{-\infty}^{x_1} p(x_1, x_2 \ldots, x_d) dx_1 \cdots dx_d. \quad (A.1)$$

Let $f : \mathbb{R}^d \to \mathbb{R}$ be a square integrable function over $\mathbf{x} = (x_1, \ldots, x_d)$. The expected value and the variance of the function $f(\mathbf{x})$ with respect to the probability density function $p(\mathbf{x})$ are defined as

$$\mathrm{E}[f(\mathbf{x})] = \int_{\mathbb{R}^d} f(\mathbf{x}) p(\mathbf{x}) d\mathbf{x},$$
$$\mathrm{Var}[f(\mathbf{x})] = \int_{\mathbb{R}^d} \left(f(\mathbf{x}) - \mathrm{E}[f(\mathbf{x})]^2 \right) p(\mathbf{x}) d\mathbf{x}. \quad (A.2)$$

The *functional ANOVA decomposition* is a representation of the function $f(\mathbf{x})$ as a sum of terms of increasing dimensionality:

$$f(\mathbf{x}) = \sum_u f_u(\mathbf{x}_u) = f_\varnothing + \sum_i f_i(x_i) + \sum_{i<j} f_{ij}(x_i, x_j)$$

$$+ \cdots + f_{12\ldots d}(x_1, x_2, \ldots, x_d), \quad (A.3)$$

where the sum is assumed over 2^d subsets $u \subseteq \{1, 2, \ldots, d\}$ and $f_u(\mathbf{x}_u)$ is a function that depends on $\mathbf{x}$ only through x_i with $i \in u$. Here, $\mathbf{x}_u$ is a subset of variables whose indices are in u, whereas $\mathbf{x}_{-u}$ are the variables with indices not in u, and $|u|$ is the cardinality of the set u.

According to Sobol's definition, for the representation given by (A.3) to be a functional ANOVA decomposition it has to satisfy the so-called *zero means* and *orthogonality* properties [2, 3]. Let random variables X_i ($i = 1, 2, \ldots, d$) be mutually independent with a joint probablity density function $p(\mathbf{x}) = p_1(x_1) p_2(x_2) \cdots p_d(x_d)$. Using an analogy with the case of uniformly distributed input variables, one can demonstrate that the functional ANOVA can be constructed by applying the following recurrent formula:

$$f_u(\mathbf{x}_u) = \int_{\mathbb{R}^{d-|u|}} \left(f(\mathbf{x}) - \sum_{v \subset u} f_v(\mathbf{x}_v) \right) p(\mathbf{x}_{-u}) d\mathbf{x}_{-u}. \quad (A.4)$$

The constant mean term, $f_\varnothing$, is thus obtained by calculating $f_\varnothing = \int_{\mathbb{R}^d} f(\mathbf{x}) p(\mathbf{x}) d\mathbf{x}$, first order effects $f_i(x_i)$ (where $i = 1, \ldots, d$) are obtained from $f_i(x_i) = \int_{\mathbb{R}^{(d-1)}} (f(\mathbf{x}) - f_\varnothing)[p(\mathbf{x})/p(x_i)] d\mathbf{x}/dx_i$ and so on. For functions $f_u(\mathbf{x}_u)$ obtained with recurrence (A.4), the zero means property becomes

$$E[f_u(\mathbf{x}_u)] = \int_{\mathbb{R}^{|u|}} f_u(\mathbf{x}_u) p(\mathbf{x}_u) d\mathbf{x}_u = \int_{\mathbb{R}^d} f_u(\mathbf{x}_u) p(\mathbf{x}) d\mathbf{x} = 0. \quad (A.5)$$

The orthogonality property holds in the weighted form

$$\int_{\mathbb{R}^d} f_u(\mathbf{x}_u) f_v(\mathbf{x}_v) p(\mathbf{x}) d\mathbf{x} = 0. \quad (A.6)$$

Properties (A.5) and (A.6) are crucial for the functional ANOVA method because they lead to the variance decomposition formula:

$$\text{Var}[f(\mathbf{x})] = \sum_{u \subseteq \{1, 2, \ldots, d\}} \text{Var}[f_u(\mathbf{x}_u)]. \quad (A.7)$$

Let us assume that the function $f(\mathbf{x})$ allows order-wise decomposition over subsets of variables (A.3). Applying the variance operator (A.2) to the left-hand side and right-hand side of (A.3) and using a standard statistical formula, we can write:

$$\text{Var}[f(\mathbf{x})] = \text{Var}\left[\sum_u f_u(\mathbf{x}_u) \right] = \sum_u \text{Var}[f_u(\mathbf{x}_u)]$$

$$+ 2 \sum_{u, v \neq u} \text{Cov}[f_u(\mathbf{x}_u), f_v(\mathbf{x}_v)]. \quad (A.8)$$

By definition,

$$\text{Cov}[f_u(\mathbf{x}_u), f_v(\mathbf{x}_v)] = \int_{\mathbb{R}^d} (f_u(\mathbf{x}_u) - E[f_u(\mathbf{x}_u)])(f_v(\mathbf{x}_v)$$

$$- E[f_v(\mathbf{x}_v)]) p(\mathbf{x}) d\mathbf{x}, \quad (A.9)$$

and it can be observed from (A.9) that properties (A.5) and (A.6) lead to the zero-covariance condition $\text{Cov}[f_u(\mathbf{x}_u), f_v(\mathbf{x}_v)] = 0$ for $u \neq v$ and hence to the variance decomposition in the form of (A.7).

B. Approximation of Multidimensional Integrals with Sparse Grid Quadratures

Let $\varphi : \Omega \rightarrow \mathbb{R}$ be a continuous function of its arguments and with bounded mixed derivatives of order r:

$$\left\| \frac{\partial^{\|k\|_1} \varphi(x_1, \ldots, x_d)}{\partial x_1^{k_1} \cdots \partial x_d^{k_d}} \right\|_\infty < \infty, \quad k_i \leq r, \quad (B.1)$$

where $\Omega = \Omega_1 \cdots \Omega_d$, d is the dimensionality of the problem and $\Omega_i \subset \mathbb{R}$ ($i = 1, 2, \ldots, d$) are bounded or unbounded intervals. We consider an approximation to the integral

$$I[\varphi(\mathbf{x})] = \int_\Omega \varphi(\mathbf{x}) \varrho(\mathbf{x}) d\mathbf{x}, \quad (B.2)$$

where $\mathbf{x} = (x_1, \ldots, x_d)$, with the tensor product form $\varrho(x) = \varrho_1(x_1) \cdots \varrho_{d_{\text{eff}}}(x_d)$ of the weight function ϱ.

In order to construct a multidimensional sparse grid quadrature, let us consider a sequence of univariate quadrature formulas

$$Q_{l_i}[\psi(x_i)] = \sum_{j=1}^{m_{l_i}} w_{j_i}^{l_i} \psi\left(x_{j_i}^{l_i}\right), \quad (B.3)$$

which approximate one-dimensional integrals

$$\int_{\Omega_i} \psi(x_i) \varrho_i(x_i) dx_i, \quad i = 1, 2, \ldots, d. \quad (B.4)$$

Here, $\psi : \Omega_i \rightarrow \mathbb{R}$ is a continuous function of its argument, $l_i \in \mathbb{Z}$, $l_i \geq 0$ is the accuracy level of the quadrature formula, m_{l_i} is the number of abscissas (knots) $x_{j_i}^{l_i}$ of the quadrature, and $w_{j_i}^{l_i}$ is the corresponding weight. The index l_i is written explicitly over abscissas and weights in order to remind that they may change for different levels. $\mathcal{H}_{l_i} = \{x_{j_i}^{l_i} : 1 \leq j_i \leq m_{l_i}\}$ will be used to denote the set of knots of the one-dimensional quadrature formula.

In the sparse grid method, the integral (B.2) is approximated via the Smolyak formula [14, 15], defined for an accuracy level $\ell \in \mathbb{Z}$ ($\ell \geq 0$) of the sparse grid as follows:

$$Q_{\ell, d}[\varphi(\mathbf{x})] = \sum_{\ell-d+1 \leq \|\mathbf{l}\|_1 \leq \ell} (-1)^{\ell - \|\mathbf{l}\|_1} \binom{d-1}{\ell - \|\mathbf{l}\|_1} \bigotimes_{i=1}^d Q_{l_i}[\varphi(\mathbf{x})], \quad (B.5)$$

where $\|\mathbf{l}\|_1 = \sum_{i=1}^{d} l_i$, and the multi-index $\mathbf{l} = (l_1, l_2, \ldots, l_d) \in \mathbb{Z}^d$ contains the accuracy level of the one-dimensional quadrature (B.4) for each dimension. The tensor product $\otimes$ in (B.5) can be calculated as

$$\bigotimes_{i=1}^{d} Q_{l_i}[\varphi(\mathbf{x})] = \sum_{j_1=1}^{m_{l_1}} \cdots \sum_{j_d=1}^{m_{l_d}} \varphi\left(x_{j_1}^{l_1}, \ldots, x_{j_d}^{l_d}\right) \prod_{i=1}^{d} w_{j_i}^{l_i}, \quad (\text{B.6})$$

where the tensor product of quadrature weights $w_{j_i}^{l_i}$ is replaced with the ordinary product, since they are real numbers. As one can see from the structure of (B.5) and (B.6), quadrature $Q_{\ell,d}[\varphi(\mathbf{x})]$ is a linear functional that depends on φ through function values at a finite set of points. This set of points is called a "sparse grid" and is denoted by $\mathcal{H}_{\ell,d}$. A sparse grid is defined as the union

$$\mathcal{H}_{\ell,d} = \bigcup_{\ell-d+1 \leq \|\mathbf{l}\|_1 \leq \ell} \left(\mathcal{H}_{l_1} \times \cdots \times \mathcal{H}_{l_d}\right). \quad (\text{B.7})$$

For nested one-dimensional sets ($\mathcal{H}_{l_i} \subset \mathcal{H}_{l_i+1}$), the corresponding sparse grids are also nested $\mathcal{H}_{\ell,d} \subset \mathcal{H}_{\ell+1,d}$ and can be simplified, yielding

$$\mathcal{H}_{\ell,d} = \bigcup_{\|\mathbf{l}\|_1 = \ell} \left(\mathcal{H}_{l_1} \times \cdots \times \mathcal{H}_{l_d}\right). \quad (\text{B.8})$$

The integral (B.2) can now be approximated by the sum:

$$Q_{\ell,d}[\varphi(\mathbf{x})] = \sum_{\mathbf{x}_{\mathbf{j}}^{\mathbf{l}} \in \mathcal{H}_{\ell,d}} w_{\mathbf{j}}^{\mathbf{l}} \varphi\left(\mathbf{x}_{\mathbf{j}}^{\mathbf{l}}\right), \quad (\text{B.9})$$

where multidimensional knots $\mathbf{x}_{\mathbf{j}}^{\mathbf{l}} = (x_{j_1}^{l_1}, x_{j_2}^{l_2}, \ldots, x_{j_d}^{l_d})$ can be constructed based on (B.7) and (B.8). The formulae for the quadrature weights $w_{\mathbf{j}}^{\mathbf{l}}$ in (B.9) can be obtained in an analytical form only in a few particular cases; in all the other cases weights can be either precalculated or calculated online using (B.5) and (B.6).

Acknowledgments

The authors would like to thank the South African National Research Foundation (NRF) for their financial support and Professor Kostadin N. Ivanov from Pennsylvania State University for his valuable suggestions.

References

[1] A. Saltelli, "Sensitivity analysis for importance assessment," *Risk Analysis*, vol. 22, no. 3, pp. 579–590, 2002.

[2] I. M. Sobol', "Sensitivity estimates for nonlinear mathematical models," *Mathematical Modelling and Computational Experiment*, vol. 1, pp. 407–414, 1993.

[3] I. M. Sobol, "Global sensitivity indices for nonlinear mathematical models and their Monte Carlo estimates," *Mathematics and Computers in Simulation*, vol. 55, no. 1–3, pp. 271–280, 2001.

[4] N. Lüdtke, S. Panzeri, M. Brown et al., "Information-theoretic sensitivity analysis: a general method for credit assignment in complex networks," *Journal of the Royal Society Interface*, vol. 5, no. 19, pp. 223–235, 2008.

[5] B. Auder and B. Iooss, "Global sensitivity analysis based on entropy," in *Safety, Reliability and Risk Analysis: Theory, Methods and Applications*, S. Martorell, C. G. Soares, and J. Barnett, Eds., pp. 2107–2115, Taylor & Francis Group, London, UK, 2009.

[6] U. Reuter and M. Liebscher, "Global sensitivity analysis in view of nonlinear structural behaviour," in *German Conference Organized by DYNAmore GmbH*, Anwenderforum, Bamberg, SC, USA, October 2008.

[7] B. Sudret, "Global sensitivity analysis using polynomial chaos expansions," *Reliability Engineering and System Safety*, vol. 93, no. 7, pp. 964–979, 2008.

[8] K. Ivanov, M. Avramova, I. Kodeli, and E. Sartori, "Benchmark for uncertainty analysis in modeling (UAM) for design, operation and safety analysis of LWRs," NEA/NSC/DOC 23, OECD Nuclear Energy Agency, 2007.

[9] R. A. Johnson and D. W. Wichern, *Applied Multivariate Statistical Analysis*, Pearson Prentice Hall, Upper Saddle River, NJ, USA, 6th edition, 2007.

[10] A. Saltelli and I. M. Sobol', "Sensitivity analysis for nonlinear mathematical models: numerical experience," *Matematicheskoe Modelirovanie*, vol. 7, no. 11, pp. 16–28, 1995.

[11] J. Jacques, C. Lavergne, and N. Devictor, "Sensitivity analysis in presence of model uncertainty and correlated inputs," *Reliability Engineering and System Safety*, vol. 91, no. 10-11, pp. 1126–1134, 2006.

[12] S. Kucherenko and N. Shah, "The importance of being Global. Application of Global sensitivity analysis in Monte Carlo option pricing," *Wilmott Magazine*, vol. 4, pp. 2–10, 2007.

[13] A. B. Owen, "Monte Carlo extension of Quasi-Monte Carlo," in *Proceedings of the 30th Conference on Winter Simulation*, pp. 571–577, Washington, DC, USA, December 1998.

[14] S. Smolyak, "Quadrature and interpolation formulas for tensor products of certain classes of functions," *Doklady Akademii Nauk SSSR*, vol. 4, pp. 240–243, 1963.

[15] E. Novak and K. Ritter, "High dimensional integration of smooth functions over cubes," *Numerische Mathematik*, vol. 75, no. 1, pp. 79–97, 1996.

[16] E. Novak, "Simple cubature formulas with high polynomial exactness," *Constructive Approximation*, vol. 15, no. 4, pp. 499–522, 1999.

[17] G. E. P. Box and M. E. Muller, "A note on the generation of random normal deviates," *The Annals of Mathematical Statistics*, vol. 29, no. 2, pp. 610–611, 1958.

[18] I. Krykova, *Evaluating of path-dependent securities with low discrepancy methods [M.S. thesis]*, Worcester Polytechnic Institute, 2003.

[19] B. Moro, "The Full Monte," *Risk*, vol. 8, no. 2, pp. 57–58, 1995.

[20] B. T. Rearden, M. L. Williams, M. A. Jessee, D. E. Mueller, and D. A. Wiarda, "Sensitivity and uncertainty analysis capabilities and data in scale," *Nuclear Technology*, vol. 174, no. 2, pp. 236–288, 2011.

[21] M. Williams, M. Jessee, R. Ellis, and B. Rearden, "Sensitivity/uncertainty analysis for OECD UAM benchmark of peach bottom BWR," in *Uncertainty Analysis and Modelling (UAM-4) Workshop*, Pisa, Italy, April 2010.

[22] W. M. Stacey, *Nuclear Reactor Physics*, John Wiley & Sons, New York, NY, USA, 2001.

[23] J. Burkardt, Source Codes in Fortran90, 2010, http://people.sc.fsu.edu/~jburkardt/f_src/f_src.html.

[24] P. M. Bokov, "Asymptotic analysis for the variance-based global sensitivity indices," *Science and Technology of Nuclear Installations*, vol. 2012, Article ID 253045, 8 pages, 2012.

3

Analysis of Subchannel and Rod Bundle PSBT Experiments with CATHARE 3

M. Valette

CEA Grenoble, Commissariat à l'Énergie Atomique et aux Énergies Alternatives, DEN, DM2S/SMTH, 38054 Grenoble, France

Correspondence should be addressed to M. Valette, michel.valette@cea.fr

Academic Editor: Maria Avramova

This paper presents the assessment of CATHARE 3 against PWR subchannel and rod bundle tests of the PSBT benchmark. Noticeable measurements were the following: void fraction in single subchannel and rod bundle, multiple liquid temperatures at subchannel exit in rod bundle, and DNB power and location in rod bundle. All these results were obtained both in steady and transient conditions. Void fraction values are satisfactory predicted by CATHARE 3 in single subchannels with the pipe module. More dispersed predictions of void values are obtained in rod bundles with the CATHARE 3 3D module at subchannel scale. Single-phase liquid mixing tests and DNB tests in rod bundle are also analyzed. After calibrating the mixing in liquid single phase with specific tests, DNB tests using void mixing give mitigated results, perhaps linked to inappropriate use of CHF lookup tables in such rod bundles with many spacers.

1. Introduction

CATHARE 3 is a new two-phase thermalhydraulics system code developed at CEA Grenoble [1]. It has been designed to expand the capabilities of CATHARE 2 and to improve the simulation accuracy of light water reactor accidents. New features include additional field, like a droplet field or a bubble field, and coupled equations of turbulence transport for a continuous field or interfacial area transport for a dispersed field. Beside the unchanged choices for numerical schemes for time and space discretization, a numerical solver gathering the different modules of a circuit has been rewritten and improved compared to CATHARE 2 in order to allow new capabilities of coupling with external codes, for example, for neutronics, or detailed CFD. A preliminary version V1 needs a wide validation program. This paper deals with the 1D and 3D module validation of the code against various boiling experiments at subchannel scale.

Following the BWR Full-size Fine-Mesh Bundle tests (BFBT) benchmark, the PWR subchannel and bundle tests (PSBT) benchmark [2] is proposed by OECD/NRC. Both are based upon a NUPEC database obtained in full-scale subchannels and rod bundles and include detailed measurements of fluid temperature, void fraction, and critical power or DNB power in steady and transient conditions. These experiments are useful to check and validate the code closure laws in rod bundles, especially the turbulence dispersion coefficients for heat in single-phase flow and void in two-phase flow, the wall and interfacial friction coefficients, and the wall-to-fluid heat transfer models. The PSBT phase I exercises are devoted to the void fraction measurements, performed in single subchannels and in 5×5 rod bundles in steady and transient conditions. In the first exercise, phase II features liquid temperature measurements in all subchannels of a heterogeneously heated rod bundle in steady conditions, and, in the following exercises, DNB measurements taken in various rod bundles in steady and transient conditions, that is, power value and location of first detected DNB.

Single subchannel experiments are simulated by the CATHARE 3 pipe 1D module while rod bundle cases are simulated with the CATHARE 3 3D module meshed at a subchannel scale, that is, one cell per subchannel in a horizontal cross cut. The 3D module for the rod bundle has been coupled with a 1D module in order to improve the inlet flow simulation along the downcomer.

Useful balance equations and closure laws are briefly presented in the following Section 2. Then, results of comparisons between simulations and measurements of void fraction for Phase I exercises are presented in Section 3. In Section 4 are presented the results of temperature and DNB simulations of Phase II exercises.

2. CATHARE 3 Balance Equations and Closure Laws

Both 1D and 3D modules of CATHARE 3 solve the same set of balance equations, except that the energy balances are written using enthalpy in the 3D module and internal energy in the pipe module. The closure laws remain identical as far as possible. A first-order donor cell scheme is used in both modules as far as space discretization is concerned. For time discretization, the pipe module calls a fully implicit scheme, while the 3D module uses a semi-implicit scheme.

Contrary to the preceding BFBT simulations [3], featuring high void two phase flow, most of the PSBT benchmark database remain in the low- or medium-void range and hence, simulations do not need an additional droplet field beside the standard 6-equation model because the boiling flow regime never changes towards an annular dispersed flow.

For a given generation of steam along a single heated channel, the local void fraction is governed by wall and interfacial friction. In a 3D flow inside a rod bundle, cross-flows between adjacent subchannels lead to void dispersion. Also turbulent dispersion or diffusion may affect the temperature map in the single phase region. The void dispersion phenomena can be modelled by a mixing term in the momentum balance equations. The temperature dispersion (caused by nonrandom flow from one subchannel to a neighbour) and diffusion (caused by random fluctuations of flow between adjacent subchannels) are modelled by a single term in the liquid energy balance equation. The velocity diffusion is presently neglected in the momentum equation; its implementation had no effect on results of several tests, either in single phase or two phase flow in bundles.

2.1. Momentum Balance. Consider the following equation:

$$\alpha_k \rho_k \left[\frac{\partial}{\partial t} V_k + V_k \cdot \nabla V_k \right]$$
$$= -\alpha_k \nabla P + p_i \nabla \alpha_k + (-1)^k \tau_i + \tau_{pk} + \alpha_k \rho_k g. \quad (1)$$

Wall frictions τ_p of both phases are calculated using the Blasius friction coefficient f_k multiplied by a phase-dependent multiplier c_k

$$\tau_{pk} = \chi c_k f_k \rho_k \frac{|V_k| V_k}{2},$$
$$\text{with } c_g = \alpha^{1.25}, \quad c_l = \frac{(1-\alpha)\rho_l}{(1-\alpha)\rho_l + \alpha\rho_g}. \quad (2)$$

Interfacial friction in bubbly, slug, and churn vertical flow is given by

$$\tau_i = \frac{K_l \rho_l + K_g \rho_g}{L} \alpha (1-\alpha)^{3.6} \left[V_g - V_l \right]^2. \quad (3)$$

L is the maximum bubble size, limited by the Laplace length ℓ and the hydraulic diameter $K_g = 29$

$$K_l = \left(F_\mu \right)^{0.25} f_l$$

$$\text{with } F_\mu = \frac{\mu_l}{\sqrt{\rho_l \sigma \ell}}, \quad f_l = 2.81 + 34 \left(\frac{L}{D_h} \right)^5 \left(6 - \frac{5L}{D_h} \right). \quad (4)$$

τ_i and τ_p are unchanged compared to CATHARE 2 6-equation model.

The mixing term p_i is calculated from an assessment of the turbulent kinetic energy.

For a single-phase flow in a tube or a subchannel (far from a spacer grid), the turbulent kinetic energy k_l can be assessed by

$$k_l = 0.0367 V_l^2 \, \text{Re}^{-1/6} \quad (5)$$

(see [4]). The associated turbulent viscosity can be assessed as

$$\nu_t = 0.5 D_H \sqrt{k_l} \quad (6)$$

and the dispersion term p_i:

$$p_i = 0.4 \, \mu_t \frac{V_l}{D_H}. \quad (7)$$

The coefficient 0.4 comes from an order of magnitude for the velocity gradients between subchannels; the velocity difference is evaluated at 40% of the axial velocity, which is close to the velocity module.

At the end, it comes

$$p_i = 0.038 \rho_l V_l^2 \text{Re}^{-1/12}. \quad (8)$$

The coefficient 0.5 in the ν_t formula has been adjusted so as to better match the void fraction measurements in the PSBT Phase I tests, and the temperature measurements in the Phase II, given a Pr_t equal to 1 (see (9) below). This coefficient appears to be several orders of magnitude above the figure calculated using simple turbulence [4, 5]; the void dispersion due to cross flows (and not only diffusion) seems to be the main driving phenomenon (see [6]).

2.2. Continuous Liquid Energy Balance. It is written using internal energy e_l for the 3D module as follows:

$$\frac{\partial}{\partial t} (\alpha_l \rho_l e_l) + \nabla \cdot (\alpha_l \rho_l e_l V_l)$$
$$= q_{li} + \chi_c q_{pl} - \Gamma H_{lc} - P \left[\frac{\partial \alpha_l}{\partial t} + \nabla \cdot (\alpha_l V_l) \right] \quad (9)$$
$$+ \nabla \cdot \left[\alpha_k \left(\lambda_l T_l + \frac{\rho_l \nu_{tl}}{\text{Pr}_t} \nabla e_l \right) \right].$$

The molecular diffusion is neglected compared to the turbulent diffusion term.

To summarize, the turbulence is modelled here by two different algebraic terms: one described just above in the energy balance and the p_i grad α in the momentum equations. No additional transport equation of turbulence quantity was solved in this study.

The Departure from Nucleate Boiling appears on a hot wall when the heat flux towards the fluid exceeds the so-called "Critical Heat Flux," which is assessed in six-equation model of CATHARE 2 and CATHARE 3 using home-made polynomials interpolating CHF lookup tables, given the local values of mass flux, pressure, and steam quality. The tables are based on the 1995 Groeneveld tables [7], using a simple rod bundle coefficient but no effect of the spacer grids.

3. Overview of the PSBT Benchmark

The PWR Subchannel and Bundle Tests (PSBT) benchmark is proposed by OECD/NRC. Pennsylvania State University (PSU) under the sponsorship of U.S. Nuclear Regulatory Commission (NRC) prepared the specification and organized the benchmark with the Japan Nuclear Energy Safety (JNES) Organization. The Nuclear Power Engineering Corporation (NUPEC) released a database including various single subchannel and full-scale rod bundle tests in boiling conditions, with detailed void distribution and DNB measurements. Both system codes and CFD codes can match their results against the averaged (macroscopic data at subchannel scale) or fine experimental results.

Two phases are proposed:

The first one is devoted to the void distribution and includes four exercises.

Exercise 1—steady-state single subchannel benchmark,
Exercise 2—steady-state bundle benchmark,
Exercise 3—transient bundle benchmark,
Exercise 4—pressure drop benchmark.

The second is devoted to DNB prediction and includes three exercises.

Exercise 1—steady-state fluid temperature benchmark,
Exercise 2—steady-state DNB benchmark,
Exercise 3—transient DNB benchmark.

This benchmark, especially through its accurate measurements, is a very good opportunity to assess the capabilities of system codes such as CATHARE 3 to simulate boiling flows in PWR core geometry.

4. Void Fraction in PSBT Phase I Exercises

4.1. Exercise I.1 Steady State in Single Subchannels. Four different geometries of single subchannels, for central, side, and corner locations (side and corner are relative to a rod

bundle in a square box) have been tested, resulting in void fraction measurements at 1400 mm level inside a 1555 mm long heated subchannel (Figure 1). They fit to standard PWR rod bundle subchannel geometry (except the heated length shorter than a real reactor core), with an additional test section corresponding to a central subchannel heated by 3 and not 4 contributing rods, one rod being replaced by a thimble. The axial power distribution is uniform. The CT scanner gives for every steady run a detailed void fraction array through the measuring section. Results can be compared with CFD simulations or averaged over the cross section for comparisons with 1D module simulation by system codes.

A set of 39 tests is proposed in the benchmark among a large database of 126 tests. The range of flow pressure is from 50 to 170 Bars and the range of mass flux is 500 to 4200 kg m^{-2} s^{-1}. We calculated the whole database and compared the simulation results versus the measurement data.

Calculations were performed using a quasiuniform 31 cell meshing.

The void fraction has been measured by a γ-ray attenuation CT scanner measurement in 1 mm wide beams, giving an array of values throughout the cross section. These values can be integrated in the whole cross section, giving a single void fraction figure associated to an experimental error bar of +/−3%.

The results are gathered in Figure 2. One can see a good coherence, but yet a slight bias, negative in any series, and a medium dispersion of the predicted values. Some rare points are outside the range +/−10%. The result statistics is presented in Table 1, given in % of void for the difference "predicted minus measured fraction."

Some examples of void axial profiles simulated by CATHARE 3 in various flow conditions in the standard central subchannel are shown on Figure 3, as well as the measured value at 1.4 m elevation.

4.2. Exercise I.2 Steady State in Rod Bundles. Several types of rod bundles were tested, most of them including a 5 × 5 matrix of rods settled with 17 spacer grids of 3 different types, with uniform or cosine power profile and with or without a central thimble instead of a heated rod. The heated length was 3658 mm; the rod diameter and pitch were 9.5 and 12.6 mm (Figure 4).

The heated part of the rod bundle was modelled by a 3D grid, with 6 × 6 cells in the *x-y* directions at the subchannel scale, and 66 cells in the *z* direction, taking one axial short cell for every 17 spacer grid, with 3 axial cells between two adjacent spacers. The 3 different kinds of spacer grids (Mixing Vane spacer, NonMixing Vane spacer, simple spacer) are described by their porosities, hydraulic diameter, and pressure loss coefficient in every subchannel.

An array of void fraction values in the different subchannels was measured at 3 different levels along the upper part of the heated length, reconstructed by 6 chordal averaged values in *x* and 6 in *y* directions. Here on Figure 5 is shown an example of the sensitivity of the void fraction distribution to the p_i grad α term.

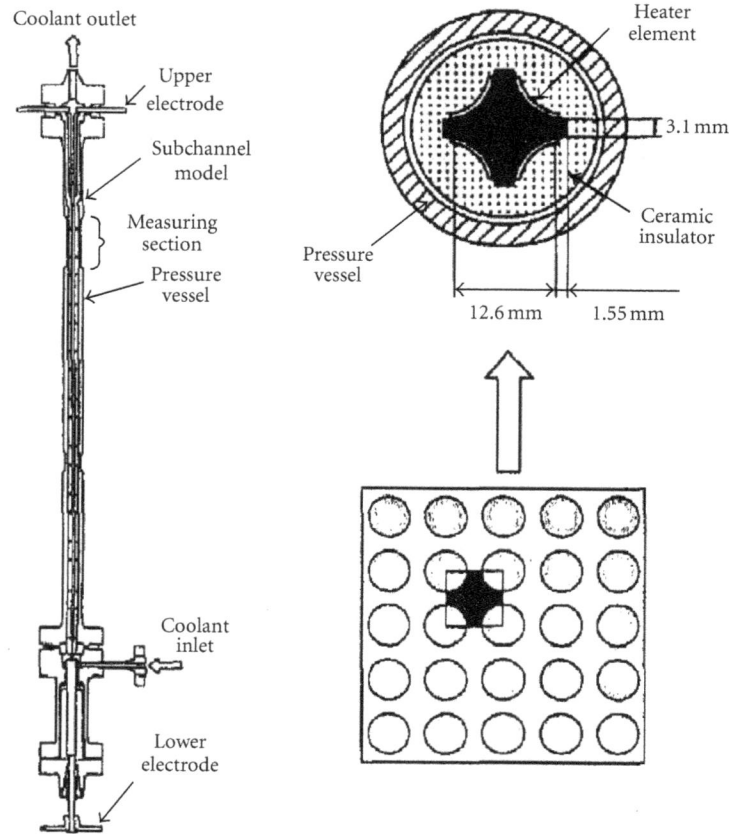

FIGURE 1: Test section for void measurement in a central subchannel.

TABLE 1: Distribution of deviations: "calculated minus measured void fraction" in the different series of single subchannel PSBT tests.

	Test number	Average	Standard dev.
Series 1: standard central subchannel	43	−2.3%	4.8%
Series 2: central subchannel close to a thimble	43	−1.8%	5.3%
Series 3: lateral subchannel	20	−3.0%	6.0%
Series 4: corner subchannel	20	−5.4%	3.2%
All series	126	−2.7%	5.0%

Only the averaged value of the 4 central subchannel void fractions was available in the benchmark database, and this is the compared data versus the void calculated by CATHARE 3 hereafter on Figure 6 for all the tests of the benchmark.

The points are more dispersed than for the single subchannel tests. The statistics of the results (difference: computed minus measured void fraction given in absolute %) is presented in Table 2.

The series number correspond to 3 different bundles; the series 8 is tested with the same bundle as series 5 as repeated cases, which appear to be less satisfying.

4.3. Exercise I.3 Transient in Rod Bundles for Void Fraction Prediction. A set of 12 transient tests is proposed in the benchmark, including power increase (PI), flow reduction, temperature increase (TI), and depressurization in each of the 3 same tested bundles as in the steady tests of the exercise 2. The void fraction was also measured at the same 3 elevations during the transient.

The flow parameters (pressure, flow rate, and inlet temperature) were measured outside of the main vessel, near the inlet nozzle (referred as Coolant Inlet on Figure 7) and should be imposed as boundary conditions in the simulation. As the inlet conditions remain in quasi-incompressible liquid subcooled conditions, pressure and flow values remain more or less uniform all along the inlet pipes and devices, and the parameter variations are not delayed between the inlet nozzle and the bottom of the heated length in the bundle.

Hence, except for the "Temperature Increase" transients, as the inlet temperature remains quasiconstant during the transient, the same computational domain was considered as in the previous steady tests, that is, the heated length in the rod bundle only and the requested inlet conditions were applied at the bottom of the heated length.

TABLE 2: PSBT rod bundle comparison statistics for void fraction tests.

Series	5	6	7	8	All gathered
Power profile	Uniform	Cosine	Cosine	Uniform	
Central thimble	No	No	Yes	No	
Average	−0.22%	−2.39%	1.13%	−6.65%	−1.96%
Standard deviation	4.27%	5.43%	5.64%	6.73%	6.32%

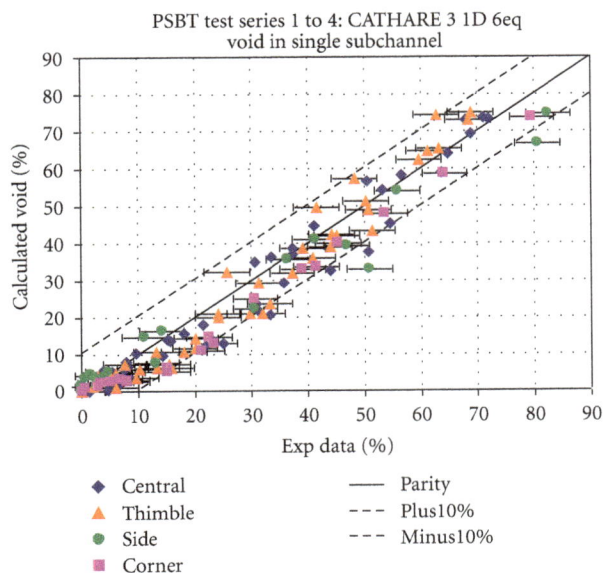

FIGURE 2: PSBT single subchannel test comparison (thimble means a central subchannel close to an unheated thimble).

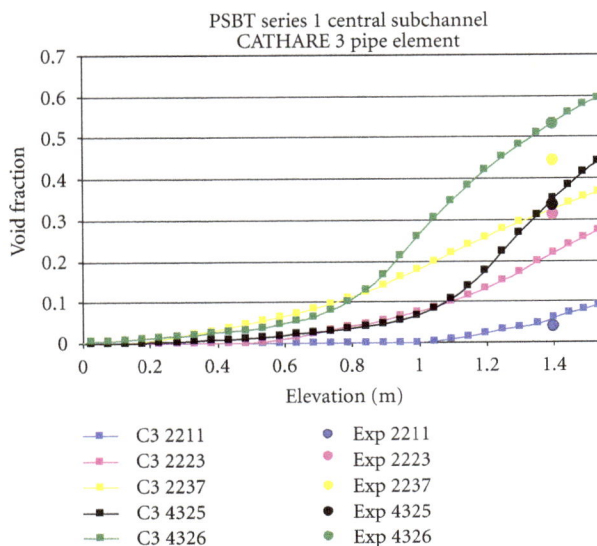

FIGURE 3: PSBT single subchannel test comparison: axial void profiles of 5 tests in a standard central subchannel.

An example of comparison is given below (Figure 8) for the test 7TPI, in the bundle B7 (with a central thimble and a cosine power profile) and given a linear increase of power, keeping constant the flow rate, pressure, and inlet temperature. The void fraction is slightly underpredicted at the upper location but is satisfactory in front of the 2 lower void measurement elevations. The other tests show less satisfactory results, the upper and medium level void often remaining underpredicted. This behaviour is consistent with the conclusions of the preceding exercise for steady tests, where the higher void fractions are underpredicted while the lower void fractions are more satisfactory.

For this 7TPI test, the location of the inlet temperature measurement is not sensitive because the temperature remains more or less constant during the transient. However, for Temperature Increase tests, this location must be at the boundary of the computational domain. Otherwise, a temperature delay would induce a bias in the simulation. Hence, for these TI transients, another domain has been set up, adding the downcomer as an axial module upstream the 3D module simulating the whole rod bundle.

An example of simulation is presented on Figure 9.

The comparison is less satisfying than for the other test 7TPI. All maximum void fractions are underpredicted and while the time of void take off is well predicted, the time of maximum void is delayed and the void curves seem to be

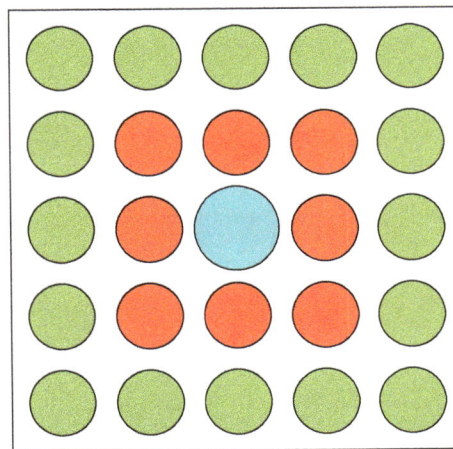

FIGURE 4: Power distribution in the bundle B7: red: 100%, green: 85%, blue: unheated thimble.

widened. This seems to be the consequence of a too large axial diffusion of void and perhaps also of the temperature step in the inlet part of the domain.

4.4. Exercise I.4 Pressure Drop in Rod Bundle. No data is available for code-to-data comparison for this exercise, except a single value given at the beginning of one transient test

PSBT test 51121; no P_i term
void fraction at 3, 156 m

PSBT test 51121
void fraction at 3, 156 m

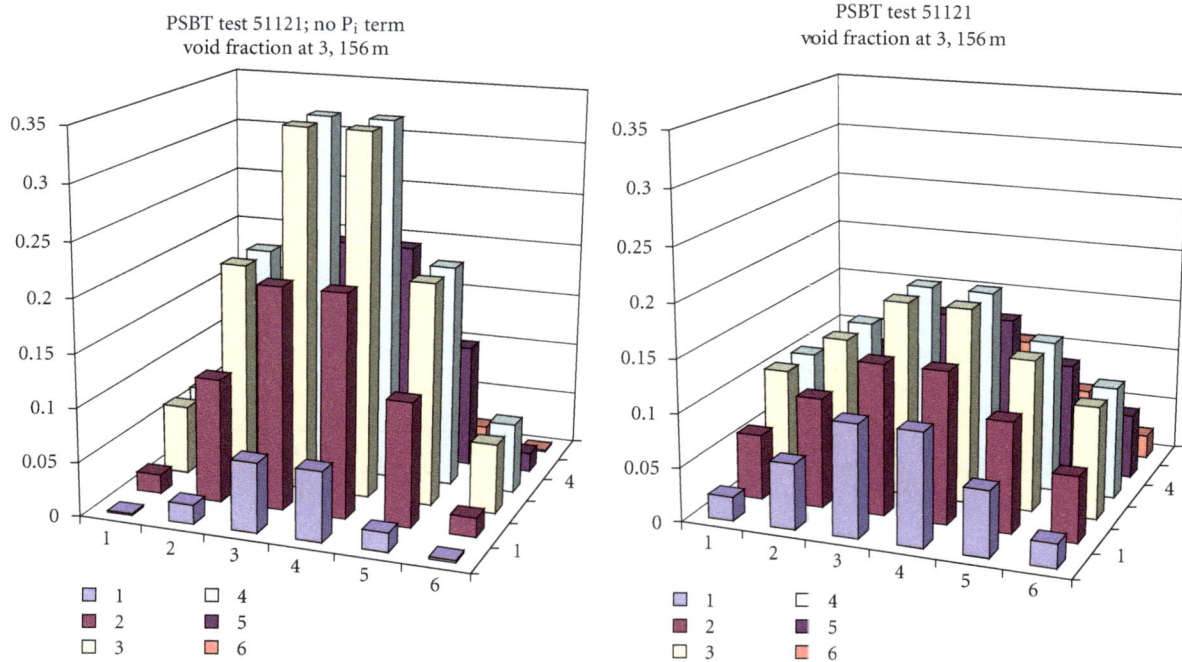

FIGURE 5: Sensitivity to the void diffusion term.

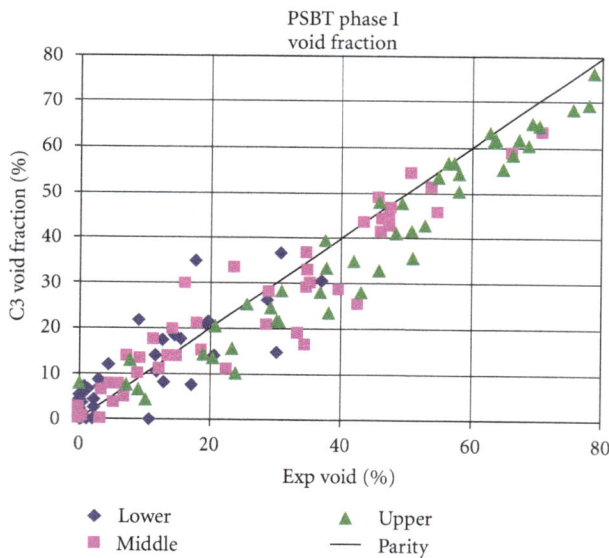

PSBT phase I
void fraction

FIGURE 6: PSBT rod bundle test comparison of predicted and measured void fraction in central region at 3 different elevations along the heated length; lower: 2216 mm; medium: 2669 mm; upper: 3177 mm.

(7TPI) in nonboiling steady-state condition. The benchmark specifications recommended the pressure loss coefficients to be used for every type of spacer grid. Using these values, CATHARE 3 predicted the overall bundle pressure drop at 1.85 kg cm^{-2}, while the measured pressure drop is 1.6 kg cm^{-2}. This can be considered as a satisfying bias, considering the constant pressure loss coefficient with no dependency on the flow Reynolds number.

5. Departure from Nucleate Boiling in PSBT Phase II Exercises

5.1. Exercise II.1 Steady-State Fluid Temperature in Rod Bundles. This exercise is particularly useful to assess the code capabilities for turbulent dispersion and diffusion in single-phase flow.

In a 5 × 5 rod bundle featuring a heterogeneous power distribution (Figure 10), a set of 36 thermocouples measure fluid temperature in every subchannel 50 cm above the top of the heated length.

Nine tests at high pressure (from 50 to 170 bars) are proposed for simulations in a wide range of mass fluxes (between 500 and 4700 kg/m^{2}s). In Figure 11 are presented the compared W'E profiles of temperatures, averaged in the N/S direction. The x-axis numbers correspond to the subchannel columns 1 to 6. The shown temperature gradient is due to the power distribution and is governed by the diffusion and dispersion across the subchannels. The actual profile is more complex due to mixing vanes, which tend to swirl the flow but this effect is not modeled by CATHARE 3.

A first step of analysis allowed us to calibrate the turbulent viscosity used in the liquid energy balance (also in the void mixing term which is not useful in this exercise). Figure 11 shows the sensitivity of the turbulent viscosity value.

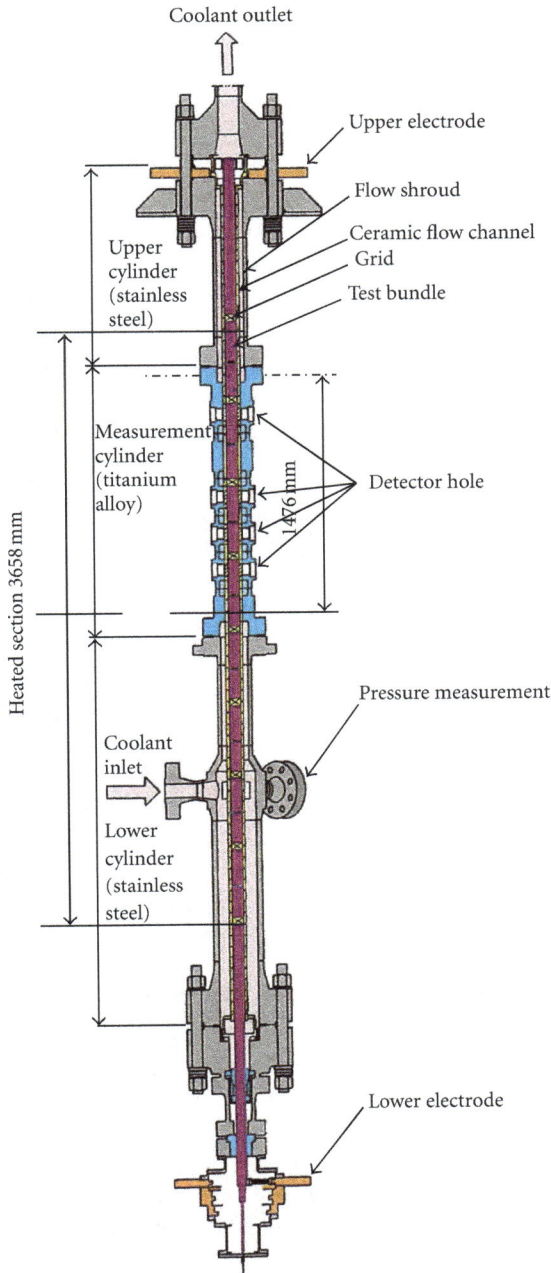

FIGURE 7: PSBT test vessel and flow channel structures.

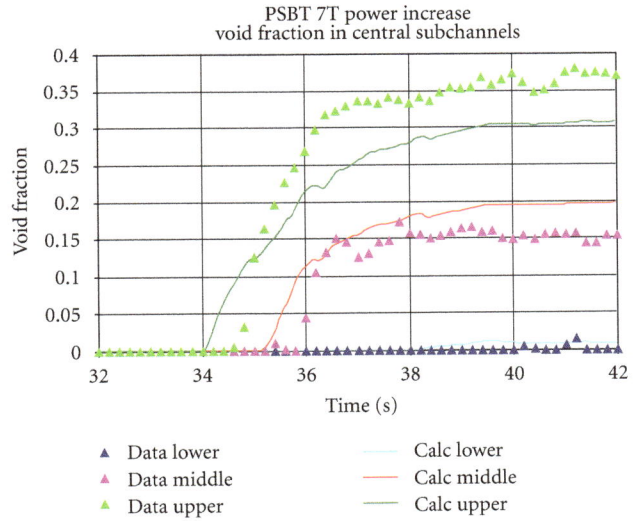

FIGURE 8: Comparison of void fraction transient predicted by CATHARE 3 and measured in PSBT 7TPI test.

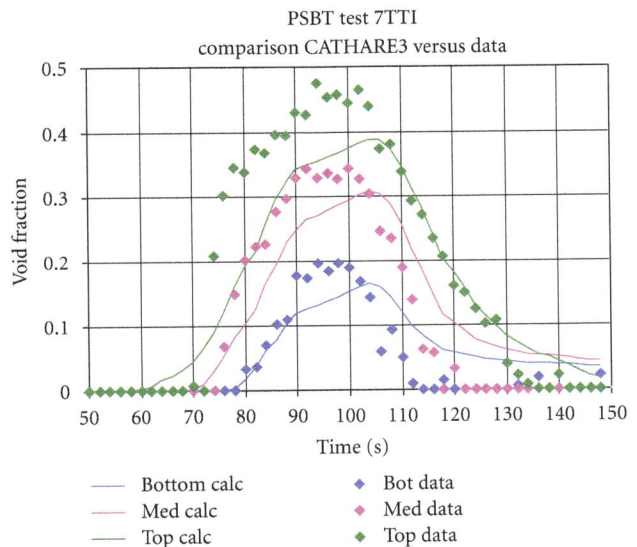

FIGURE 9: Comparison of void fraction transient predicted by CATHARE 3 and measured in PSBT 7TTI test.

The profiles of half of the proposed tests are correctly predicted. The other tests show unbalanced measured temperatures at the outlet compared to inlet flow parameters and bundle power and hence, comparisons are not significant.

In the Table 3, one can see that the parameters of the two correct tests presented on Figure 12 are very close except the inlet temperature (which obviously may have a slight effect on the Reynolds number). However, the temperature profiles show unexplained different behaviours, which are not predicted by CATHARE 3.

Using the available correct tests, a satisfying value of the temperature dispersion parameter has been selected, and

implemented in CATHARE 3 for the other void and DNB simulations.

5.2. Exercise II.2 Steady State DNB in Rod Bundles. As for the void fraction tests proposed in phase I, different bundles were tested. The DNB is detected both in experiments and simulation by a significant rise of the wall temperature (more than 11°C) when the bundle power is slowly increased.

We calculated 6 test series in 5 different bundles corresponding to several geometries and power profiles, as shown in Table 4. The DNB location is given only in the A4 and A8 bundles. All calculated tests of bundles 4 and 8 were run at the same pressure of 150 bars, while the range in series 0,2,3 spreads from 50 to 170 bars.

TABLE 3: Test parameters for temperature measurements.

Test number	Pressure (kg/cm^{2}a)	Mass flux (10^6 kg/m^2 hr)	Inlet temperature (°C)	Power (MW)
01-6232	169.1	2.10	251.5	0.42
01-5252	150.0	1.95	113.9	0.41

TABLE 4: Results of 6 series of DNB simulations in rod bundles.

Bundle	Rods	Spacers	Radial power	Axial power	Calculated tests	Predicted power	Std dev	Small flow rate tests
A0	5×5	**13**	A	**Uniform**	9	96.80%	6.24%	0
A2	5×5	17	A	**Uniform**	11	87.70%	17.04%	2, overpredicted
A3	**6×6**	17	D	**Uniform**	8	79.46%	3.90%	0
A4	5×5	17	A	Cosine	20	78.44%	3.07%	2, over 5 std dev
A4	5×5	17	A	Cosine	27	78.50%	3.72%	0
A8	5×5	17	**B (thimble)**	Cosine	24	81.79%	8.70%	2, overpredicted

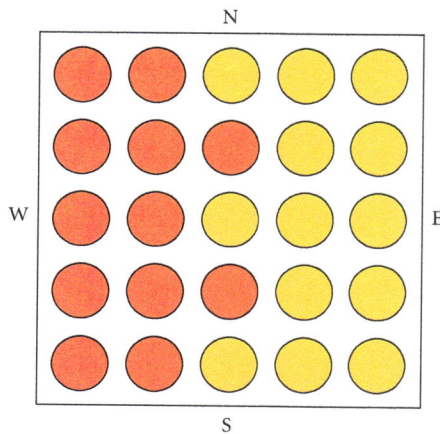

FIGURE 10: Rod power distribution in fluid temperature measurement tests; red rod power 100%, yellow 25%.

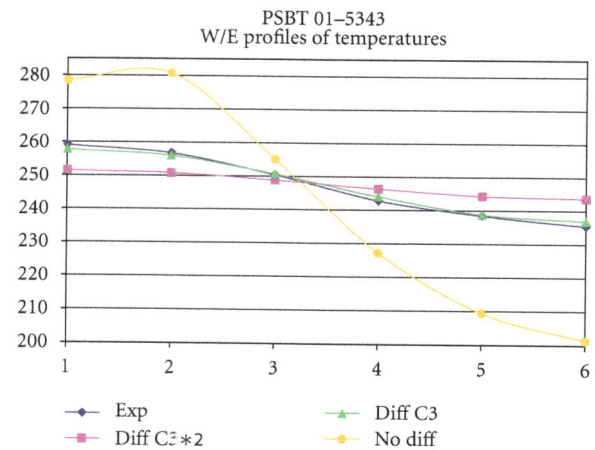

FIGURE 11: Sensitivity of the temperature profile to the turbulent diffusion term.

Some statistics of the simulation results (relative power: computed over data in % of experimental data) are given in Table 4.

One can see that the result for bundle A0 is satisfying results while the others show a significant bias.

The A3 bundle, featuring a 6×6 rod array, is not better than the A2 bundle, while the side effects are weaker.

The results in the bundle A2 are weakened by two tests at very low flow rate (330 kg m^{-2} s^{-1}), which are overpredicted contrary to the 9 other tests; this enlightens the large value of the standard deviation for this bundle. A similar behaviour exists in series 4 and series 8 where the 2 tests at very low flow rate show significant differences (larger DNB power) compared to the other tests. Generally speaking, the location of the first detected DNB matches better in the series 4 than in the series 8.

The general underprediction of the DNB power in rod bundles may be linked to the use of lookup tables in a 3D analysis; such tables can predict CHF or DNB given 3 parameters: mass flux, pressure, and steam quality. These tables were built using 1D analysis of numerous tests. But in a 3D analysis, the steam quality and the mass flux must obey a local definition with local void fraction and velocities and may display wrong values. As a consequence, the code computation of the local CHF may deviate from the recommended value. Better results can be expected when this point is improved.

Moreover, the better results for the A0 bundle seem point out that the CHF is better predicted with 13 spacers than with 17. The number of 13 is closer to the usual number of spacers in industrial bundles and CHF experiments used to build up the lookup tables. So, the DNB predictions would depend strongly on the spacer number, either through the CHF calculated with the lookup tables, or through the mixing effects simulated in the 3D computation. From this point of view, CATHARE 3 is not able now to predict the DNB power at subchannel scale with accuracy better than 20% in a new bundle.

Given the underprediction of the high-void fraction noticed in the preceding sections, the predicted DNB power should have been overpredicted because the local CHF increases when the local void fraction decreases. This also shows that these lookup tables are not convenient for

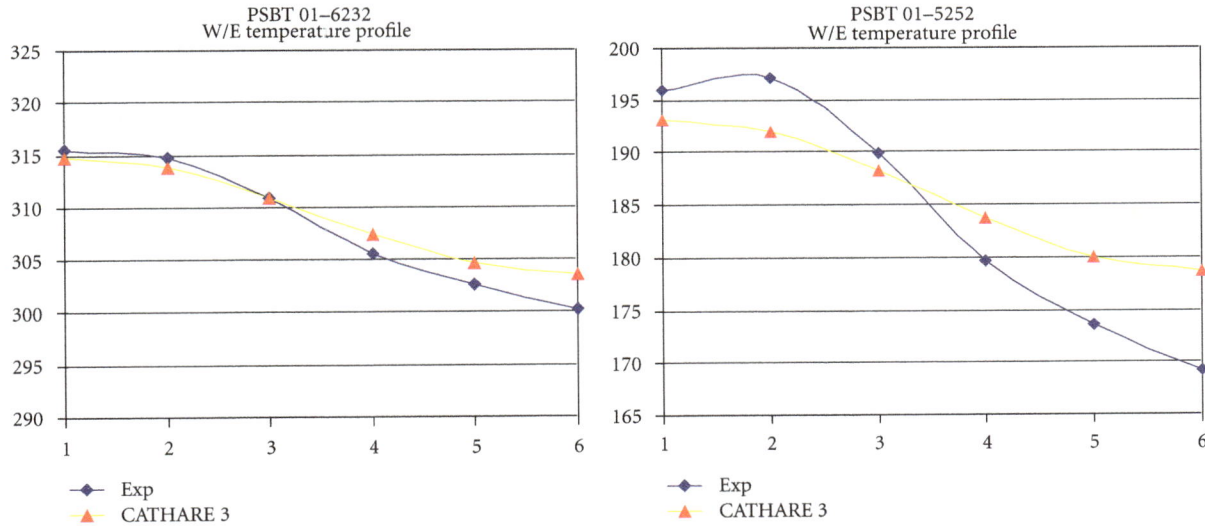

FIGURE 12: Temperature profiles in 2 fluid temperature measurement tests.

the analysis of the PSBT tests. It has to be noted that the original purpose of the 3D module of CATHARE2 and 3 was an improvement of the code behaviour in very large volumes within the reactor vessel such as the downcomer and the lower plenum. Applying this module in core subassemblies at subchannel scale is beyond the usual scope of the module, and the BFBT and PSBT benchmarks were just opportunities to check the module capabilities.

5.3. Exercise II.3 Transient DNB in Rod Bundles. Several tests were proposed in the benchmark specification report, in the rod bundles A4 and A8 (see Table 4 for details). As an example, the Temperature Increase transients prescribed a linear increase of the inlet temperature while the three other flow parameters (flow rate, pressure, and power) remained unchanged. The simulation results show the same behaviour as in the steady tests: the DNB power occurred at 86% of the experimental value. This behaviour is also seen in other transients: Flow Reduction, Power Increase, and Depressurization.

6. Conclusion

The 1D and 3D modules of the CATHARE 3 system code were used for the simulations of PSBT benchmark tests. Results of void fraction in phase I and temperature measurements and DNB power measurements in phase II have been compared to calculation results. The void comparisons show that our models of wall and interfacial friction, coupled with void dispersion, lead to satisfactory results, with a slight bias towards void underprediction, for both single subchannels and full rod bundles.

The exercise 1 of phase II, devoted to single-phase mixing and cross flows in liquid phase, shows good results as far as the experimental heat balance of the tests remains satisfactory.

In the exercise 2 of benchmark phase II, steady DNB simulations in 5 different rod bundles show significant underprediction of the critical power in the whole bundle (20% bias). The main reason should be a poor local CHF assessment, more than a rough mixing model. The analysis of transient tests for exercise 3 confirmed this behaviour. The CHF lookup tables in CATHARE 3 underpredict the CHF values in the PSBT rod bundles, which have 17 spacer grids. Improving these features would require improving the modelling of the interaction between spacer grids, turbulence, and local CHF or using CHF correlations designed for specific bundles.

Nomenclature

Roman Letters

c: Wall friction phase multiplier
e: Internal energy
f: Wall friction coefficient
g: Gravity
H: Enthalpy
k: Turbulent kinetic energy
K: Interfacial friction coefficient
ℓ: Laplace length
P: Pressure
p_i: Void dispersion coefficient
Pr: Prandtl number
q: Heat flux
t: Time
T: Temperature
V: Velocity

Greek Letters

α: Phase volume fraction
Γ: Boiling/condensation rate
χ: Friction area (or heated area) over control volume

λ: Molecular heat conductivity
μ: Dynamic viscosity
ν: Kinematic viscosity
ρ: Phase density
σ: Surface tension
τ_i: Interfacial stress
τ_p: Wall stress

Indexes

i: Interface
k: Any phase
l: Liquid phase
p: Wall
t: Turbulent.

Acknowledgments

This work was made in the frame of the development of the NEPTUNE project, which is jointly developed by the Commissariat à l'Energie Atomique et aux énergies alternatives (CEA, France) and Electricité de France (EDF) and also supported by the Institut de Radioprotection et de Sûreté Nucléaire (IRSN, France) and AREVA-NP.

References

[1] P. Emonot, A. Souyri, J. L. Gandrille, and F. Barré, "CATHARE-3: a new system code for thermal-hydraulics in the context of the NEPTUNE project," *Nuclear Engineering and Design*, vol. 241, no. 11, pp. 4476–4481, 2011.

[2] A. Rubin, A. Schoedel, M. Avramova, H. Utsuno, S. Bajorek, and A. Velazquez-Lozada, "OECD/NRC benchmark based on NUPEC PWR subchannel and bundle tests (PSBT)," Volume I: Experimental Database and Final Problem Specifications NEA/NSC/DOC (2010)1, November 2010.

[3] M. Valette, "Analysis of boiling two phase flow in rod bundle for NUPEC BFBT benchmark with 3-field neptune system code," in *Proceedings of the 12th International Topical Meeting on Nuclear Reactor Thermal Hydraulics (NURETH '07)*, Pittsburgh, Pa, USA, September-October 2007.

[4] M. Chandesris, G. Serre, and P. Sagaut, "A macroscopic turbulence model for flow in porous media suited for channel, pipe and rod bundle flows," *International Journal of Heat and Mass Transfer*, vol. 49, no. 15-16, pp. 2739–2750, 2006.

[5] M. Chandesris and D. Jamet, "Derivation of jump conditions for the turbulence $k - \varepsilon$ model at a fluid/porous interface," *International Journal of Heat and Fluid Flow*, vol. 30, no. 2, pp. 306–318, 2009.

[6] M. Drouin, O. Grégoire, O. Simonin, and A. Chanoine, "Macroscopic modeling of thermal dispersion for turbulent flows in channels," *International Journal of Heat and Mass Transfer*, vol. 53, no. 9-10, pp. 2206–2217, 2010.

[7] D. C. Groeneveld, L. K. H. Leung, P. L. Kirillov et al., "The 1995 look-up table for critical heat flux in tubes," *Nuclear Engineering and Design*, vol. 163, no. 1-2, pp. 1–23, 1996.

4

Modeling Forced Flow Chemical Vapor Infiltration Fabrication of SiC-SiC Composites for Advanced Nuclear Reactors

Christian P. Deck, H. E. Khalifa, B. Sammuli, and C. A. Back

General Atomics, P.O. Box 85608, San Diego, CA 92186-5608, USA

Correspondence should be addressed to Christian P. Deck; christian.deck@ga.com

Academic Editor: Hangbok Choi

Silicon carbide fiber/silicon carbide matrix (SiC-SiC) composites exhibit remarkable material properties, including high temperature strength and stability under irradiation. These qualities have made SiC-SiC composites extremely desirable for use in advanced nuclear reactor concepts, where higher operating temperatures and longer lives require performance improvements over conventional metal alloys. However, fabrication efficiency advances need to be achieved. SiC composites are typically produced using chemical vapor infiltration (CVI), where gas phase precursors flow into the fiber preform and react to form a solid SiC matrix. Forced flow CVI utilizes a pressure gradient to more effectively transport reactants into the composite, reducing fabrication time. The fabrication parameters must be well understood to ensure that the resulting composite has a high density and good performance. To help optimize this process, a computer model was developed. This model simulates the transport of the SiC precursors, the deposition of SiC matrix on the fiber surfaces, and the effects of byproducts on the process. Critical process parameters, such as the temperature and reactant concentration, were simulated to identify infiltration conditions which maximize composite density while minimizing the fabrication time.

1. Introduction

Advanced nuclear reactor concepts promise significant improvements over current technology, including increased efficiency, higher fuel burn-up, and longer core life. However, these features put increasing demands on the performance of fuel cladding and other reactor components, and materials must be developed for these reactors that are both resistant to high levels of irradiation damage and offer accident tolerant behavior. Silicon carbide fiber/silicon carbide matrix (SiC-SiC) composites offer many desirable properties, and are being considered for use in advanced nuclear reactor designs, such as the General Atomics Energy Multiplier Module (EM^2) concept. Experiments on monolithic silicon carbide have shown that it maintains excellent mechanical performance in harsh, high temperature, and high irradiation rate environments, but its low toughness limits its application [1–3]. High purity and high quality SiC fiber-reinforced composites have shown similar performance under harsh

conditions but offer improved toughness to address this limitation. In these composites, a silicon carbide matrix is deposited within a preform composed of high purity, near-stoichiometric silicon carbide fibers, such as Tyranno-SA fibers (Ube Industries, Ube, Japan) or Hi-Nicalon type S fibers (Nippon Carbon Co., Ltd., Tokyo, Japan). The performance of these composites has the potential to enable the development and construction of high temperature, long-life advanced reactor concepts.

Several techniques have been developed to fabricate SiC matrix composite materials, including melt infiltration, polymer infiltration and pyrolysis, and chemical vapor infiltration (CVI) [4–6]. However, in order to achieve good irradiation resistance, very high purity material is required, and CVI is the most reliable approach to produce a sufficiently pure matrix for nuclear applications. In CVI, a silicon carbide precursor (or precursors) is introduced into a high temperature chamber in the gas phase. This is commonly done under vacuum, and the precursors are allowed to diffuse into the preform and chemically react, forming a silicon carbide matrix within the sample. The most commonly used precursor is methyltrichlorosilane (CH_3SiCl_3), which is mixed with hydrogen and decomposes to form silicon carbide according to (1) [1]

$$CH_3SiCl_{3(g)} + H_{2(g)} \longrightarrow SiC_{(s)} + 3HCl_{(g)} + H_{2(g)}. \quad (1)$$

While CVI produces a very high purity matrix, the deposition process is dependent on the diffusion of reactants into the fiber preform, and a slow reaction rate can be desirable to ensure uniform transport of reactants throughout the fiber preform. The reaction rate can be controlled through selection of the deposition parameters; for example, higher process temperatures generally lead to more rapid reaction and higher deposition rates. Uniform matrix deposition is essential to achieve high composite density and good material properties, but by slowing the reaction rate, the fabrication process may require long infiltration times. For fiber-reinforced composite materials, small voids between fibers within the tows are vulnerable to being closed off by matrix deposits on the tow surface. Because of this, density uniformity is especially important for thicker samples, and reduced density can lead to reduced material performance. Often, a density gradient will exist in the composite, where the densest regions are located near the surface (where the precursors first reach the composite), and the least dense areas are located towards the center, especially inside fiber tows. A schematic of this conventional CVI is shown in Figure 1(a).

Several approaches to reduce fabrication time have been reported in the literature. Two of the more promising routes are thermal gradient chemical vapor infiltration and pressure gradient chemical vapor infiltration (also called forced flow CVI). In thermal gradient CVI, modifications to both the CVI chamber and fixtures are designed to establish a temperature gradient in the opposite direction of the diffusion-related reactant concentration gradient. This allows for faster reaction rates near the center of the composite, which helps offset the reduction in precursor concentration, and reduces the overall infiltration time needed to make the composite. In forced flow CVI (FFCVI), a pressure gradient is established to enhance reactant transport into the fiber preform (compared to transport via diffusion only). By improving transport of reactants into the preform, the depletion of the precursors is reduced, and conditions allowing for faster deposition rates can be used while still achieving acceptable final composite densities. However, this approach must be carefully controlled, as variations in pressure in different regions of the composite can have detrimental effects on the matrix infiltration uniformity. In addition, it is necessary to hold the sample in more complicated fixtures to direct the flow of reactants (Figure 1(b)).

In order to help understand the effects of different parameters on the composite infiltration process, the diffusion and chemical reaction of the precursors can be modeled. Several approaches have been reported in the literature, including modeling chemical compositions of the gas and solid phases [7], modeling diffusion through a fibrous preform [8], and

simulating precursor concentrations on a sample in a typical reactor [9]. These models can be used to consider the effects of different reaction parameters, although the trends predicted by some models have not always been consistent with experimentally observed results, and simplified models cannot account for all the phenomena occurring during deposition.

In this work, SiC-SiC composites were fabricated by chemical vapor infiltration, and an empirical model was developed to simulate forced-flow CVI. This work expands on a model which we had previously developed [10] to simulate the infiltration process at the fiber scale (~10–100 μm). In these current results, the effects of forced flow are included in the simulation. The infiltration simulated by the model was qualitatively similar to that observed in the experiments, and the effects of different fabrication parameters were investigated.

2. Sample Fabrication and Characterization

SiC matrix composites were fabricated under a range of conditions, including forced flow CVI and conventional CVI (with no pressure gradient). Various composite geometries can be fabricated; however, for most of this work, the material was produced as larger planar sheets that were then cut into samples appropriate for different characterization techniques.

The sample fabrication process has been described previously [10]. Briefly, SiC fiber fabric is cut to shape and stacked to achieve a nominally 1 mm thick preform of ~35% fiber volume fraction. This fiber preform is then processed under vacuum and at elevated temperatures (900–1100°C) to form a composite using chemical vapor infiltration. This infiltration process is used to first deposit a thin pyrolytic carbon interface layer over the fibers and then infiltrate the SiC matrix. The silicon carbide matrix was deposited from a dilute reactant mixture composed of methyltrichlorosilane (MTS) evaporated into a hydrogen flow.

Samples were infiltrated using both conventional (isobaric) CVI and forced flow CVI, and all samples were weighed prior to infiltration and at regular intervals during infiltration to monitor weight gain as a function of processing time. Forced flow samples were processed until the pressure differential through the sample increased approximately tenfold from the starting pressure drop, which varied with the fiber preform size and structure. Depending on the deposition conditions, this process took up to 3 days. In order to allow a more direct comparison, the durations of the conventional CVI runs were selected to yield samples with similar densities to the forced flow samples, and these runs took significantly longer than the forced flow runs. The relative sample density was calculated from the measured sample volume and mass and was compared to the maximum theoretical density of 3.2 g/cm^3 for SiC fiber-reinforced samples.

3. Modeling SiC Matrix Infiltration

In order to help optimize infiltration process parameters, a two-dimensional computer simulation was developed to

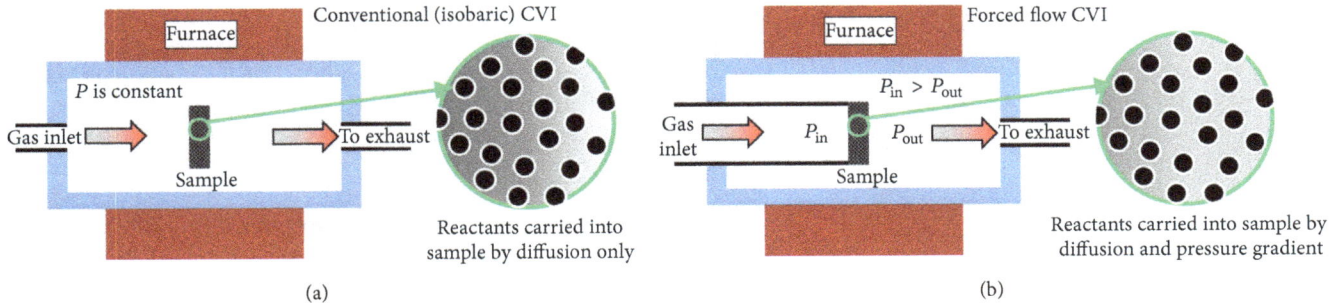

FIGURE 1: (a) Schematic of conventional (isobaric) CVI and (b) schematic of forced flow or pressure gradient CVI.

model the effects of different deposition conditions on the infiltration process. To accurately simulate the matrix infiltration process, the transport of the reactants through the fiber preform must be followed. However, due to the size of the fibers (7–10 μm diameter), a system including several fibers would be much too large (approximately 100 μm) to practically model and track the motions of individual reactant molecules. Typical Knudsen numbers (Kn = mean free path λ, divided by characteristic length L) for deposition conditions are also too large ($\lambda/L \gg 1$) to use standard computational fluid flow codes to model the infiltration. In order to address these restrictions and accommodate the larger system size, the model uses a direct simulation Monte Carlo (DSMC) solver with the OpenFOAM toolbox.

With the DSMC approach, statistically representative parcels are tracked, rather than individual molecules, and each parcel contains many particles. During the simulation of conventional (isobaric) infiltration, the initial velocity of each parcel is obtained from a thermal velocity distribution function. In the simulation of forced flow infiltration, a net velocity is applied to the parcels in the model, with this velocity representing the effect of a pressure gradient on the reactant gas flow. In both cases, the subsequent motions of the parcels are tracked ballistically, accounting for collisions with walls and other particles.

In addition to tracking the motion of different parcels, the DSMC solver also tracks the reaction of MTS on surfaces to produce SiC. The general growth rate equation used in the model is $G = S_{MTS}F_{MTS}$, where G is the growth rate, S_{MTS} is the sticking coefficient, or probability that MTS will react to form silicon carbide on a surface, and F_{MTS} is the flux of MTS on that surface. The sticking coefficient of MTS is given by $S^0_{MTS} = A \exp(-E_{aMTS}/RT)\sqrt{P_{H_2}}$, where P_{H_2} is the partial pressure of hydrogen in the system, E_{aMTS} is the activation energy for MTS decomposition, R is the natural gas constant, T is the temperature, and A is a constant [11]. This SiC deposition is modeled by depleting the number of MTS particles in each parcel in contact with a surface. The simulation also accounts for the presence of HCl byproducts, which have been shown both experimentally and in other models to inhibit the deposition process [11–14]. HCl can occupy surface sites, and the probability of MTS conversion to silicon carbide is reduced by $1 - \theta/\theta_m$, in proportion to the surface area covered by adsorbed HCl (θ/θ_m).

As MTS decomposes on the surface to form SiC, γ is the fraction of HCl that remains on the surface, and the HCl adsorption rate is given by (2), where S^0_{HCl} is the sticking coefficient of HCl on the surface and F_{HCl} is the flux of HCl particles on the surface:

$$\frac{d\theta}{dt} = \left(S^0_{MTS}\gamma F_{MTS} + S^0_{HCl}F_{HCl}\right)\left(1 - \frac{\theta}{\theta_m}\right). \quad (2)$$

HCl will also desorb from the surface according to $d\theta/dt = -\nu\theta \exp(-E_{aHCl}/RT)$, where E_{aHCl} is the activation energy for the desorption of HCl and ν is a rate constant. The steady state growth rate is given by (3), where C is a constant:

$$G = \frac{S^0_{MTS}F_{MTS}C}{1 + \left(\gamma S^0_{MTS}F_{MTS} + S^0_{HCl}F_{HCl}\right)/\theta_m \nu \exp\left(-E_{aHCl}/RT\right)}. \quad (3)$$

In this equation, the activation energies are taken from the literature, with E_{aMTS} = 188 kJ/mol [12] and E_{aHCl} = 268 kJ/mol [13]. SiC deposition experiments were performed across a range of different process conditions, and coating rates were measured and compared to the simulation in order to empirically fit values for the MTS and HCl sticking coefficients and γ. For each collision between a given parcel and a surface, the modified DSMC solver uses the growth rate equation to determine the amount of MTS that reacts to form SiC. The solver also tracks the HCl generated by the MTS decomposition, of which a fraction remains on the surface (given by γ), and the remainder is added to the parcel. The solver represents the deposited SiC by adding elements to fiber surfaces upon which SiC has been deposited, increasing the fiber diameter and reducing the open pore volume.

4. Results and Discussion

4.1. Forced Flow Chemical Vapor Infiltration. Forced flow chemical vapor infiltration offers a means to increase transport of reactants through the fiber preform during the matrix infiltration process. As a result, process conditions that allow for more rapid deposition can be used, such as higher temperatures and MTS partial pressures. This can reduce process time, while the pressure gradient helps ensure matrix deposition uniformity.

(a)

(b)

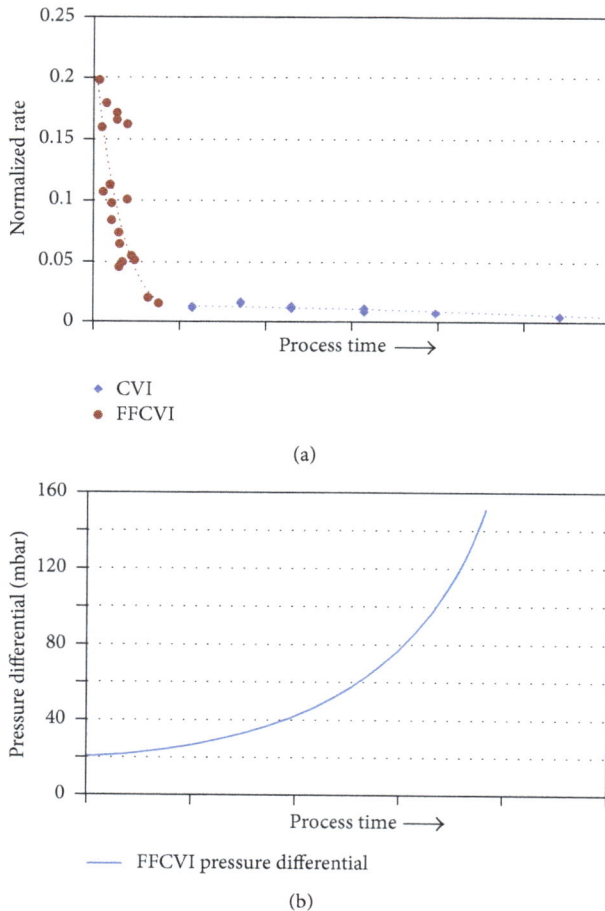

Figure 2: (a) Plot of normalized rate of mass gain as a function of process time for conventional CVI and FFCVI and (b) plot of increasing pressure differential across sample observed during FFCVI.

The infiltration process was interrupted at several intervals in order to measure the mass gain of the samples and remove portions for characterization. It was found that the infiltration process could be started and stopped in this manner without causing discontinuities in the matrix. Figure 2(a) shows a typical rate of mass gain for conventional CVI and forced flow CVI. Also plotted is the pressure differential between the inlet and exhaust sides of the sample in the forced flow case (Figure 2(b)). As a result of the improved reactant transport into the sample during FFCVI, higher deposition rates can be used while still achieving reasonable infiltration uniformity.

For both FFCVI and isobaric CVI, the rate of mass gain typically starts out high, and then drops as the infiltration process continues. The amount of SiC deposited on the samples is a function of CVI conditions but also of the available surface area of the substrate. Initially, the fiber preform has a very high surface area, and consequently, a high rate of mass gain is observed. However, as the matrix fills in the voids within the preform, some regions become closed off, trapping voids within the sample. This reduces the available surface area as well as the rate of mass gain.

In conventional CVI, very low deposition rates are used to allow sufficient reactant diffusion into the preform. The added pressure gradient in FFCVI allows for higher deposition rates to be achieved while maintaining uniformity through the sample. If process conditions are changed in a conventional CVI process to achieve deposition rates comparable to FFCVI, a significant SiC deposition gradient develops from the surface of the sample (the surface exposed to the reactant flow) towards the center of the sample (Figure 3(a)). The gradient is much reduced when a forced flow configuration is used to achieve similar deposition rates (Figures 3(b) and 3(c)).

In both conventional and forced flow CVI, the rate of mass gain drops with time, as pores are closed off and the available surface area for deposition is reduced. In conventional CVI, the reactants are flowing around the sample, so SiC deposition on the sample and this eventual pore closure do not affect the process pressure. However, in forced flow CVI, all reactants are forced to flow through the sample, and as the open porosity begins to be filled or closed off, the pathways for gas flow are reduced, and the pressure differential across the samples increases. This pressure difference increases rapidly towards the end of the process and can result in the upstream side of the sample seeing significantly higher pressures than the exhaust. As the deposition rate can be strongly dependent on reactant partial pressures, this increasing pressure differential can lead to nonuniform deposition (Figure 4). This can result in excess SiC build-up on the inlet side of the composite and can also result in deposition conditions that produce nonstoichiometric, silicon-rich SiC.

4.2. Simulation of SiC Matrix Deposition. Although forced flow CVI can be used to potentially provide increased deposition rates, the conditions must be understood to ensure uniform infiltration. To more rapidly investigate the effects of various process parameters and sample geometries on the fabrication process, a computer model was used to simulate SiC infiltration into an idealized fiber arrangement. The influence of different process parameters was explored in the context of the maximum simulated composite density achieved and the infiltration duration required to reach that density. The end of the simulation was defined as the time at which the reactant transport pathways to the inside fibers became completely blocked off; after this point, any additional deposition would occur only on the surface, and the composite density and pore volume would no longer change.

Prior to the start of infiltration, the structure of the composite consists of bundles or tows of hundreds or thousands of SiC fibers, and these tows are woven into the desired macroscopic component geometry. The simulation developed in this work focuses on the deposition at the microscale, and the geometry and boundary conditions were selected to provide an approximation of an actual fiber bundle. Although transport of reactants within the CVD chamber and between adjacent tows (on the macroscopic scale) is also important to the overall composite uniformity, running the simulation

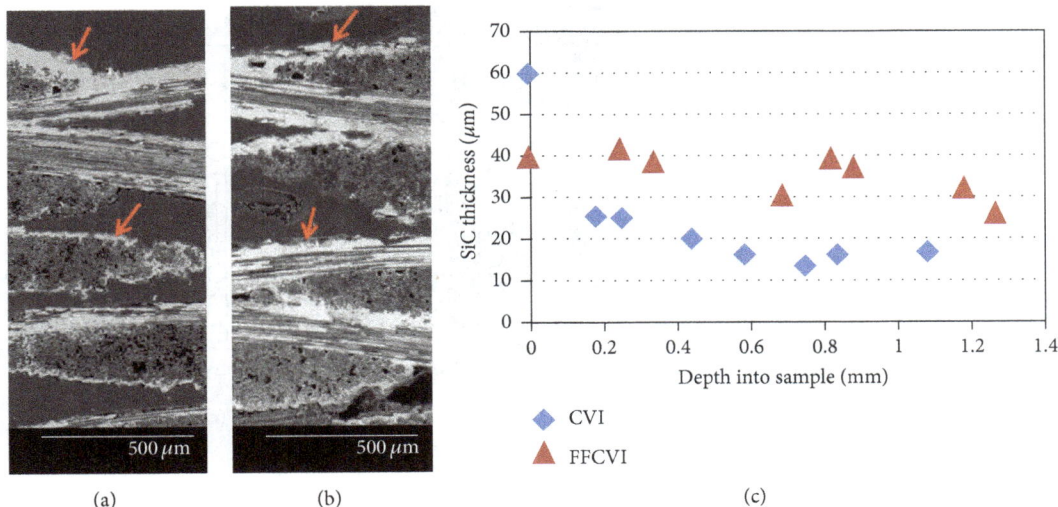

(a)   (b)   (c)

FIGURE 3: (a) Cross-section of fiber preform partially infiltrated with higher rate isobaric CVI, (b) cross-section of fiber preform partially infiltrated with forced flow CVI, and (c) SiC deposit thickness as a function of depth into the sample for CVI and FFCVI.

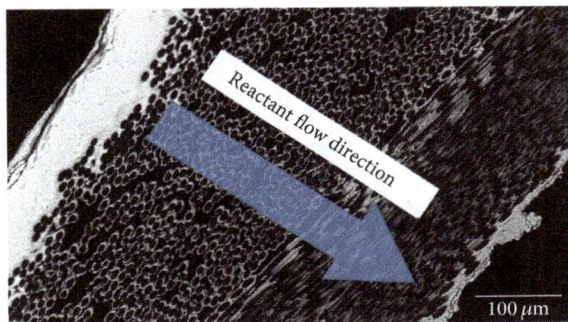

FIGURE 4: SiC build-up on the upstream side of the sample caused by increasing reactant pressure and varying deposition conditions on the upstream side of the sample during FFCVI.

at this scale would be too computationally intensive to be practical.

Infiltration into the fiber tows was simulated with a narrow column of fiber cross-sections. Fiber spacing was kept constant, although the simulation approach could also be used to model a more realistic system or randomly arranged fibers. The fiber tows used in these composites have elliptical cross-sections, and the column of fibers used in the simulation represents approximately half the yarn minor axis, as shown in Figure 5. Reactant gases were introduced from one side, and a net velocity could be applied to the reactants to simulate forced flow conditions. The boundary conditions used for the other sides of the system are shown in Figure 5(c).

With the empirically determined reactant sticking coefficients, the simulated SiC infiltration is qualitatively similar to the experimentally observed deposition gradient, as shown in Figure 5(d). For this comparison, the coating thickness was measured as a function of position within the fiber bundle, and the model reproduced the variation in coating thickness away from the tow surface that was observed in

the actual sample. The simulation was then used to model the infiltration for both isobaric and forced flow conditions and to investigate the effects of different coating parameters on the deposition process. Certain parameters (including temperature, pressure, and MTS concentration) were varied within ranges that coincided with the typical conditions used in the experiments.

4.2.1. Reactant Velocity Effects. In the model, reactants can be introduced with a net velocity which can be set for different forced flow conditions or reduced to zero to represent isobaric (conventional) CVI. At very low velocities, reactant transport into the fiber preform occurs due to diffusion. If this transport is slow relative to the reaction rate of the precursors, SiC will be deposited towards the outer surface of the sample, and a large deposition gradient will lead to a composite with low final density and high internal porosity. Slow deposition rates are needed to achieve higher final composite densities.

If the reactants have a positive net velocity, this provides additional means to facilitate their transport into the fiber preform. At moderate velocities, the time scale for transport through the sample is comparable to the reaction rate of the precursors, and deposition uniformity is improved. This allows a faster reaction rate to be used which reduces composite fabrication time while maintaining improved uniformity. However, the simulation indicated that if the reactant velocities are very high, transport through the fiber preform can be too fast, and the short residence time can reduce the deposition uniformity. This effect could be exaggerated in our simulation, as the system volume is very small relative to an actual sample (which would be thicker and contain many thousands of fibers). Thus, a given velocity in the model could be sufficient to transport the precursors out of the system before they can completely react, whereas in an actual sample, these unreacted precursors would react and deposit elsewhere.

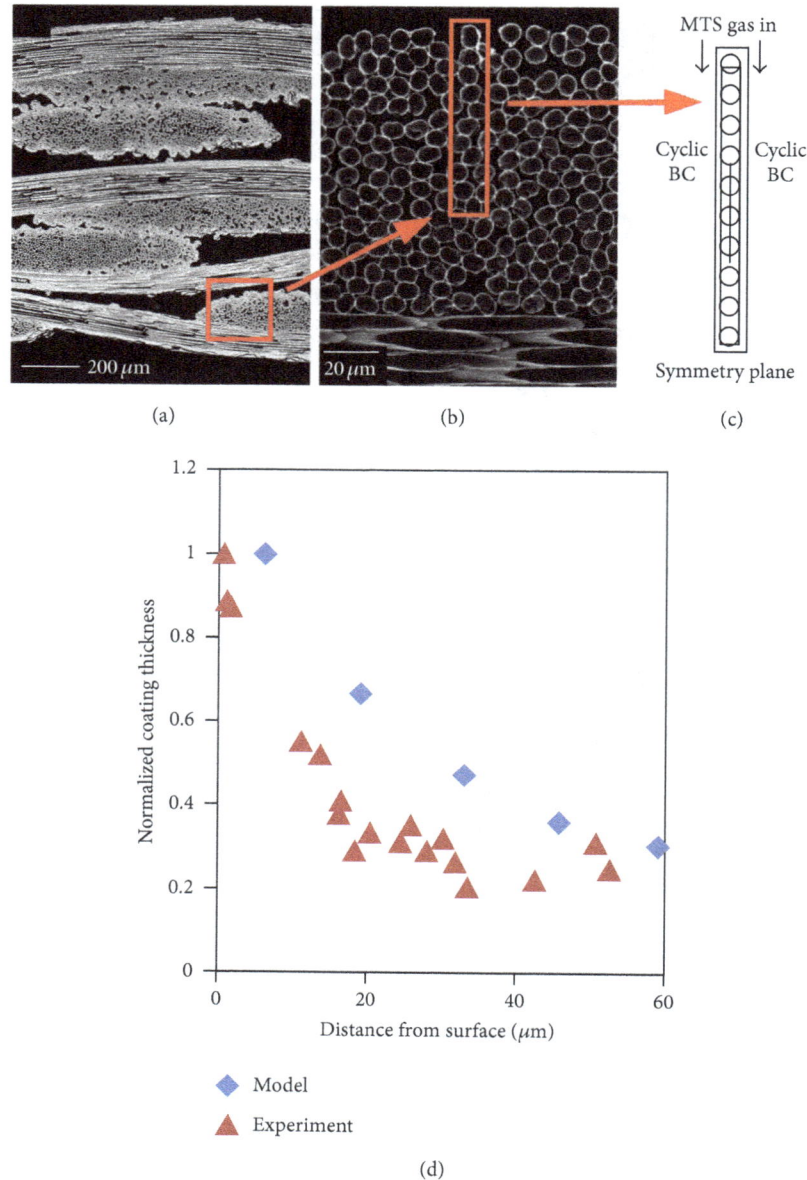

FIGURE 5: (a) Cross-section of stacked fabric layers, (b) cross-section of fiber bundle, (c) model geometry showing MTS introduction and boundary conditions, and (d) comparison between simulated and experimentally observed SiC deposition gradients into the fiber preform.

To examine the influence of reactant velocity, the model was run with zero, moderate, and fast reactant velocities. The precursor partial pressure was held constant, and infiltration was simulated at two different SiC deposition temperatures (900°C and 1100°C). These temperatures are in the range of infiltration temperatures commonly reported in the literature [5]. Reactant velocity was found to have a significant effect on the deposition uniformity and the corresponding final composite relative density. For both slow and moderate reactant velocities, similar deposition was seen at the surface. However, the improved reactant transport achieved by forced flow CVI was apparent in the simulated deposition inside the fiber preform. Moderate velocities (which could be obtained under appropriate FFCVI conditions) provided the highest

relative density and most uniform infiltration, as shown in Figure 6(a). The effects of reactant velocity were similar for both temperatures modeled (Figure 6(b)).

4.2.2. Reactant Concentration Effects. The reactant partial pressure also plays an important role in the infiltration process. The MTS flux at the fiber surface is proportional to the MTS partial pressure, and the growth rate increases with the increasing flux (according to (3)). In the model, the reactant concentration is varied by setting the MTS partial pressure (a function of the MTS concentration and the overall pressure). Unlike isobaric infiltration, in FFCVI, the overall pressure cannot be simply controlled. Typically, the downstream pressure is fixed, but unless a bypass is used,

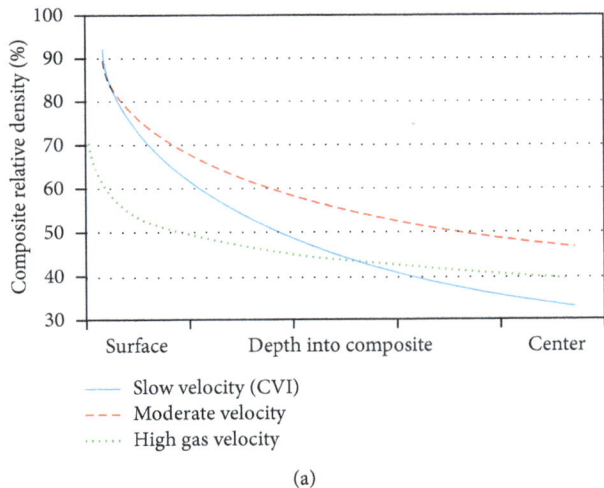

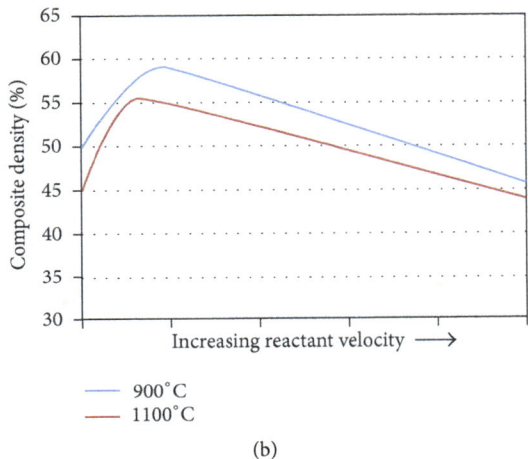

FIGURE 6: (a) Simulated deposition gradient for varying reactant velocities and (b) variation in simulated composite density as a function of temperature and reactant velocity.

the upstream pressure increases as the transport pathways through the fiber preform close during infiltration. As the process pressure then varies as a function of time and position within the sample, control over the reactant concentration becomes the means by which the reactant partial pressure can be set.

Simulations and experiments were carried out over a range of MTS concentrations, with temperature and reactant velocity held constant. For low MTS concentrations, SiC deposition can be limited by reactant availability. With increasing reactant concentration, the reaction becomes limited by surface site availability, and the deposition rate is no longer strongly influenced by reactant concentration. Both the simulation and experimental results show this effect, and the infiltration process becomes less sensitive to the MTS partial pressure at high MTS concentrations. The simulated infiltration as a function of depth into the composite is shown in Figure 7(a), and for moderate and higher MTS concentrations, there is a minimal effect on the infiltration.

Model results were also compared with experimental results for a range of MTS partial pressures (Figure 7(c)). These experiments were carried out in both CVI and FFCVI conditions, and the infiltration duration was limited (which limited the densities that could be achieved). At higher MTS partial pressures, there was good agreement between the simulated and experimental densities, but at lower partial pressures, the experimental densities were much lower than the simulation. This is likely caused by reactant depletion in the experiment before the precursors reach the fiber preform. Precursors can react on other surfaces inside the furnace, and with very low reactant partial pressures, this can deplete the reactant concentration enough to impact the sample density achieved. At higher reactant partial pressures, the relative amount of depletion is greatly reduced, and in the simulation, the reactants are introduced directly at the fiber preform surface, so the depletion in other regions of the furnace is not considered.

4.2.3. Fiber Spacing Effects. The final parameter modeled in this work is the average fiber spacing in the preform. This variable is governed by the fiber architecture and sample preparation and is unlike reactant pressure or velocity, in that it cannot be set during the actual infiltration. However, the same simulation can be used to model the effects of different fiber spacing by increasing the system size. Coupled with the cyclic boundary conditions, this effectively increases the width of the gaps between adjacent fibers, which allows for more infiltration to occur before the reactant pathways become blocked.

In this study, the width of the simulated system was increased from 15 to 30 μm. This effectively tripled the gap between adjacent fibers from a 7.5 μm gap to a 22.5 μm gap. More SiC deposition is required before these larger reactant transport pathways are closed off. As a result, prolonged infiltration is required for this closure to occur, and a 3.1x increase in processing time was observed in the simulation. This is roughly proportional to the 3x increase in gap width used in this study. While increased fabrication time is not desirable in terms of rapid processing, if this is accompanied by an increase in composite density and performance, then these gains could offset the long infiltration required. At the end of infiltration, the simulated composite with the larger gap spacing showed an increase of 57% and 39% in density compared to the narrow gap model for the 900°C and 1100°C simulations, respectively (Figure 8).

Composite density has been shown to have a very significant effect on mechanical and thermal performance, and careful control of the average fiber spacing within the composite could be a means to optimize the tradeoff between fabrication time and improved material properties. However, in increasing the average gap between fibers, the overall fiber volume fraction of fibers in the composite is reduced as well. Maintaining a sufficient fiber loading in the composite is essential to obtain increased fracture toughness and more graceful failure characteristic of a composite, so any adjustments to the fiber architecture would need to be evaluated in the context of composite behavior as well.

(a)   (b)

(c)

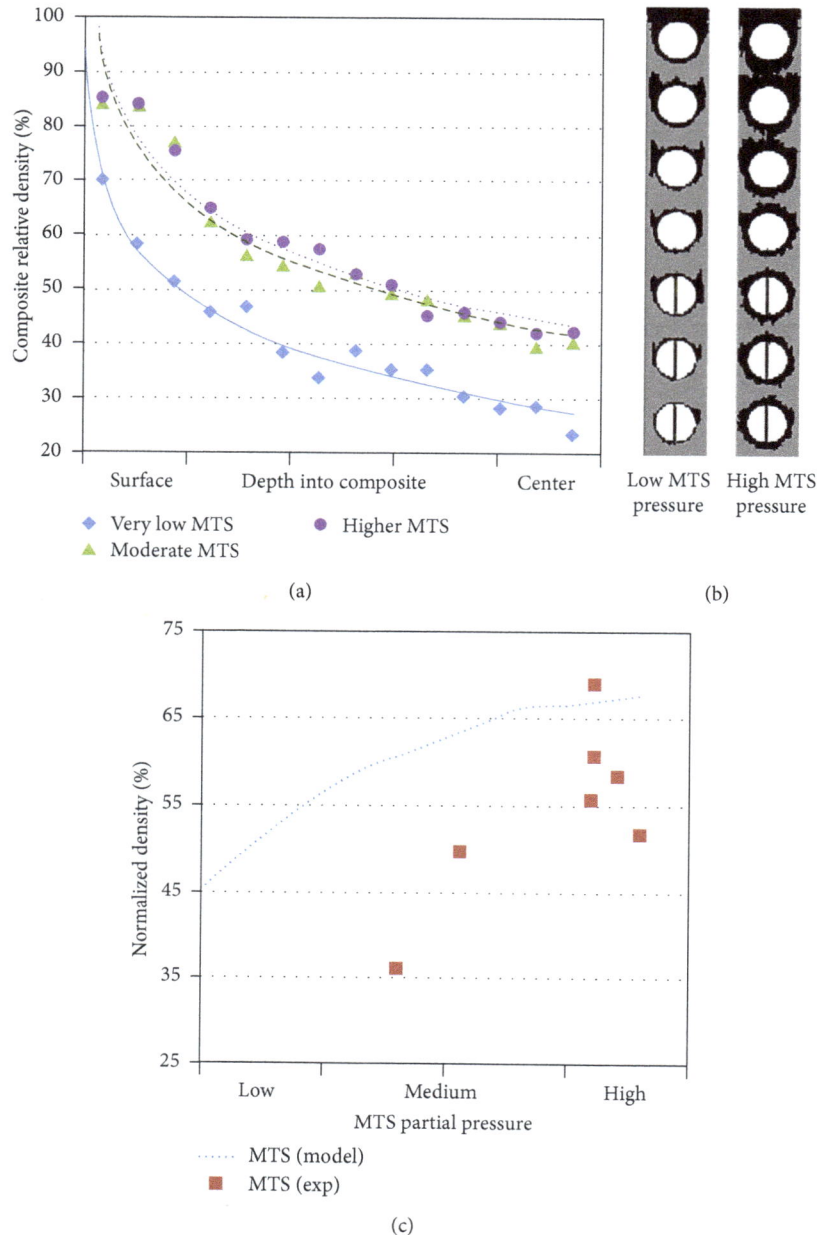

FIGURE 7: (a) Simulated deposition gradient for varying MTS partial pressures, (b) simulated infiltration for low and high MTS partial pressures, where black indicates deposited SiC matrix, and (c) comparison between simulation and experimental results for MTS concentration effects on density.

Simulations of the SiC infiltration process have also been performed by other groups and trends that are consistent with the results of this work have been reported. Roman et al. [15, 16] developed an extensive model of the reaction and deposition process for forced flow CVI and observed similar trends with increasing precursor velocity through the sample. In that work, reactant introduction was given as a volumetric flow, rather than a net velocity, and this makes it difficult to make an exact comparison with this work. However, in both that model and ours, increasing reactant flow from very low to moderate velocities resulted in reduced processing time and increased density and matrix uniformities (Figure 9).

5. Conclusions

SiC-SiC composites exhibit many properties that could make them very desirable for use in advanced nuclear reactor concepts, including strength at high temperatures and exceptional stability under irradiation. However, one of the challenges that must be overcome before these materials are widely adopted is the time-consuming and expensive fabrication process required to produce very high purity material. Forced flow CVI is a modification to the conventional isobaric infiltration process, which allows for increased densification rates while maintaining infiltration

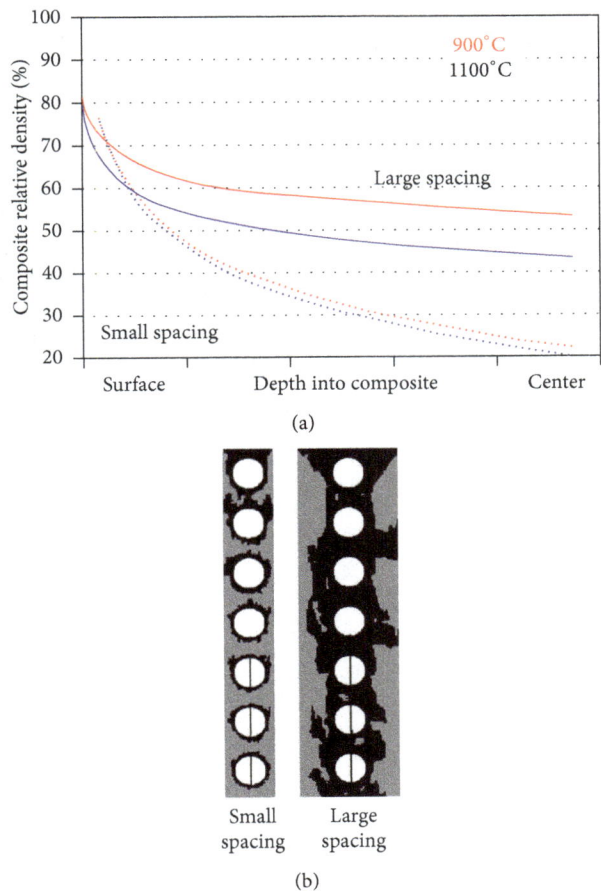

FIGURE 8: (a) Simulated deposition gradient for varying fiber spacings and temperatures, (b) simulated infiltration for small and large fiber spacing, where black indicates deposited SiC matrix around white fiber cross-sections.

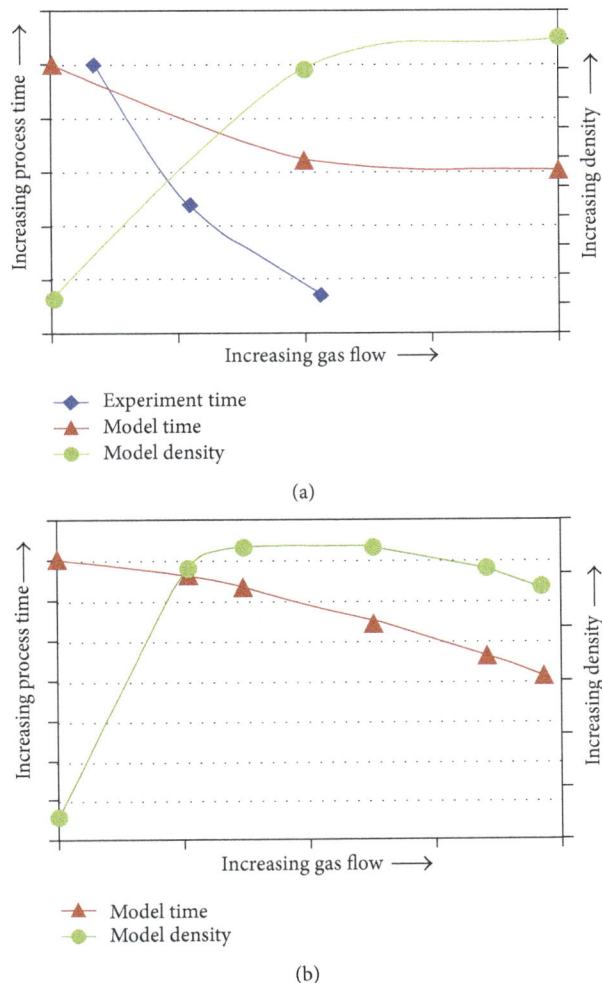

FIGURE 9: (a) Simulation and experimental results from the current work, showing effects of reactant velocity and (b) simulation results from Roman et al., [15] showing similar trends with increasing gas flow.

uniformity. In this work, a model was developed to simulate the chemical vapor infiltration process for the fabrication of SiC-SiC composites under both isobaric and forced flow conditions. This model uses a direct simulation Monte Carlo approach and accounts for MTS deposition, depletion, and the inhibition of this reaction due to the HCl byproducts.

The effects of different fabrication parameters were simulated to investigate ways to optimize forced flow CVI. Increasing reactant flow through the sample increased both the densification rate (reducing the required processing time) and the final composite density. However, very high reactant flow rates were found to reduce uniformity in the simulation, which could be due to insufficient reactant residence time. Very high reactant velocities are also associated with larger pressure gradients across the sample, and in the supporting experimental work, high pressures were found to lead to non-uniform, nonstoichiometric deposition on the sample. The trends of reactant velocity were similar between the model developed in this work and the experimental results and were also consistent with simulations reported in the literature. The concentration of the MTS precursor (and the corresponding partial pressure) was also found to influence the infiltration process, although the effect was most pronounced at lower concentrations. Finally, an increase in the average spacing between fibers led to a composite with a higher final density, but it increased the required processing time as well. Any infiltration improvements achieved by a modification to the fiber architecture would need to be carefully balanced with any impacts the altered structure had on the mechanical performance of the composite.

Further improvements to the model could allow simulation of reactant transport through larger intertow porosity and also account for randomized fiber positioning and the more complex two- or three-dimensional structure found in woven preforms. The model provides means to explore the effects of different microstructures and infiltration process parameters on the composite densification process. Using this model, these effects can be simulated in far less time than would be required to complete a corresponding experimental test matrix. This allows for a much faster approach towards optimization of the fabrication process, but it needs to be coupled with, and confirmed by, experimental results. More efficient composite fabrication can lead to improved consistency and reduced fabrication cost and time, which will bring SiC-SiC composites closer to the deployment in actual advanced nuclear reactor applications.

Acknowledgments

The authors would like to acknowledge the contributions of the EM^2 lab personnel to the operation of the SiC composite fabrication facility at General Atomics. The work presented here was supported by General Atomics internal funding.

References

[1] L. L. Snead, T. Nozawa, Y. Katoh, T. S. Byun, S. Kondo, and D. A. Petti, "Handbook of SiC properties for fuel performance modeling," *Journal of Nuclear Materials*, vol. 371, no. 1–3, pp. 329–377, 2007.

[2] Y. Katoh, T. Nozawa, L. L. Snead, K. Ozawa, and H. Tanigawa, "Stability of SiC and its composites at high neutron fluence," *Journal of Nuclear Materials*, vol. 417, no. 1–3, pp. 400–405, 2011.

[3] K. Hironaka, T. Nozawa, T. Hinoki et al., "High-temperature tensile strength of near-stoichiometric SiC/SiC composites," *Journal of Nuclear Materials*, vol. 307–311, no. 2, pp. 1093–1097, 2002.

[4] M. Kotani, A. Kohyama, and Y. Katoh, "Development of SiC/SiC composites by PIP in combination with RS," *Journal of Nuclear Materials*, vol. 289, no. 1-2, pp. 37–41, 2001.

[5] R. Naslain, "Design, preparation and properties of non-oxide CMCs for application in engines and nuclear reactors: an overview," *Composites Science and Technology*, vol. 64, no. 2, pp. 155–170, 2004.

[6] K. Shimoda, J. S. Park, T. Hinoki, and A. Kohyama, "Microstructural optimization of high-temperature SiC/SiC composites by NITE process," *Journal of Nuclear Materials*, vol. 386–388, pp. 634–638, 2009.

[7] W. G. Zhang and K. J. Hüttinger, "CVD of SiC from methyl-trichlorosilane—part II: composition of the gas phase and the deposit," *Chemical Vapor Deposition*, vol. 7, no. 4, pp. 173–181, 2001.

[8] G. Y. Chung and B. J. McCoy, "Modeling of chemical vapor infiltration for ceramic composites reinforced with layered, woven fabrics," *Journal of the American Ceramic Society*, vol. 74, no. 4, pp. 746–751, 1991.

[9] X. Wei, L. Cheng, L. Zhang, Y. Xu, and Q. Zeng, "Numerical simulation of effect of methyltrichlorosilane flux on isothermal chemical vapor infiltration process of C/SiC composites," *Journal of the American Ceramic Society*, vol. 89, no. 9, pp. 2762–2768, 2006.

[10] C. P. Deck, H. E. Khalifa, B. Sammuli, T. Hilsabeck, and C. A. Back, "Fabrication of SiC-SiC composites for fuel cladding in advanced reactor designs," *Progress in Nuclear Energy*, vol. 57, pp. 38–45, 2012.

[11] T. M. Besmann, B. W. Sheldon, T. S. Moss III, and M. D. Kaster, "Depletion effects of silicon carbide deposition from methyltrichlorosilane," *Journal of the American Ceramic Society*, vol. 75, no. 10, pp. 2899–2903, 1992.

[12] C. Lu, L. Cheng, C. Zhao, L. Zhang, and Y. Xu, "Kinetics of chemical vapor deposition of SiC from methyltrichlorosilane and hydrogen," *Applied Surface Science*, vol. 255, no. 17, pp. 7495–7499, 2009.

[13] M. T. Schulberg, M. D. Allendorf, and D. A. Outka, "The adsorption of hydrogen chloride on polycrystalline β-silicon carbide," *Surface Science*, vol. 341, no. 3, pp. 262–272, 1995.

[14] H. C. Chang, T. F. Morse, and B. W. Sheldon, "Minimizing infiltration times during isothermal chemical vapor infiltration with methyltrichlorosilane," *Journal of the American Ceramic Society*, vol. 80, no. 7, pp. 1805–1811, 1997.

[15] Y. G. Roman, J. F. A. K. Kotte, and M. H. J. M. de Croon, "Analysis of the isothermal forced flow chemical vapour infiltration process—part I: theoretical aspects," *Journal of the European Ceramic Society*, vol. 15, pp. 875–886, 1995.

[16] Y. G. Roman, M. H. J. M. de Croon, and R. Metselaar, "Analysis of the isothermal forced flow chemical vapour infiltration process—part II: experimental study," *Journal of the European Ceramic Society*, vol. 15, no. 9, pp. 887–898, 1995.

5

Solutions without Space-Time Separation for ADS Experiments: Overview on Developments and Applications

B. Merk and V. Glivici-Cotruţă

Helmholtz-Zentrum Dresden-Rossendorf, Institut für Sicherheitsforschung, Postfach 51 01 19, 01314 Dresden, Germany

Correspondence should be addressed to B. Merk, b.merk@hzdr.de

Academic Editor: Alberto Talamo

The different analytical solutions without space-time separation foreseen for the analysis of ADS experiments are described. The SC3A experiment in the YALINA-Booster facility is described and investigated. For this investigation the very special configuration of YALINA-Booster is analyzed based on HELIOS calculations. The results for the time dependent diffusion and the time dependent P_1 equation are compared with the experimental results for the SC3A configuration. A comparison is given for the deviation between the analytical solution and the experimental results versus the different transport approximations. To improve the representation to the special configuration of YALINA- Booster, a new analytical solution for two energy groups with two sources (central external and boundary source) has been developed starting form the Green's function solution. Very good agreement has been found for these improved analytical solutions.

1. Introduction

Different current and planned experiments (MUSE [1], YALINA [2], GUINEVERE [3]) are designed to study the zero power neutron physical behavior of accelerator-driven systems (ADSs). The detailed analysis of the kinetic space-time behavior of the neutron flux is important for the evaluation of these ADS experiments. Current analysis for all these experiments is based on the standard methods [4]—Sjöstrand method [5] and Slope method [6]—both are based on the point kinetics equations [7]. The point kinetics equations are developed from different approximations to the transport equation. Nevertheless, all these partial differential equations of the transport equation are solved by the separation of space and time to derive the point kinetics equations. The separation of space and time does not provide useful results for cases with strongly space-time dependent external source [8]. In recent years, two big projects have been launched to solve the problem of subcriticality determination in ADS experiments and during ADS operation for the future. In the 6th European Framework program in the integrated project EUROTRANS, the domain 2, ECATS [9] has been dedicated to ADS experiments and the analysis

methods for the experiments. In the same time frame, the IAEA has launched the coordinated research project "Analytical and Experimental Benchmark Analyses of Accelerator Driven Systems (ADS)" [10]. Different correction methods based on Monte Carlo Results for the YALINA-Booster system have been derived for the Sjöstrand method as well as for the Slope method in EUROTRANS/ECATS [11, 12]. Good results have been achieved with this correction method for the analysis of the detectors in the thermal zone [13]. Nevertheless, there is still a problem that exists in the fast zone, which is the really important zone, since the follow-up experiment GUINEVERE will be a pure fast system. The results for the analysis of the fast zone of YALINA-Booster are still not convincing, even with the use of correction factors. Good results, but once more only for the thermal and the reflector positions, have been shown in the IAEA CRP by the US American group, especially. These results rely on the use of correction factors from deterministic calculations [14]. The conclusion of the IAEA meeting suggests for further activities, maybe a further CRP among others, the following two topics: "Online Reactivity Monitoring and Control of Sub-critical System"; "Determination of Sub-criticality Level and Uncertainties Analyses" [15].

2. Developed Analytical Solutions for the Analysis of ADS Experiments

To overcome the problems in the analysis of the fast zone in YALINA-Booster and in GUINEVERE as completely fast system, it has been proposed from Helmholtz-Zentrum Dresden-Rossendorf to solve a more elaborate approximation of the transport equation—either the diffusion equation itself (1) or even the time-dependent Telegrapher's equation (2) [16] without separation of space and time. The Telegrapher's equation and the diffusion equation without delayed neutron production are solved completely analytically by using the Green's function method [17, 18]:

$$\frac{1}{v}\frac{\partial \phi}{\partial t} = D\frac{\partial^2 \phi}{\partial x^2}\phi - \Sigma_a \phi + S, \tag{1}$$

$$\frac{3D}{v^2}\frac{\partial^2 \phi}{\partial t^2} + \frac{1}{v}(1 + 3D\Sigma_a)\frac{\partial \phi}{\partial t} = D\frac{\partial^2 \phi}{\partial x^2}\phi - \Sigma_a \phi + S. \tag{2}$$

Solutions for the Telegrapher's equation have already been provided for a Dirac type pulsed external source [19], for the start-up [20, 21], and for the switch-off [8] of an external source, even with consideration of the delayed neutron production. For the comparison with the experimental results, obtained at the YALINA-Booster facility, a special external source (switch-on followed by a switch-off after a finite time period) has been used for the determination of the analytical solution for the neutron flux [22–25]. The derived solutions for the space-time dependent neutron flux were compared to the detector responses at different locations in the fast area of the YALINA-Booster core [26, 27]. Major results of these comparisons were a good agreement for the spatial distributions during the pulse and only small differences between time-dependent diffusion and time-dependent P_1 transport using identical cross-sections and coefficients [23, 25]. The analysis of the specifics of the YALINA facility forced to extend the analytical solution of the diffusion equation using two energy groups, (3) [28]:

$$\frac{1}{v_1}\frac{\partial \phi_1}{\partial t} = D_1\frac{\partial^2 \phi_1}{\partial x^2} - \Sigma_{a1}\phi_1 + \chi_1\left(\nu\Sigma_{f1}\phi_1 + \nu\Sigma_{f2}\phi_2\right)$$
$$- \Sigma_{1\to 2}\phi_1 + \Sigma_{2\to 1}\phi_2 + \chi_{s1}S,$$
$$\frac{1}{v_2}\frac{\partial \phi_2}{\partial t} = D_2\frac{\partial^2 \phi_2}{\partial x^2} - \Sigma_{a2}\phi_2 + \chi_2\left(\nu\Sigma_{f1}\phi_1 + \nu\Sigma_{f2}\phi_2\right)$$
$$+ \Sigma_{1\to 2}\phi_1 - \Sigma_{2\to 1}\phi_2 + \chi_{s2}S. \tag{3}$$

For a better representation, especially for the experiments in the GUINEVERE facility [3, 29], analytical solutions for one energy group for a two-region system, consisting of a multiplicative core with source and a reflective, only absorbing surrounding (reflector), (4) [30], were calculated:

$$\frac{1}{v_1}\frac{\partial \phi_1}{\partial t} = D_1\frac{\partial^2 \phi_1}{\partial x^2} - \Sigma_{a1}\phi_1 + \nu\Sigma_{f1}\phi_1 + S,$$
$$\frac{1}{v_2}\frac{\partial \phi_2}{\partial t} = D_2\frac{\partial^2 \phi_2}{\partial x^2} - \Sigma_{a1}\phi_2. \tag{4}$$

For all problems symmetry of the system was assumed and the solutions were derived for a half of the region. Reflective boundary conditions are used in the center of the system, as well as at the outer boundary. In order to connect two regions, continuous neutron flux and neutron current were used for the two-region solution.

The derived Green's functions $G_i(\xi, \tau \mid \xi_0, \tau_0)$ are universal solutions for the solved equation, which are still independent of the definition of the external source in space and time.

The analytical solution for the time-dependent one- and two-group diffusion equation with initial and boundary conditions (1) to (4) is expressed by a double integral in terms of the corresponding Green's functions and the external source, which has to be defined:

$$\Phi_i(\xi, \tau) = \int_{\tau_a}^{\tau}\int_{\xi_0=0}^{R} G_i(\xi, \tau \mid \xi_0, \tau_0)[S(\xi_0, \tau_0)]d\xi_0\, d\tau_0,$$
$$i = I \text{ or } I, II. \tag{5}$$

$G_i(\xi, \tau \mid \xi_0, \tau_0)$, $i = I$ or I, II are the Green's functions. A Green's function is a solution for the problem associated with the given problem (one-group diffusion, one-group P_1, two-group diffusion, or two-region one-group diffusion) with the same boundary and initial conditions, in which the nonhomogeneous contribution—in our case the external neutron source $S(\xi_0, \tau_0)$—is replaced by the unit impulse function $\delta(\xi - \xi_0)\delta(\tau - \tau_0)$. In a following step after the determination of the Green's functions, the different kinds of external sources have to be defined. In this way, solutions for a Dirac type pulsed external source [19], for the start-up [20, 21], and for the switch-off [8] of an external source for the time-dependent one-group diffusion and the time-dependent one-group P_1 equation were determined. For the comparison with the YALINA-Booster experiment the external neutron source like it is used in the experiment has to be included and the flux has to be determined by solving (5).

For the described YALINA-Booster experiment with the switch on of an external source followed by a switch off after a finite time step, a Heaviside function **H** was used to model a source function. This Heaviside function is locally concentrated at the center part of the system (see Figure 1). In the figure the normalized spatial coordinate points to the front, the normalized time coordinate points to the right, and the source amplitude points upwards. The mathematical definition of the described external source is given by

$$S(\xi_0, \tau_0) \equiv \begin{cases} \dfrac{s(\mathbf{H}(\tau_0 - \tau_i) - \mathbf{H}(\tau_0 - \tau_a))}{\Delta\xi}, & 0 \le \xi_0 \le \Delta\xi, \\ 0, & \Delta\xi \le \xi_0 \le R. \end{cases} \tag{6}$$

This source was used for the already mentioned comparisons of the one energy group time-dependent diffusion and P_1 solution with the YALINA-Booster experiments [22, 23, 25].

To improve the reproduction of the special experimental conditions with the influences described in Specificity of

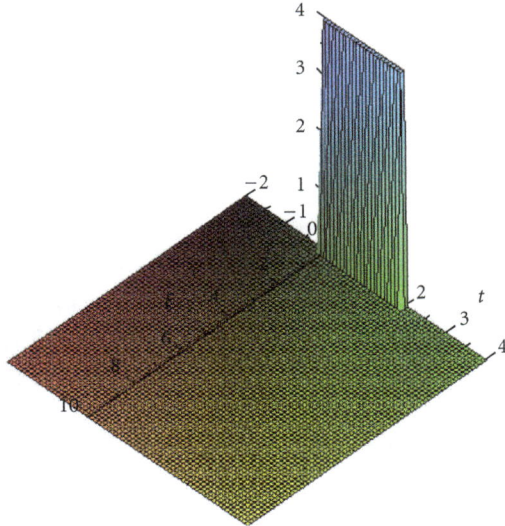

FIGURE 1: Illustrative sketch of the external neutron source used for the one-group analysis.

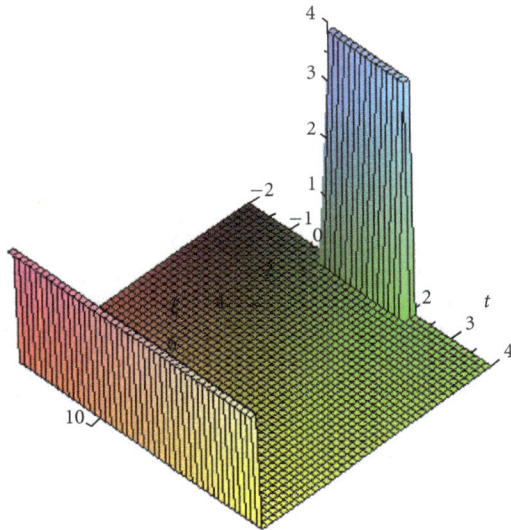

FIGURE 2: Illustrative sketch for the combination of the space-time dependent external neutron sources for the two-energy group solution.

YALINA-Booster section an additional source has been introduced into the two-group solution to represent the streaming of neutrons from the fast area into the thermal area:

$$S_b(\xi_0, \tau_0) \equiv \begin{cases} 0, & 0 \le \xi_0 \le R - \Delta\xi_b, \\ \dfrac{s_b \mathbf{H}(\tau_0 - \tau_b)}{\Delta\xi_b}, & R - \Delta\xi_b \le \xi_0 \le R. \end{cases} \quad (7)$$

Using the sources, defined in (6) and (7), a combined source is created for the two-energy group analytic solution, as it is sketched in Figure 2. It has to be mentioned that the different contributions of the central source and the boundary source can vary independently for the location and

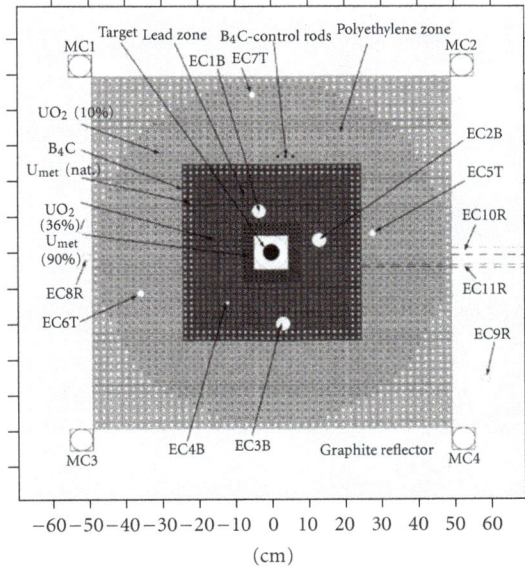

FIGURE 3: General configuration of the YALINA-Booster core.

for both energy groups. Thus for the experiments the central neutron source in the thermal group is set to zero, since the external source produced via the D-T reaction provides mono-energetic neutrons with 14 MeV. For the time behavior of the boundary source, a simple approximation has been used. The source has been used as constant, since the start-up of the source is not of interest for the analysis and the decay of the neutron flux in the thermal area is significantly slower than in the fast area, due to the strong difference in the neutron generation time.

3. Specificity of YALINA-Booster

The core of the YALINA-Booster facility, located in Belarus, consists of the central target region, surrounded by the region with 90% enriched uranium metal (U_{met} 90%) or 36% enriched uranium oxide (UO_2 36%) rods in a lead matrix. This central region is surrounded by another fast region consisting of a lead matrix with 36% enriched uranium oxide (UO_2 36%) rods. In this region the three experimental channels (EC1B, EC2B, and EC3B) are located (see Figure 3). The fast region is decoupled from the surrounding thermal region by the so-called "valve," consisting of one row of natural uranium metal (U_{met}) rods and one row of boron carbide (B_4C) rods. The thermal region consists of a polyethylene matrix with holes, which are filled either with 10% enriched UO_2 fuel or with air. The core is surrounded by a graphite reflector.

The SC3A configuration (see the HELIOS model in Figure 4) of the YALINA-Booster facility is used for the comparison with the different analytical solutions. The configuration has been modeled in all details in HELIOS for the determination of the one- and two-group cross-section sets [23, 24]. In the SC3A configuration the 90% enriched uranium metal fuel of the inner fast zone is replaced by 36%

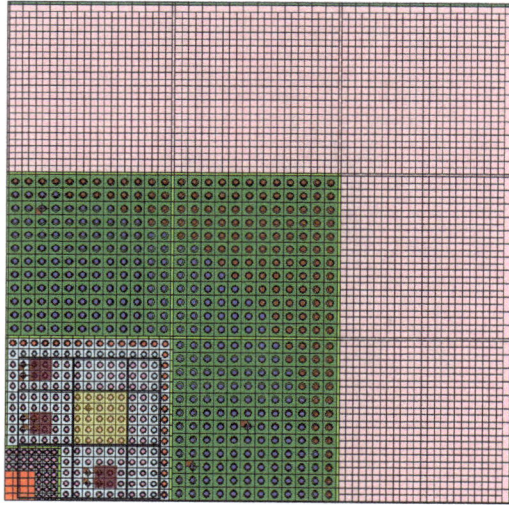

FIGURE 4: SC3A configuration of the YALINA-Booster core in the HELIOS model for XS preparation.

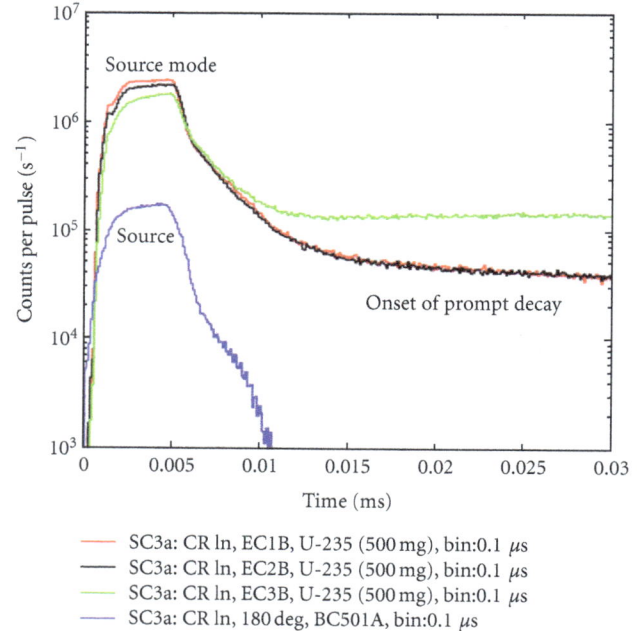

— SC3a: CR ln, EC1B, U-235 (500 mg), bin:0.1 μs
— SC3a: CR ln, EC2B, U-235 (500 mg), bin:0.1 μs
— SC3a: CR ln, EC3B, U-235 (500 mg), bin:0.1 μs
— SC3a: CR ln, 180 deg, BC501A, bin:0.1 μs

FIGURE 5: Detector responses in the pulsed neutron source experiment at three different detector positions in the fast area of the SC3A configuration of YALINA-Booster [31].

enriched uranium oxide fuel (tight rows of bright pink rods, which surround the red target zone). The thermal region in green contains the rods in deep blue and the empty positions in brown, respectively. The graphite reflector is shown in rose. The critical 2D HELIOS calculation has been corrected with a leakage term (defined in the HELIOS input by using the input buckling B^2 option) in the third dimension to reach a comparable result to a 3D MCNP calculation for a critical problem [31]. This is used to correct for the really small depth of the YALINA-Booster core ~0.5 m. The k_{eff} of the system with the above-mentioned leakage correction is 0.949090.

For the comparison of the space-time dependent analytical solutions with the real experiment a cross-section set is needed. This cross-section set is calculated with the licensing grade code module HELIOS 1.9 [26]. One-quarter of the YALINA-Booster facility is reproduced in a two-dimensional model in all details in unstructured mesh. The experimental channels are all relocated into the modeled core quarter, but only with the weight of 25%. The microscopic cross-sections are taken from the HELIOS internal library with 190 energy groups. A two-dimensional 190 energy group neutron flux solution is calculated using this library and the above-given geometry with the SC3A material configuration. These neutron flux solutions are used to produce condensed one- and two-group cross-section sets for the yellow overlaid reference element for the YALINA-Booster core in SC3A configuration.

The external source is not taken into account in the HELIOS calculation for the cross-section preparation. The absence of the external source neutrons with 14 MeV from the D-T reaction has the potential to influence the neutron spectrum slightly in the very center of the system. Nevertheless it has to be kept in mind that the amount of external source neutrons is small <5% and the slowing down to "reactor energies" occurs after some collisions in the lead matrix.

The YALINA-Booster facility has a very specific and unusual design. The basic idea was to produce a small fast subcritical reactor experiment for ADS study. The thermal part around the fast system in the lead matrix was introduced to reduce the leakage out of the fast system and to reach an acceptable criticality level in a small facility; the overall core size is below one by one meter. To create the possibility of doing fast system measurements the "valve" was inserted. In theory, the rows consisting of B_4C and natural uranium should prevent the thermal neutrons from the thermal zone from entering the fast zone.

Figure 5 shows the detector responses at the three detector positions in the fast zone of the YALINA-Booster experiment induced by a finite source pulse. The neutron generator was operated in pulse mode. The detector's signals were registered after each pulse. In Figure 5 the histograms of pulses were produced by adding data from all pulses to each other. For a more detailed analysis, the response of the source monitor (blue line) is added to create an insight into the time behavior of the external neutron source. The responses at the two innermost detectors show the expected results, but the response of the outer detector (EC3B in green) shows an unexpected behavior. After the prompt exponential die out of the neutron pulse, all curves should end at approximately the same delayed neutron level. The stable level of the detector response, which is significantly higher at the EC3B even after 0.03 ms, has to be explained. The most convenient explanation for the high detector response after the die out of the prompt pulse in the fast area is the inflow of neutrons from the thermal system. Since the neutron generation time in the thermal area is significantly longer, there is still a high

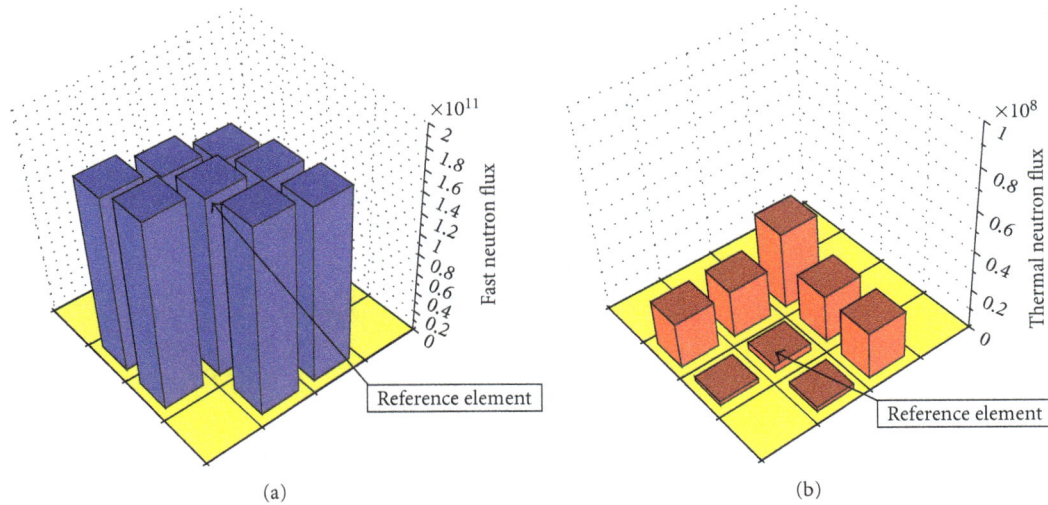

FIGURE 6: Fast neutron flux (a) and thermal neutron flux, (b) in the different relevant fuel elements in the fast part of YALINA-Booster for SC3A.

neutron flux available. The inflow of neutrons from the thermal area should be suppressed by the design using the "valve," the combination of one row of B_4C and one row of natural uranium. The "valve" is more efficient for thermal neutrons, since the absorption coefficient of B-10 for thermal neutrons is high, but fission neutrons can penetrate the "valve" since they are definitely not all well thermalized before they appear at the "valve." A conclusion of the described effect of the insufficient die out of the detector response is that fast fission neutrons born in the thermal area can influence at least the outermost detector in the fast area. Thus, "EC3B is affected by the thermal zone" [31].

A detailed analysis has been performed on the basis of the HELIOS model, developed for the cross-section preparation to explain the unexpected results in the cross-section sets [22] and to confirm the thesis of the inflow of neutrons from the thermal area.

The 190 energy group neutron flux calculated with HELIOS is condensed for the detailed analysis of the spatial neutron flux distributions on the basis of the fuel element-like structures, marked in Figure 4. The relevant fast and the thermal neutron flux (cutoff energy is $6.2506 \cdot 10^{-1}$ eV) distributions are shown in Figure 6. The spatial distribution of the fast neutron flux (Figure 6(a)) shows the expected cosine-like distribution. An influence of the neutrons penetrating through the "valve" is not explicitly visible here. The result for the thermal neutron flux (Figure 6(b)) is in strong contrast to the result for the fast neutron flux. In the fast area the thermal flux has to be very low due to the lead matrix in the fast zone of the YALINA-Booster core. The thermal neutron flux shows a distribution which is neither expected nor typical for a reactor core. In contrast to the cosine distribution of the fast neutron flux, the thermal neutron flux grows exponentially with increasing distance from the center. The thermal neutron flux is the highest in the corner of the fast zone, which is surrounded by the thermal zone on both sides. This kind of distribution can be explained only by

the ingress of thermal neutrons from the moderated outer, thermal area through the "valve" into the fast area. The comparison of the thermal neutron flux shows an 18 times higher thermal neutron flux in the corner element compared to the elements in the inner row. This significantly higher thermal neutron flux in the outer fuel elements has strong influence on the production of the cross-section set for all kinds of deterministic calculations for YALINA-Booster. The influence of the thermal neutrons on the cross-section preparation is very strong, since the microscopic cross-section, for example, fission of U-235, is more than hundred times higher than for fast neutrons. The macroscopic production cross-section $\nu\Sigma_f$ is roughly a factor of 3 higher in the corner element than in the elements of the inner row. The effect is only a spectral effect, since the materials in all elements are completely identical. In the standard cross-section preparation a fuel element is calculated in infinite grid, where this spectral effect is lost. Thus an adequate flux weighting of the cross-sections requires a simulation of the full YALINA-Booster core. Due to the strong differences in the thermal flux distribution, the cross-sections for identical material at different positions will vary significantly [22].

4. Comparison with the Experiment

4.1. One Group P_1 and Diffusion Solution. The experimental results, shown in Figure 5, are compared with the developed analytical approximation solutions for the first 0.01 ms on a logarithmic scale for a qualitative comparison in Figure 7. The results for the analytical solution for the different detector positions are given as lines, and the experimental results are given as diamonds in the identical color. This graph already shows one of the major problems. The mathematical representation of the external neutron source by two Heaviside functions is only an approximation of the detected pulse of the external source. In the experiment the external neutron source is created by a D-T reaction in the target

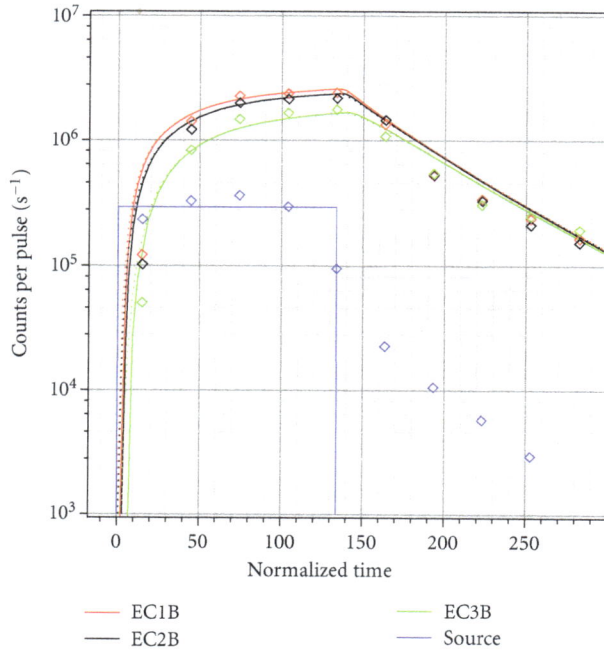

FIGURE 7: Comparison of the analytical results for the space-time dependent neutron flux (full line—P_1 transport; dotted line—diffusion) with the results for the SC3A experiment at the YALINA-Booster facility on logarithmic scale.

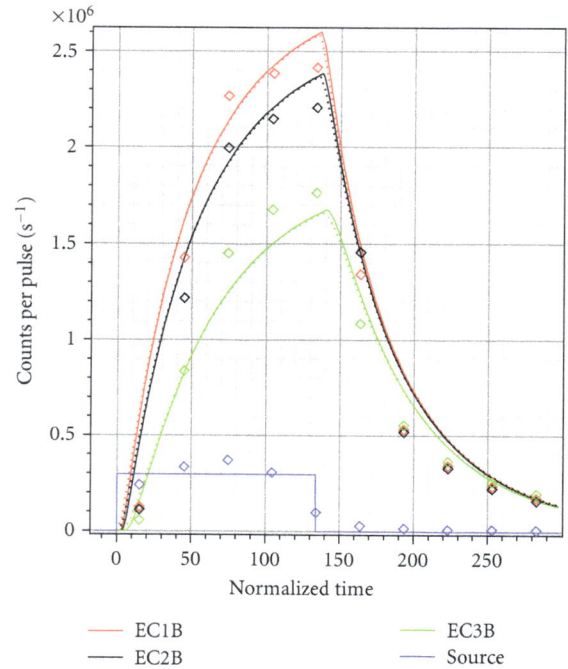

FIGURE 8: Comparison of the analytical results for the space-time dependent neutron flux (full line—P_1 transport; dotted line—diffusion) with the results for the SC3A experiment at the YALINA-Booster facility on linear scale.

in the center of the YALINA-Booster facility. Unfortunately, switch on as well as switch off of the accelerator, which provides the source with deuterium ions, takes some finite time. With this limitation it is impossible to create really sharp pulse fronts for the external neutron source.

Additionally, the source detector is located in another room with a distance of roughly 4 meters from the target, and the signal, taken from the source monitor, follows another chain of electronics which might have other time properties. Both facts cause some time delay in the source counts compared to the counts in the core. These effects have been taken into account by an estimated time correction of $2\,\mu s$ for the source.

Nevertheless, a good qualitative agreement between the analytic approximation solutions (P_1 transport—solid line, diffusion—dotted line) and the experimental data is observable. Especially, the spatial distribution in the plateau phase shows a good agreement. This exactly confirms the need and demonstrates the progress of the developed analytic approximation solutions derived without separation of space and time in contrast to the currently used methods based on point kinetics.

The detailed comparison of the developed analytical solutions with the SC3A experimental results at the YALINA-Booster facility on linear scale is shown in Figure 8. It is once more observable that the response of the source detector in the experiment (blue diamonds) does not have the sharp rectangular time behavior like it is used in the development of the analytical solutions. The switch on of the accelerator as well as the switch off of the accelerator cannot be performed

in sharp way like it is defined in the mathematical way. This fact causes observable differences in the results for the experiment and the analytical solution on the linear scale. Nevertheless, there is a good agreement between the analytical results with the experimental results. The detector response for the three different detector positions EC1B red, EC2B black, and EC3B green is given at different time points of the experiment by diamonds. The time behavior of the neutron flux at the different positions of the detectors calculated with the analytical solution for the time-dependent P_1 transport solution is given by the full-lines and the results obtained with the analytical solution for the time-dependent diffusion solution with the dotted line: EC1B-red; EC2B-black; EC3B-green. Both analytical solutions show similar behavior with only minor differences in a short time period after the step change in the external source, for the switch on as well as for the switch off. The comparison with the experiment demonstrates that the developed solutions without separation of space and time can reproduce the behavior of the neutron flux at the different detector positions. A small overestimation for the diffusion solution in comparison with the time-dependent P_1 solution is observable in the initiation phase of the transient. The neutron flux grows rapidly after the switch on of the source. After roughly 10 neutron generations the neutron production exceeds due to multiplication of the external neutron source. The calculated neutron flux rises as long as the external source is in operation and would reach a steady state value if the external source would be operated long enough.

The neutron flux decays very rapidly in an exponential manner after the switch off of the external source. The response at the outermost detector is somewhat lower in the calculation. This difference can be once more explained by the ingress of neutrons from the thermal area.

The effect of the fast neutrons, which travel from the thermal into the fast area, is the starting point for the discussion of another important fact. Currently, the analytic approximation is only for one region and one energy group and it has only been applied to the fast area of the system. However, the k_{eff} for the fast area is not known and even a prediction is very problematic, since, on the one hand, the fast area is heavily influenced by the thermal area and the "neutron valve," and on the other hand, the model for the cross-sections is only two-dimensional. These two facts make it nearly impossible to draw a reliable balance between neutron gains and losses for the fast region. Additional problems are caused by the definition of the boundary, since including or excluding of the strong absorber in the "valve" influences the result significantly. A rough estimation of the k_{eff} has given a value around 0.6 for the fast region. This value has finally been used for the calculations with the analytic approximation solutions. This means an analytic solution for two or more regions would be needed to overcome this problem at least partly. Ideally the solution for the thermal area should be additionally expanded to two-energy group. Summing up all mentioned above, we can conclude that a really complicated experiment like YALINA-Booster is not the ideal test case for the development of a new analysis method. A system like Guinevere [3], consisting of only two regions, a pure lead region containing the fuel, and another pure lead region, acting as reflector, will simplify the problem. Nevertheless, at least an analytical solution for two regions would be needed even for this kind of system to determine the real multiplication factor for the system.

4.2. Diffusion versus P_1 Transport. The decision for either the time-dependent P_1 transport equation or the time-dependent diffusion equation for the further development of the analytic approximation solutions is very important. It is required to solve a second-order equation for the development of the analytic approximation solution for the P_1 transport equation. To solve only the time-dependent diffusion equation would lead to a significantly reduced complexity in the solution, since there is only a first-order partial differential equation in time to solve. This reduced complexity offers the possibility to invest more into the spatial or energetic domain of the problem by tracking a system with two or more regions or more than one energy group [28].

It has already been demonstrated that there is a visible deviation between the results for the time-dependent P_1 transport equation and the time-dependent diffusion equation [8, 20, 21]. An evaluation of the deviation Figure 9(a) and the difference Figure 9(b) between the P_1 transport and the diffusion solution for the three different detector positions in YALINA-Booster is shown in Figure 9. The structure of the deviations is identical for all three detector positions. The major deviations occur certainly in the moments directly after the change in the external neutron source. The

deviation in the beginning is rather high since the neutron flux used for normalization is comparably low. This deviation dies out after roughly 50 neutron generations for the case of a constant operating source. Following the switch off of the source the deviation rises once more but only to roughly 3% and dies out rapidly during the fade away of the neutron flux. The deviation in the beginning is strongly dependent on the distance from the source. The reason for this deviation is the infinite velocity of a perturbation in the diffusion equation compared to the finite velocity of the spreading of the perturbation in the P_1 transport solution. Due to this fact, the time delay of the reduction of the deviation increases with increasing distance from the external source.

The difference between the time-dependent P_1 result and the time-dependent diffusion result is normalized on an average plateau value of $2 \cdot 10^6$ counts per pulse. This way of evaluation avoids the weighting of the difference by different flux values, which leads, especially in cases of very low neutron flux, to tremendously high deviation. The evaluation of the difference gives a good overview on the quality of the result independent of the actual flux level. The difference during the whole transient is below 4% with peaks at the points of the changes of the external source. Negative difference indicates overestimation by the diffusion result.

Both evaluations show significant differences between the diffusion and the P_1 result, but these differences are low in comparison with the differences between the experimental results and the calculations. The major reasons for the differences between experiment and calculation are the limited representation of the real external source in the experiment and the difficulties to represent the complicated geometric structure (three different regions, three dimensions) of the experimental facility by one energy group, one region, and one-dimensional analytical solution.

The comparison of the time-dependent diffusion and the time-dependent P_1 solutions, both without separation of space and time, shows that already the diffusion solution provides good results for the evaluated finite source pulse experiment. The limitation for the recalculation of the experiment is given by the incomplete representation of the character of the source pulse. Unfortunately, the pulses are not as sharp as it would be needed for a detailed analysis of pulsed source experiments based on a Heaviside step function.

4.3. Two-Energy Group Solution. The results of the analytic approximation solutions for the time-dependent one- and two-group diffusion equation for three different detector positions (EC1B, EC2B, and EC3B) are compared to the detector response in the experiment SC3A in Figure 10. First of all, the same conditions (difference in the temporal source shape, influence of the thermal zone...) exist, but the observed time period has been extended for the analysis of the delayed neutron level. The detector response for the three different detector positions—EC1B (red), EC2B (black), and EC3B (green)—is given once more by diamonds at different time points of the experiment. The time behavior of the neutron flux is calculated in this case with the analytical time-dependent one group diffusion solution (Figure 10(a)). The neutron flux at the different positions of the detector is

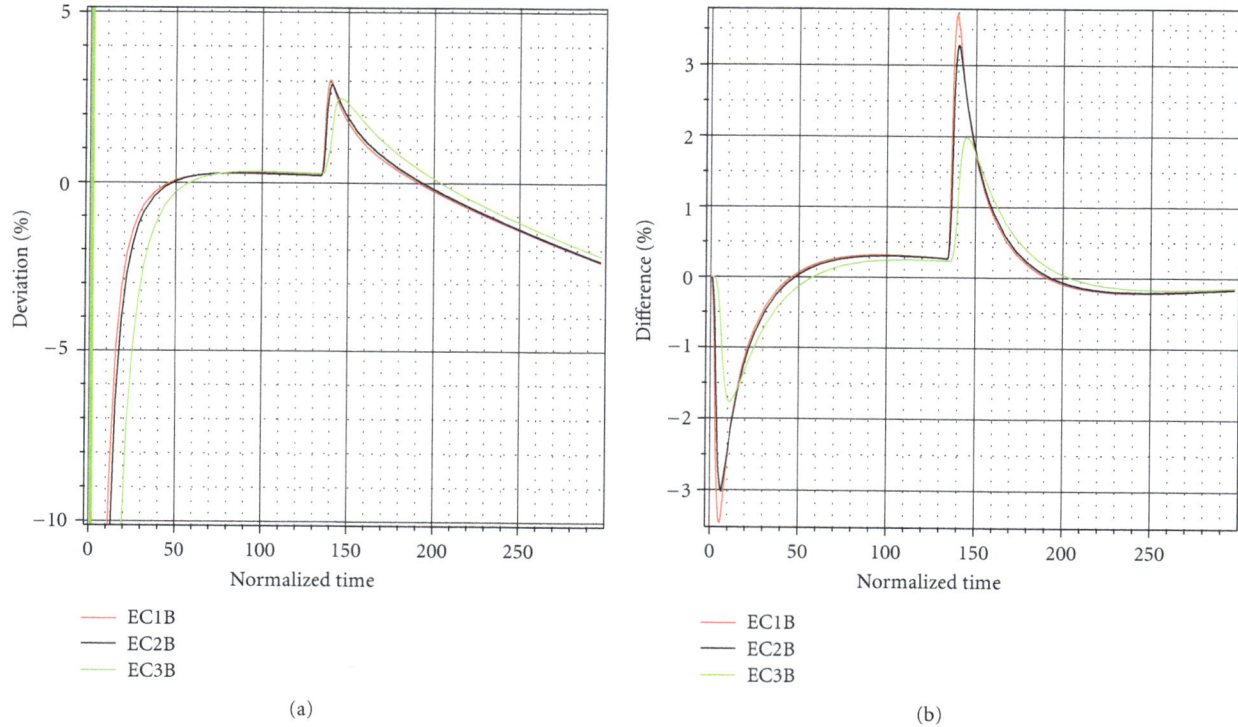

FIGURE 9: Evaluation of the deviation (a) (deviation = $(\Phi_{P_1}/\Phi_{\text{diffusion}} - 1) * 100$) and difference, (b) (difference = $\Phi_{P_1} - \Phi_{\text{diffusion}}$) between the time dependent P_1 transport solution and the time dependent diffusion solution at the different detector positions.

given by the full lines: EC1B-red; EC2B-black; EC3B-green. The comparison with the experiment demonstrates that the developed one group solution without space and time separation can reproduce the behavior of the neutron flux at the different positions of the detectors, like it was demonstrated before. The neutron flux decays very rapidly in an exponential manner after the switch off of the external source. Two points show deviations which have to be discussed: first, the response at the outermost detector (green) is somewhat lower in the calculation. This difference can be explained by the location close to the thermal area position. Second, the steady state value of the neutron flux after the switch off of the source is in the experiment significantly higher at the outermost position than in the calculation. This difference indicates once more that there is an influence from the thermal area to the fast area. The neutrons have a significantly longer lifetime in the thermal area; thus there is still a high neutron density available in the thermal area, while the neutron density in the fast area has already decreased significantly. This gradient between the fast and the thermal area leads to the possibility of a neutron inflow to the fast area.

The results of the improved mathematical model, which uses the time-dependent two-group diffusion equation and an adopted source, are shown in the right part of Figure 10. The external source is a combination of the central external source pulse and the time independent boundary source (to reproduce the incoming neutrons from the thermal area),

like it is shown in Figure 2. The first significant difference can be seen in the values after the die out of the pulse. The adopted solution shows a very good agreement when the effect of the pulse has died out and the neutrons are still streaming from the thermal area into the fast area due to the significantly slower decay of the prompt flux in the thermal area. Thus the influence of the thermal area on the fast area can be represented with the new model. With the knowledge of having captured this effect, a close look on the peaks is of our interest. In the one group analysis without caring for the neutrons streaming into the fast area, an increased deviation was visible in the outermost detector EC3B. Now this deviation has vanished nearly completely. The representation of all peaks is good and reflects a good reproduction of the spatial distribution of the neutron flux during the pulse. Additionally, the time shape of the pulses at all detector positions has improved and comes closer to the experiment. This improvement can be explained with the use of two different neutron velocities, or neutron generation times, in the two energy group solution, which is simply a better approximation of the real continuous energy flux. The change in the horizontal axis is caused by the different, in the two-group case averaged, neutron generation times.

Overall, the comparison of the results of the different analytic approximations with experimental ones from the YALINA-Booster facility has been successful for the SC3A. The extended solution has only been required due to the very special design of the YALINA-Booster facility.

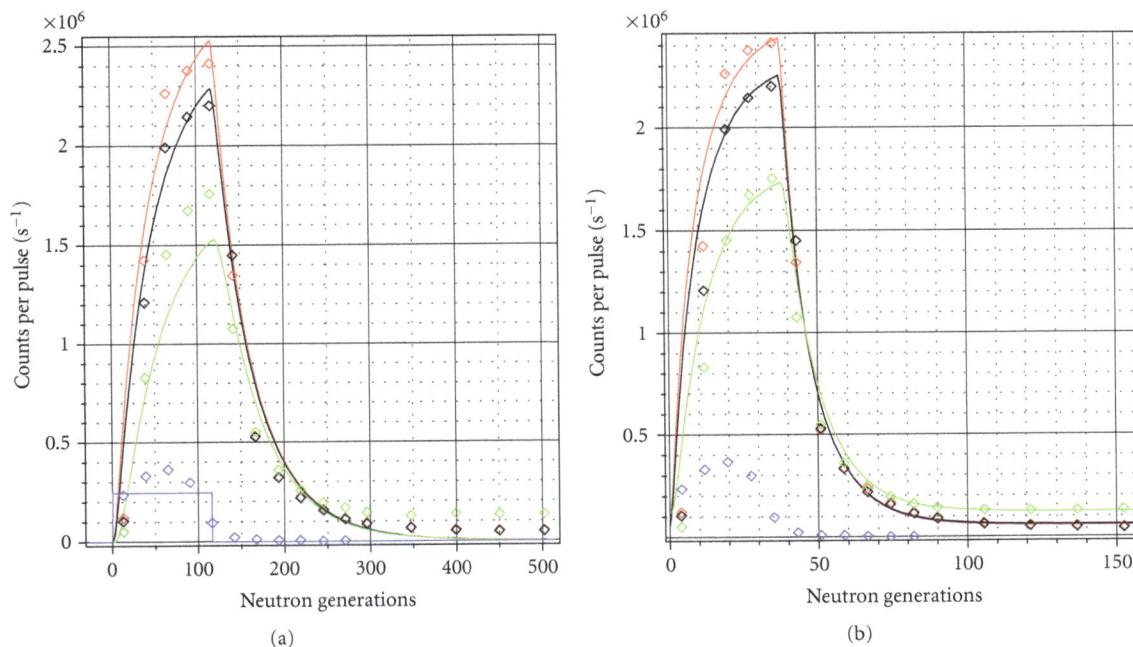

FIGURE 10: Comparison of the analytical results gained with the analytical solution of the time dependent diffusion equation (one-group—(a), two-group—(b)) with the experimental results for the SC3A configuration of the YALINA-Booster facility.

4.4. Two-Spatial Region-Solution.

The final idea is to use a solution with two spatial regions, especially for the analysis of the experiments in the GUINEVERE facility [3, 29]. The two-region solution without separation of space and time for the coupled system of a multiplying core with external source and a reflecting surrounding has been derived recently. The flux and the current density are continuous across the interface between two media, and the outer boundaries are reflecting. The exact analytical solution is expressed in terms of a Green's function. The solution is developed by the application of the Laplace transformation [30]. The problems that occur for the one-region solution in YALINA, where the multiplication factor has been calculated only for the region, which is simulated, should be eliminated with this solution for GUIEVERE, where only two regions exist.

The solution will not be applied to YALINA, since two regions would not be sufficient for this complicated configuration. The experiments in the GUINEVERE facility are planned in the beginning of 2012. The successful coupling of the GENEPI accelerator with subcritical VENUS reactor has been announced on October 28, 2011. Early in December an official start-up is scheduled and the first experiments start.

5. Conclusion

A newly developed methodology for reaching a deeper understanding of ADS experiments has been presented as basis for a new method for the analysis of ADS experiments. In the last years at the Helmholtz-Zentrum Dresden-Rossendorf the developed method for the solution of the time-dependent P_1 transport equation, avoiding separation of space and time and based on Green's functions, is promising. Further solutions for the time-dependent diffusion equation for one region and one- and two-energy groups have been used and compared with the YALINA experiments. A special solution for a finite pulse of the localized external neutron source has been developed for all analytical solutions. As an input for the new methodology, a cross-section set has to be created. This is performed with the HELIOS code with a detailed two-dimensional model for the whole core of the YALINA-Booster facility. This detailed two-dimensional transport model has been used to get a deeper insight into the specificity of the YALINA-Booster experimental setup. The detailed analysis of the fast and the thermal neutron flux distribution in the fast area has shown a strong influence of the thermal area on the fast area. This strong influence requires a significantly increased effort for the cross-section preparation. A full model should be used instead of the standard method, which uses a fuel element in reflective surrounding, to catch the effect of the neutron ingress from the thermal to the fast area.

The first comparison of the results of the analytic approximation solutions with experiments from the YALINA-Booster facility has been successful. Good agreement between the experiment and the calculation was obtained in space as well as in time. The comparison of the time-dependent P_1 results with the time-dependent diffusion results has shown that the differences between the modeling and the experiment cause in the case of YALINA significantly stronger differences than the use of different transport approximations. For a better representation of the specifics of the YALINA-Booster setup, a two-energy group solution with a special arrangement of the sources has been developed which leads to improved results. Since the complicated YALINA-Booster system cannot be represented by a one-region solution the analysis had to be concentrated on the fast zone only. A drawback of this reduction is the limitation of the k_{eff} calculation only to the fast zone. For a prediction of the k_{eff} for the SC3A configuration of the YALINA-Booster facility a calculation for the full system would be essential (at least 3 regions). To overcome this problem for the future experiment GUINEVERE a two region solution has already been developed for the full representation of the GUINEVERE configuration.

Generally, it has to be mentioned that the advantage of the analytical solution over the numerical method lies in the following: the ansalytical solution is an exact solution, which gives dependences on variables; the numerical method is only an approximation to the problem, which causes an inaccuracy, if a large region is considered. Additionally, the numerical method does not give a continuous solution and, in some cases, can be time and resource consuming due to the number of iterations.

Overall, very promising results have been obtained, and a good agreement between the experiment and the calculation has been met in space, as well as in time by using analytical solutions developed without separation of space and time. The flexibility of the analytical solutions for the complicated experimental settings has been demonstrated. Thus, analytical solutions without separation of space and time are a very promising tool to develop a new method for the analysis of ADS experiments.

References

[1] R. Soule, W. Assal, P. Chaussonnet et al., "Neutronic studies in support of accelerator-driven systems: the MUSE experiments in the MASURCA facility," *Nuclear Science and Engineering*, vol. 148, no. 1, pp. 124–152, 2004.

[2] A. I. Kievitskaia et al., "Experimental and theoretical research on transmutation of long-lived fission products and minor actinides in a subcritical assembly driven by a neutron generator," in *Proceedings of the 3rd International Conference on ADTTA*, Praha, Czech Republic, 1999.

[3] H. Aït Abderrahim and P. Baeten, "The GUINEVERE-project at VENUS, project status," in *ECATS Meeting*, Cadarache, France, Januray 2008.

[4] C. M. Persson, P. Seltborg, A. Åhlander et al., "Analysis of reactivity determination methods in the subcritical experiment Yalina," *Nuclear Instruments and Methods in Physics Research, Section A*, vol. 554, no. 1–3, pp. 374–383, 2005.

[5] N. G. Sjöstrand, "Measurements on a subcritical reactor using a pulsed neutron source," *Arkiv för Fysik*, vol. 11, p. 13, 1956.

[6] B. E. Simmons and J. S. King, "A pulsed neutron technique for reactivity determination," *Nuclear Science and Engineering*, vol. 3, p. 595, 1958.

[7] K. Ott and R. Neuhold, *Introductory Nuclear Reactor Dynamics*, American Nuclear Society, La Grange Park, Ill, USA, 1985.

[8] B. Merk and F. P. Weiß, "A three-scale expansion solution for a time-dependent P_1 neutron transport problem with external source," *Nuclear Science and Engineering*, vol. 163, no. 2, pp. 152–174, 2009.

[9] G. Granget et al., "EUROTRANS/ECATS or neutronic experiments for the validation of XT-ADS and EFIT monitoring: status, progress and first assessments," in *OECD Nuclear Energy Agency International Workshop on Technology and Components of Accelerator Driven Systems*, Karlsruhe, Germany, March 2010.

[10] Analytical and Experimental Benchmark Analyses of Accelerator Driven Systems (ADS), http://www.iaea.org/inisnkm/nkm/aws/fnss/crp/crp7.html.

[11] P. Gajda, "Pulsed neutron source measurement in YALINA-subcritical facility," in *Proceedings of the 3rd Environmental Compatible Air Transport System Progress Meeting (ECATS' 10)*, Mol, Belgium, March 2010.

[12] V. Bécares, "Analysis of pulsed neutron source experiments," in *Proceedings of the 3rd Environmental Compatible Air Transport System Progress Meeting (ECATS' 10)*, Mol, Belgium, March 2010.

[13] V. Bécares, D. Villamarín, M. Fernández-Ordóñez, E. M. González-Romero, and Y. Fokov, "Correction methods for reactivity monitoring techniques in Pulsed Neutron Source (PNS) measurements," in *Proceedings of the International Conference on the Physics of Reactors (PHYSOR '10)*, pp. 914–928, Pittsburgh, Pa, USA, May 2010.

[14] Y. Gohar et al., "'Argonne analyses of YALINA-booster', YALINA-thermal, and LEU YALINA-booster," in *IAEA Meeting*, Mumbai, India, February 2010.

[15] Summary of the Technical Meeting on, "'Use of Low Enriched Uranium (LEU Fuel in Accelerator Driven Sub-Critical Assembly (ADS Systems', and 3rd RCM of the CRP on 'Analytical and Experimental Benchmark Analyses of Accelerator Driven Systems'," Mumbai, India, February 2010.

[16] A. M. Weinberg and E. P. Wigner, *The Physical Theory of Neutron Chain Reactors*, University of Chicago Press, Chicago, Ill, USA, 1958.

[17] D. G. Duffy, *Green's Functions with Application*, Chapman & Hall/CRC, 2001.

[18] M. N. Özişik and B. Vick, "Propagation and reflection of thermal waves in a finite medium," *International Journal of Heat and Mass Transfer*, vol. 27, no. 10, pp. 1845–1854, 1984.

[19] B. Merk, "Time dependent analytical approximation solutions for a pulsed source problem: P_1 transport versus diffusion," in *IEEE Nuclear Science Symposium*, Dresden, Germany, 2008.

[20] B. Merk, "An analytical solution for a one dimensional time-dependent neutron transport problem with external source," *Transport Theory and Statistical Physics*, vol. 37, no. 5–7, pp. 535–549, 2008.

[21] B. Merk, "An analytical approximation solution for a time-dependent neutron transport problem with external source and delayed neutron production," *Nuclear Science and Engineering*, vol. 161, no. 1, pp. 49–67, 2009.

[22] B. Merk, V. Glivici-Cotruţă, and F. P. Weiß, "A solution for the telegrapher's equation with external source: application to Yalina—SC3A and SC3B," in *International Conference on the Physics of Reactors (PHYSOR '10)*, pp. 3053–3065, Pittsburgh, Pa, USA, May 2010.

[23] B. Merk, V. Glivici-Cotruta, and F. P. Weiß, "A solution for Telegrapher's equation with external source: development and first application," *Nuovo Cimento della Societa Italiana di Fisica B*, vol. 125, no. 12, pp. 1547–1559, 2010.

[24] B. Merk, V. Glivici-Cotruţă, and F.-P. Weiß, "Application of a two energy group analytical solutions to the YALINA experiment SC3A," in *Proceedings of the International Conference on Mathematics and Computational Methods Applied to Nuclear Science and Engineering (MC '11)*, Rio de Janeiro, Brazil, 2011.

[25] B. Merk, V. Glivici-Cotruṭă, and F.-P. Weiß, "A three scale expansion solution for the Telegrapher's equation with external source: development and first application," in *Proceedings of the 21st International Conference on Transport Theory*, Torino, Italy, 2009.

[26] HELIOS Methods, Studsvik Scandpower, 2003.

[27] V. Bécares Palacios et al., "Analysis of PNS measurements at the YALINA-Booster subcritical facility," in *Proceedings of the 2nd EUROTRANS ECATS Progress Meeting*, Aix-en-Provence, France, January 2009.

[28] B. Merk and F. P. Weiß, "A two group analytical approximation solution for an external source problem without separation of space and time," *Annals of Nuclear Energy*, vol. 37, no. 7, pp. 942–952, 2010.

[29] FREYA: Fast Reactor Experiments for hYbrid Applications, http://freya.sckcen.be/en.

[30] V. Glivici-Cotruta and B. Merk, "An analytical solution of the time dependent diffusion equation in a composite slab," submitted to: *Transport Theory and Statistical Physics*.

[31] C. Berglöf, "YALINA-booster: pulsed neutron source characterization," in *Proceedings of the 2nd EUROTRANS ECATS Progress Meeting*, Aix-en-Provence, France, 2009.

6

A Fast Numerical Method for the Calculation of the Equilibrium Isotopic Composition of a Transmutation System in an Advanced Fuel Cycle

F. Álvarez-Velarde and E. M. González-Romero

Centro de Investigaciones Energéticas, Medioambientales y Tecnológicas (CIEMAT), Avenida. Complutense 40. Ed. 17, 28040 Madrid, Spain

Correspondence should be addressed to F. Álvarez-Velarde, francisco.alvarez@ciemat.es

Academic Editor: Alberto Talamo

A fast numerical method for the calculation in a zero-dimensional approach of the equilibrium isotopic composition of an iteratively used transmutation system in an advanced fuel cycle, based on the Banach fixed point theorem, is described in this paper. The method divides the fuel cycle in successive stages: fuel fabrication, storage, irradiation inside the transmutation system, cooling, reprocessing, and incorporation of the external material into the new fresh fuel. The change of the fuel isotopic composition, represented by an isotope vector, is described in a matrix formulation. The resulting matrix equations are solved using direct methods with arbitrary precision arithmetic. The method has been successfully applied to a double-strata fuel cycle with light water reactors and accelerator-driven subcritical systems. After comparison to the results of the EVOLCODE 2.0 burn-up code, the observed differences are about a few percents in the mass estimations of the main actinides.

1. Introduction

The implementation of the advanced technologies of partitioning and transmutation (P&T) depends on the particular topics that energy policy makers of a country or region choose to optimize because of its singular constraints or motivations. Each of the different possible P&T configurations, that is, the advanced fuel cycle scenarios, will have different objectives that can impact technology choices and performance expected from such systems.

However, the full potential of any P&T policy can be exploited only if the advanced fuel cycle strategy is utilized for a minimum time period of about a hundred years [1, 2]. During this period of time, the irradiated fuels of the advanced reactors are reprocessed to obtain the useful elements for their later recycling as part of the new fresh fuel, in an iterative procedure along successive cycles. The fuel isotopic composition will evolve along a number of cycles, approaching progressively an equilibrium composition, defined as the long-term state reached after infinite cycle iterations.

The calculation of the equilibrium isotopic composition of a transmutation system in an advanced fuel cycle is a matter of special interest since this composition is often used as representative of the nuclear reactor through the whole fuel cycle [3, 4].

In this paper, we describe a fast numerical method for the calculation of the equilibrium isotopic composition in a zero-dimensional approach, taking advantage of the Banach fixed point theorem [5] for the demonstration of uniqueness and existence of a solution. The method is based on the division of the fuel cycle in successive stages: fuel fabrication, storage, irradiation inside the transmutation system, cooling, reprocessing, and incorporation of the external material into the new fresh fuel. The fuel isotopic composition is represented by an isotope vector and will be modified by each fuel cycle stage that is described in a matrix formulation. The fuel cycle is hence represented as a contractor mapping over the fuel composition vector.

With the aim of validating the methodology, a simulation of the fuel cycle with the advanced burn-up code EVOL-CODE 2.0 [6, 7] has been performed, checking that the isotopic composition has successfully reached the equilibrium and discussing the range of the possible deviations.

2. Computational Tool

EVOLCODE 2.0 is an in-house development to solve the burn-up problem, that is, the coupled problem of neutron transport and isotopic evolution. Its cycle data flow is shown in Figure 1.

The neutron transport stage is solved by the MCNPX code [8], allowing an important degree of the heterogeneity description in the reactor core model. The isotope evolution stage can be solved by the ORIGEN code [9] or by ACAB [10]. The user selects some geometrical regions or cells inside which neutronic properties and material composition are considered constant. For each of these cells, EVOLCODE 2.0 creates one-group effective cross-section libraries using the neutron flux energy spectrum provided by MCNPX in such a way that the whole calculation is faster than allowing MCNPX to do the job. Then, one ORIGEN execution is made independently for each cell providing the isotopic evolution of the materials. The ORIGEN executions are made using the neutron flux and materials at the beginning of the time step. In case that the averaged reactor power (estimated by the ORIGEN simulation) is not equal to the desired value or if its variation along the time step is large (causing a large uncertainty in the final results [7]), then the predictor/corrector method is activated to fix it. This whole procedure is repeated for several successive irradiation steps (or cycles) to reach the total burn-up specified by the user.

3. Double-Strata Fuel Cycle Scenario

One of the proposed advanced fuel cycle scenarios for waste reduction is the so-called double-strata scenario, whose scheme is shown in Figure 2. This fuel cycle has been studied in detail in the frame of different projects [11, 12] and consists in two separated strata or stages, one dedicated to energy production and the other dedicated to waste consumption.

In this particular specification of the double-strata scenario, the first stratum of the fuel cycle consists in the irradiation of UO_2 in light water reactors (LWRs) and the later advanced Purex reprocessing of the irradiated fuel for the Pu reutilization, only once, as MOX, again in LWR. Recovered MA coming from the first partitioning process and Pu and MA coming from the reprocessing of the MOX irradiated fuel are reutilized as fuel for the accelerator-driven subcritical system (ADS) in the second stratum. This second stratum is based on a fast spectrum ADS, which operates with continuous recycling of the main actinides (U, Pu, Np, Am, and Cm). This recycling is supposed to be a pyro-metallurgical process.

4. ADS Fuel Isotopic Composition

The isotopic composition of the ADS fuel depends on the LWR park since the LWR recovered material (from both UO_2 and MOX reprocessed fuels) becomes a part of the ADS fuel content in each new ADS irradiation (a 15% approximately). The other part of the ADS fuel comes from the pyro-reprocessing of the preceding ADS irradiated fuel. In different cycles, the ADS fuel isotopic composition would change (and being consumed), but maintaining similar ADS operating conditions and a constant additional supply per cycle of actinides from the first LWR-stratum, the fuel approaches an equilibrium (as it is demonstrated below) in which the fuel isotopic composition of each cycle is equal to the following one.

The iterative second stratum of this fuel cycle scenario can be divided in a succession of stages: fuel fabrication, storage, irradiation inside the ADS, cooling, reprocessing, and incorporation of the external material (from the first stratum) into the new fresh fuel. Each process can be described in a matrix formulation as an operator over the fuel composition vector V_n, obtained just before the nth ADS irradiation. Vectors and matrices length is determined by the total number of different isotopes considered in the problem. This number of isotopes is equal to $M = 26$ in this study (the detailed list is shown in Table 1), although a larger value of M has been used in other problems. If V_{n+1} is the fuel composition vector before the $(n + 1)$th ADS irradiation, it is related to V_n following the expression

$$V_{n+1} = D_s R D_c A V_n + V_{\text{ext}} = C V_n + V_{\text{ext}}, \qquad (1)$$

where A is the matrix operator representing the irradiation inside the ADS, D_c represents the cooling decay time after the irradiation (two years in this scenario), R is the reprocessing operator, D_s represents the storage time before irradiation (one year), V_{ext} is the material incorporation from the first stratum (taken at the moment just before the beginning of the ADS irradiation), and $C = D_s R D_c A$. Since the fuel composition vector can be expressed as $V = F(V) = CV + V_{\text{ext}}$, the process fulfils the requirements of the Banach fixed point theorem, and therefore there is a solution vector $V = F^n(V)$ under the requirement that $F(V)$ is a contractor mapping, that is, if there is a constant $0 < k < 1$ such as

$$\|F(V') - F(V)\| \le k\|V' - V\|, \qquad (2)$$

where the symbol $\| \cdot \|$ represents the norm of the operator. Applying $F(V) = CV + V_{\text{ext}}$, this inequality can be written as

$$\|C(V' - V)\| \le k\|V' - V\|, \qquad (3)$$

which is satisfied if $|C| < 1$. In order to prove that this condition is satisfied, we must analyze the norm of each of the matrix operators appearing in the definition of C. The norm of the decay operators D_s and D_c is equal to the unity because they only change one isotope into another leaving the total amount of material unchanged (being precise, it might be smaller than the unity as some decay products are not included in the dimensions of the composition vector).

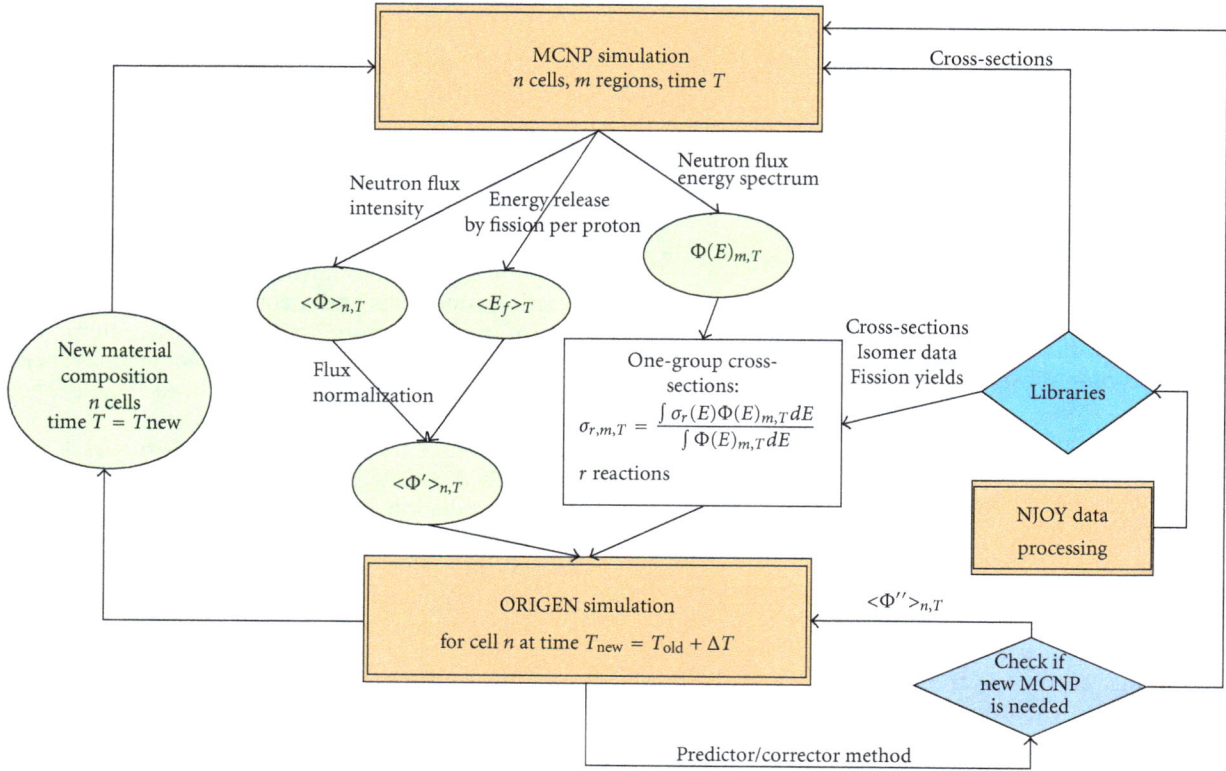

FIGURE 1: EVOLCODE 2 cycle data flow scheme.

The reprocessing operator, R, is a diagonal matrix having a constant diagonal value smaller than the unity (reprocessing recovery factor of 0.999) for the reprocessing isotopes (only the reprocessing elements U, Pu, Np, Am, and Cm have been considered), hence $|R| < 1$. Finally, the norm of the irradiation matrix A is strictly smaller than the unity because it represents consumption of heavy material (approximately a 15% for this ADS burn-up). This can be easily shown when the norm is defined as the maximum $|A_{ij}|$ and demonstrated for other definitions. For all these reasons, $|C| < 1$, therefore this process fulfils the Banach fixed point theorem and there exists a unique solution of the equation of the type $V = F^n(V_0)$, for an arbitrary V_0.

Coming back to the previous notation, we can find the solution directly from the equilibrium equation $V_{n+1} = V_n = V_{eq}$; hence the expression of the fuel composition vector can be written as

$$V_{eq} = CV_{eq} + V_{ext}. \qquad (4)$$

The solution of this equation is

$$V_{eq} = (1 - C)^{-1} V_{ext}. \qquad (5)$$

In order to reach this solution, it is necessary to calculate D_c, D_s, A, R, and V_{ext} in a matrix formulation from the data of the problem. The operator R is represented by a diagonal matrix with a constant value of 0.999, as mentioned above. V_{ext} is a vector containing the contribution of the LWR stratum so it can be provided after solving the first

noniterative stratum of the fuel cycle. The operator A can be represented by the generic evolution of a nuclear system under irradiation. The generic evolution of the isotopic composition vector $V(t)$ can be described in the zero-dimensional approximation as

$$\frac{dV(t)}{dt} = E \cdot V(t), \quad \text{with } E = XS \cdot \Phi + \text{DEC}, \qquad (6)$$

where XS is the cross-section matrix, DEC is the radioactive decay matrix, and Φ is the averaged neutron flux. Operators D_c and D_s can be obtained in case that $\Phi = 0$. The formal solution of this coupled linear system of equations of constant coefficients is

$$V(t) = e^{E \cdot t} \cdot V(0). \qquad (7)$$

In the base of eigenvectors of the matrix E, this solution can be written as

$$A = e^{E \cdot t} = U \begin{pmatrix} e^{au_1 t} & 0 & \cdots & 0 \\ 0 & e^{au_2 t} & \cdots & 0 \\ \cdots & \cdots & \cdots & \cdots \\ 0 & 0 & \cdots & e^{au_M t} \end{pmatrix} U^{-1}, \qquad (8)$$

where $U(eu_1, eu_2, \ldots, eu_M)$ is the matrix of the eigenvectors eu_i associated with the eigenvalues au_i. The calculation of the eigenvalues and eigenvectors needed to compute the exponential and the inversion of the $(1-C)$ matrix has

A Fast Numerical Method for the Calculation of the Equilibrium Isotopic Composition of a Transmutation System in an Advanced Fuel Cycle

53

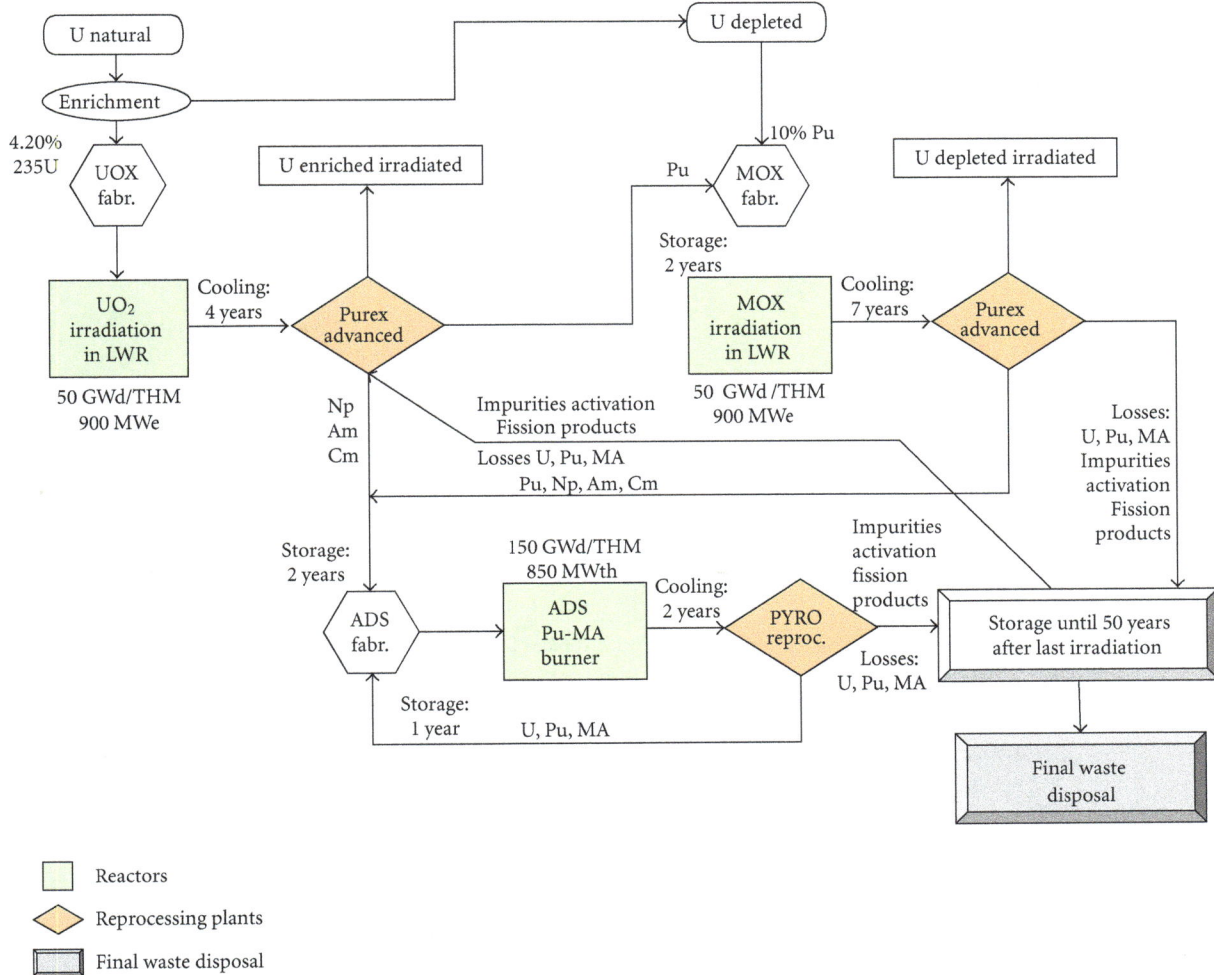

FIGURE 2: Scheme of the double-strata fuel cycle scenario with LWR and ADS.

been a difficult problem because of the large numerical instabilities involved in the manipulation of these large matrices. However, the easy access to arbitrary precision arithmetic, made recently available in popular computer codes as Maple [13] and Mathematica [14], allows using brute force and direct methods of solving these problems. For our case, it was necessary to set the arithmetic precision to 70 decimal digits to obtain the required stability. In spite of this condition, the eigenvectors computed had to be regularized by discarding very small imaginary components $(|\operatorname{Im}(eu_i)| < 10^{-25}|\operatorname{Re}(eu_i)|)$. The process was implemented with the Maple program, where efficient procedures for the calculation of eigenvalues, eigenvectors, and matrix inversions were readily available. The typical calculation time is few minutes in a modern PC.

In order to estimate the flux level and the one-group zero-dimensional equivalent cross-sections, a fully detailed transport simulation was performed. In this simulation, the exact geometry, the required total power, and an initial guess of the equilibrium composition of the reactor materials were used. Careful weighting had to be used in the procedure of averaging the cross-sections over the whole reactor volume,

which was divided in different cells with isotopic evolution. For every one of these cells, one set of one-group effective cross-sections were calculated using the EVOLCODE 2.0 simulation system. The final reactor-averaged cross-sections were calculated weighting the different one-group cross-sections of each fuel cell by its fuel mass. The initial guess of the fuel composition was set to the same vector as the vector of actinides from the external refuelling, V_{ext}.

Since the initial ADS fuel composition, chosen for the calculation of matrices XS and Φ, is different from the final equilibrium composition, this method of calculation of the equilibrium isotopic composition of the ADS fuel introduces an error in the results. An iterative process can be carried out to increase the precision of the method. Once the equilibrium isotopic composition is calculated using the previous noniterative method, the resulting isotopic composition can be used as the initial ADS fuel content and a new set of cross-sections/neutron flux could be obtained with EVOLCODE 2.0. Then, the method is repeated in order to obtain a more precise value of the equilibrium isotopic composition. The new procedure can be successively performed until the isotopic composition of a certain

TABLE 1: Equilibrium isotopic composition of the ADS fuel, before the irradiation, for cycles n and $(n + 1)$. Units are tonnes of initial heavy metal in the ADS.

Isotope*	ADS fresh fuel equilibrium composition for the nth cycle estimated with the matrix equation method (t)	ADS fresh fuel composition for the $(n + 1)$th cycle estimated with EVOLCODE 2.0 (t)	Ratio (EVOLCODE 2-Method)/ EVOLCODE 2(%)
U234	$3.677E - 02$	$3.794E - 02$	3.08%
U235	$9.303E - 03$	$9.330E - 03$	0.29%
U236	$1.296E - 02$	$1.302E - 02$	0.48%
U238	$7.171E - 05$	$7.245E - 05$	1.03%
Np237	$1.554E - 01$	$1.553E - 01$	−0.09%
Pu238	$2.791E - 01$	$2.792E - 01$	0.04%
Pu239	$4.084E - 01$	$4.067E - 01$	−0.42%
Pu240	$1.204E + 00$	$1.208E + 00$	0.36%
Pu241	$2.476E - 01$	$2.417E - 01$	−2.44%
Pu242	$5.432E - 01$	$5.441E - 01$	0.16%
Am241	$2.338E - 01$	$2.386E - 01$	2.00%
Am242m	$1.313E - 02$	$1.348E - 02$	2.59%
Am243	$2.166E - 01$	$2.166E - 01$	0.01%
Cm242	$2.183E - 04$	$2.165E - 04$	−0.83%
Cm243	$1.958E - 03$	$1.853E - 03$	−5.68%
Cm244	$1.835E - 01$	$1.801E - 01$	−1.92%
Cm245	$5.300E - 02$	$5.266E - 02$	−0.64%
Cm246	$3.430E - 02$	$3.428E - 02$	−0.04%
Cm247	$9.439E - 03$	$9.370E - 03$	−0.74%
Total	$3.643E + 00$	$3.643E + 00$	

*The short-lived isotopes U-237, U-239, Np-238, Np-239, Pu-243, Am-242, and Am-244 were also used in the calculations.

iteration is deviated less than a limit value from the previous iteration. Nevertheless, one iteration is typically enough to achieve precisions acceptable for fuel cycle studies.

5. Results

The isotopic composition of the different stages of the iterative stratum of the fuel cycle has been calculated using the matrix equation method. Table 1 shows the initial (BOL, beginning of life, before the irradiation) ADS fuel isotopic composition (in terms of actinides), calculated as the solution of the equation for V_{eq} described in the previous section. Moreover, the isotopic composition of the ADS fuel at BOL has also been calculated for the following cycle V_{n+1} by means of the advanced burn-up code EVOLCODE 2.0 for checking purposes.

The deviation between the equilibrium isotopic compositions obtained from the matrix equation method and from the irradiation (and the successive decay and reprocessing as indicated in the fuel cycle scenario) using EVOLCODE 2.0 can be seen in the fourth column in Table 1. These deviations, due to the different accuracy of both methods, are smaller than 3% for all actinides (with significant contributions to the total mass) except for Cm-243, with 5.7% difference.

The objective of the fuel cycle scenario will define the maximum allowed deviation between one cycle and the following one and also the final decision concerning the necessity of a second iteration on the matrix equation method if the equilibrium isotopic composition has not been reached. For instance, for a fuel cycle scenario where the objective is the study of a representative set of the different (primary and secondary) waste streams, the obtained equilibrium isotopic composition is a good approximation so no iterative calculations are needed.

6. Conclusions

A fast numerical method, based on an arbitrary precision arithmetic solution to the matrix equations, for the calculation in a zero-dimensional approach of the equilibrium isotopic composition of a transmutation system in an advanced fuel cycle has been developed and successfully applied to a double-strata fuel cycle with LWR and ADS.

The observed differences between the matrix equation method and the EVOLCODE 2.0 results for the successive cycle are smaller than a few percents in the mass estimations of the main actinides. The precision of the calculation can be improved, if needed, by performing a second iteration on the

A Fast Numerical Method for the Calculation of the Equilibrium Isotopic Composition of a Transmutation System
in an Advanced Fuel Cycle

55

matrix equation method, using the one-group cross-sections of the ADS with the equilibrium composition obtained in the first iteration.

References

[1] E. M. González-Romero, H. A. Abderrahim, C. Fazio et al., "Rational and Added Value of P&T for Waste Management Policies", Deliverable 1.1, PATEROS project, EU 6th FP FI6W-Contract number 036418, 2007.

[2] OECD Nuclear Energy Agency, *Advanced Nuclear Fuel Cycles and Radioactive Waste Management*, Nuclear Development, NEA/OECD No. 5990, Les Moulineaux, France, 2006.

[3] "RED-IMPACT synthesis report," Tech. Rep. 978-3-89336-538-8, Forschungszentrum Julich, Berlin, Germany, 2008.

[4] OECD Nuclear Energy Agency, "Nuclear fuel cycle transition scenario studies. status report," Tech. Rep. 6194, Nuclear Development, NEA/OECD, Les Moulineaux, France, 2009.

[5] S. Banach, "Sur les opérations dans les ensembles abstraits et leur application aux équations intégrales," *Fundamenta Mathematicae*, vol. 3, pp. 133–181, 1922.

[6] F. Álvarez-Velarde, P. T. León, and E. M. González-Romero, "EVOLCODE2, a combined neutronics and burn-up evolution simulation code," in *Proceedings of the 9th Information Exchange Meeting on Actinide and Fission Product P&T*, OECD Nuclear Energy Agency, Nîmes, France, 2007.

[7] F. Álvarez-Velarde, *Development of a computational tool for the simulation of innovative transmutation systems*, Ph.D. thesis, University of Córdoba, Andalusia, Spain, 2011.

[8] J. S. Hendricks, G. W. McKinney, H. R. Trellue et al., *MCNPX, Version 2.6.b*, LA-UR-06-3248, Los Alamos National Laboratory, Los Alamos, NM, USA, 2006.

[9] A. G. Croff, "ORIGEN2: a versatile computer code for calculating the nuclide compositions and characteristics of nuclear materials," *Nuclear Technology*, vol. 62, pp. 335–352, 1983.

[10] J. Sanz, O. Cabellos, and N. García-Herranz, ACAB Inventory Code for Nuclear Applications: User's Manual V.2008, NEA-1839/02, 2008.

[11] OECD Nuclear Energy Agency, "Accelerator-driven systems (ADS) and fast reactors (FR) in advanced nuclear fuel cycles. A comparative study," Tech. Rep. number, Nuclear Development, NEA/OECD, Les Moulineaux, France, 2002.

[12] L. Boucher, "Detailed description of selected fuel cycle scenarios," Deliverable 1.4, RED-IMPACT Project, EU 6th FP FI6W-CT-2004-002408, 2007.

[13] M. B. Monagan, K. O. Geddes, K. M. Heal et al., *Maple 13 Advanced Programming Guide*, Maplesoft, Waterloo, ON, Canada, 2009.

[14] S. Wolfram, *The Mathematica Book*, Wolfram Media, Champaign, Ill, USA, 5th edition, 2003.

7

Multiphysics Model Development and the Core Analysis for In Situ Breeding and Burning Reactor

Shengyi Si

Shanghai Nuclear Engineering Research & Design Institute, 29 Hongcao Road, Shanghai 200233, China

Correspondence should be addressed to Shengyi Si; jacksi168@gmail.com

Academic Editor: Wei Shen

The in situ breeding and burning reactor (ISBBR), which makes use of the outstanding breeding capability of metallic pellet and the excellent irradiation-resistant performance of SiC$_f$/SiC ceramic composites cladding, can approach the design purpose of ultralong cycle and ultrahigh burnup and maintain stable radial power distribution during the cycle life without refueling and shuffling. Since the characteristics of the fuel pellet and cladding are different from the traditional fuel rod of ceramic pellet and metallic cladding, the multiphysics behaviors in ISBBR are also quite different. A computer code, named TANG, to model the specific multiphysics behaviors in ISBBR has been developed. The primary calculation results provided by TANG demonstrate that ISBBR has an excellent comprehensive performance of GEN-IV and a great development potential.

1. Introduction

After 60 years of development and deployment, nuclear energy has become one of the three main energy sources supporting human society. Although it happened to three major nuclear accidents during past decades and projected a big shadow on the nuclear energy development, especially the Fukushima Nuclear Event, occurred in March, 2011, significantly twisted people's understanding to nuclear safety, but, people still could not stop 1 the prospects on nuclear energy, as the fossil energy on earth is gradually drying up and increasingly expensive, and the other renewable energy is not enough stable and reliable. By now, except for a few European countries, who claim to terminate their nuclear power projects, the main energy consumers of the world, including China and the United States, have clearly declared their positive position to their established nuclear power route, and several new projects just have been approved during the past recent years.

However, it should be pointed out that the existing nuclear power technology could not support the sustainable nuclear energy development for long-term prospects. The reason is that the existing nuclear power technology is mainly based on water-cooled reactor. It is well known that the fuel utilization of water-cooled reactor is less than 1%, which means that the existing water-cooled reactors are quickly consuming the limited natural uranium resource and producing the huge volume of depleted uranium in front end and the high radioactive spent fuel in back end. According to the 2009 edition of the IAEA red book [1], the natural uranium reserves (including proven and inference) on the earth are about 3.7 million tons (development costs < \$80/kgU) or 4.5 million tons (development costs < \$130/kgU), and the global natural uranium consumption in 2011 is about 70000 tons [2]; at this rate, the existing nuclear development way can only maintain around 50 years. Furthermore, along with the gradual expansion of global nuclear energy, natural uranium demand is expected to rise steadily to 100000 tons/year in 2015 year [2]. On the other hand, it is approximately estimated that the nuclear development during past 60 years has consumed around 2.8 million tons of natural uranium and produced about 2.5 million tons of depleted uranium and 0.25 million tons of high radioactive spent fuel. Therefore, it is clear that the existing nuclear power technology cannot guarantee a sustainable nuclear development.

In fact, the international society has long known the limitation of the existing nuclear energy technology and has been actively looking for effective ways to pursue the sustainability for nuclear development. Among them, a general consensus is that fuel breeding and closed fuel cycle are the crucial options to realize sustainable nuclear energy development. As is well known, fast reactor or harden spectrum is the necessary condition to achieve fuel breeding for current uranium-plutonium fuel cycle, whereas it seems a little optimistic if we think that the existing fast reactor technology is just the solution to the sustainable issue. Taking the existing sodium-cooled fast reactor (SFR) adopting oxide fuel as an example, the one year or so of the refueling cycle brings not only heavy burden of reprocessing and a significant increase in the fuel cycle cost, but also the 20 to 25 years of doubling time seems to give people a choking sense. In recent years, Terra Power LLC, which is founded by Bill Gates, proposed an innovative concept of traveling wave reactor (TWR), which is based on the platform of pool-type sodium-cooled fast reactor by using metallic fuel pellet and HT9 cladding; TWR can approach 30~40 years of ultralong cycle life and around 30 at % of ultrahigh burnup without refueling, but with periodical fuel shuffling during the cycle life. Theoretically, the TWR can well satisfy the need of sustainable nuclear power development, whereas, because the TWR core is composed of igniting region of medium-enriched uranium and blanket region of depleted uranium, the core radial power distribution shall become severely heterogeneous and have significant variation during the cycle life, and also the ultrahigh burnup shall pose a big challenge to the dose limitation of the HT9 cladding. Therefore, there are still a series of questions for TWR's engineering implementation.

Based on the above understanding, this paper proposed an innovative concept of in situ breeding and burning reactor (ISBBR), which is based on the platform of traditional sodium-cooled fast reactor (SFR), and can approach ultra-long cycle and ultrahigh burnup and maintain stable radial power distribution during the cycle life without refueling and shuffling. A computer code TANG modeling main multiphysics phenomenon in ISBBR has been developed for the core design balance. The primary calculation results provided by TANG demonstrate that ISBBR has an excellent comprehensive performance of GEN-IV and a great development potential.

2. The In Situ Breeding and Burning Reactor

Obviously, the prerequisites to approach in situ breeding and burning are that the fuel should have great breeding capability and the fuel reactivity should change very slowly during the cycle life. Figures 1 and 2 present the evolution of reactivity and normalized ^{239}Pu versus fuel burnup for several typical fuel materials, which are designed to have similar initial reactivity. It is clear to see that the ternary alloy of Uranium-Plutonium-Zirconium has good attributes matching the demand of ISBBR.

In addition, in order to ensure the integrity of the fuel rod under the conditions of ultralong cycle and ultra-high burnup, the ISBBR's structure material, especially the fuel cladding material, should have outstanding radiation resistant performance besides the better heat conduction

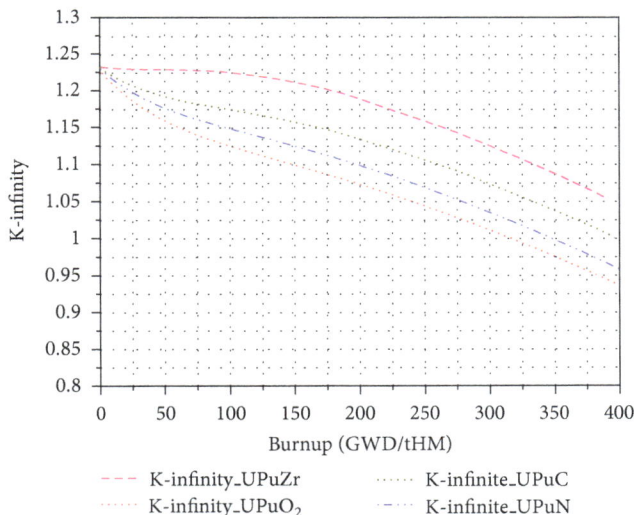

FIGURE 1: Reactivity versus burnup for typical fuel materials.

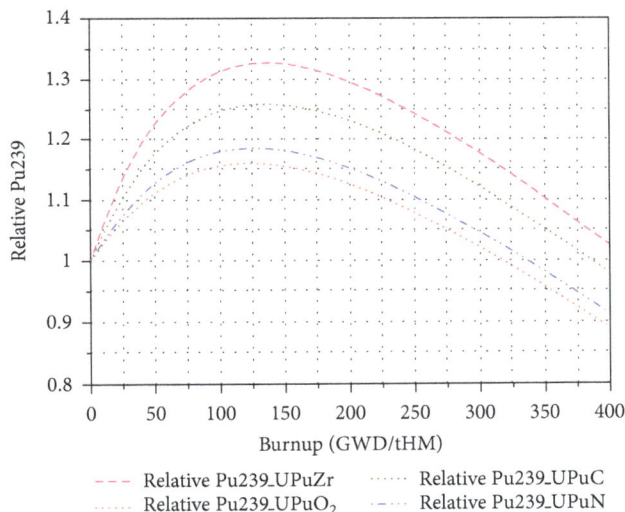

FIGURE 2: Normalized ^{239}Pu versus burnup for typical fuel materials.

capability and mechanical performance. The cladding material of the traditional sodium-cooled fast reactor (SFR) is Austenite or Martensite Stainless Steel, whose radiation-resistant performance is not enough promising. Some new structure materials under developing, such as ODS and HT9, are predicted to have good radiation-resistant capability; however, it is difficult to have a revolutionary solution because these materials are still iron based. Recently, an innovative ceramic composite material SiC_f/SiC has been causing more and more attention from the nuclear field because of its comprehensive performance of heat conduction, mechanical properties, and radiation resistant. It is said that SiC_f/SiC is the most promising for the first wall material in fusion reactor [3]. Figures 3 and 4 present the comparison of DPA cross section and radiation dose for SiC_f/SiC and stainless steel;

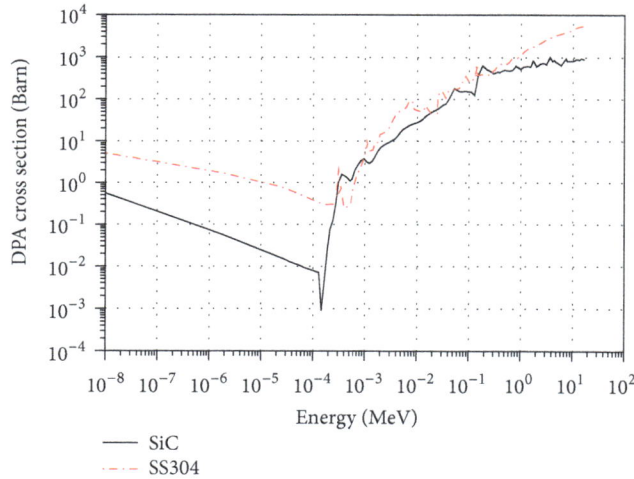

FIGURE 3: DPA cross section for stainless steel and SiC$_f$/SiC.

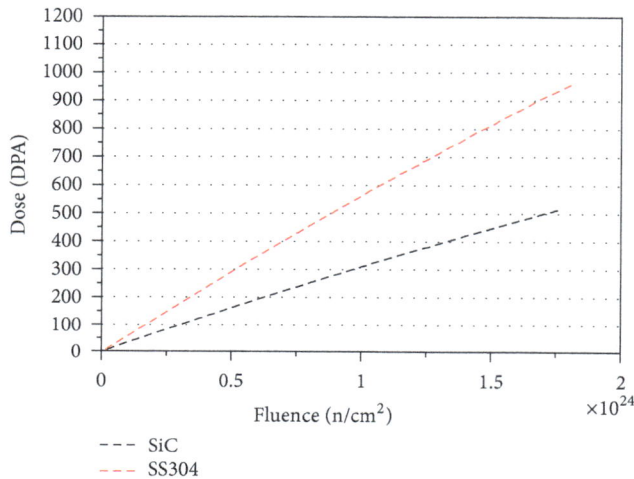

FIGURE 4: DPA dose for stainless steel and SiC$_f$/SiC.

TABLE 1: General parameters for reference core of ISBBR.

Thermal power, MWt	800
Electricity power, MWe	300
Coolant flow rate, kg/sec	5000
Inlet temperature, °C	350
System pressure, MPa	0.1
Number of fuel assembly	222
Number of control assembly	31
Number of shielding assembly	270
Flat to flat distance of FA, cm	~12.5
Number of fuel pin in a FA	60
Fuel rod pitch, cm	1.50
Fuel rod diameter, cm	1.40
Active fuel height, cm	200
Plenum height, cm	200
Cladding material	SiC$_f$/SiC
Pellet material	UPuZr
Coolant material	Sodium
Control absorber material	B$_4$C
Detail size	Reference Figure 5
Heavy metal inventory in core, ton	40.4
DU (0.3 w/o ^{235}U), ton	35.8
Reactor grade Pu, ton	4.6
Composition of Pu:	
^{238}Pu (%)	3.54
^{239}Pu (%)	50.94
^{240}Pu (%)	22.99
^{241}Pu (%)	15.15
^{242}Pu (%)	7.38

we can see that the DPA cross section of SiC is much lower than SS304 in lower energy and high energy region, and the accumulated radiation dose of SiC is much lower than SS304 for a given SFR spectrum.

In the view of the above analysis, our proposed in situ breeding and burning reactor (ISBBR) shall select the ternary alloy of Uranium-Plutonium-Zirconium as fuel pellet, SiC$_f$/SiC as cladding material, and liquid sodium as coolant. Table 1 describes the general design parameters of the reference ISBBR core, and Figure 5 illustrates the schematics of reference core, fuel assembly, and fuel rod.

The reference core is a small modular reactor. The rated thermal power is 800 MWt, the rated mass flow rate is 5000 kg/Sec, and the inlet coolant temperature is 350°C. The core is composed of 222 fuel assemblies, 30 control assemblies, 270 shielding assemblies, and a barrel. The equivalent diameter of active core is about 252 cm and the outer diameter of barrel is about 350 cm. The active core is divided into inner zone (108 assemblies, identified with 1 in Figure 6) and outer zone (104 assemblies, identified with 2 in Figure 6). The unique difference for the fuel assemblies in inner zone and outer zone is the Zirconium contents in the fuel pellet which are 20% and 10%, respectively.

The fuel assembly has an overall length of 460 cm and contains 90 fuel pins arranged in a triangular pitch array within a duct, see Figure 5. The duct thickness is 0.1 cm and the flat-to-flat distance of the duct is around 12.5 cm. Fuel pins are made of sealed SiC$_f$/SiC cladding containing a metallic fuel pellet column of 200 cm length. Just below the fuel slug is a 60 cm shield segment, with the shield being an integral part of the fuel pin in the form of an extended fuel-pin bottom end cap. Sodium is filled as the initial thermal bond between the fuel column and the cladding. And a 200 cm long fission gas plenum is located above the fuel slug and sodium bond. The fuel pin diameter and cladding thickness are 1.4 cm and 0.1 cm, respectively, and the inner diameter and outer diameter of the annular pellet are 0.4 cm and 1.1 cm, respectively. The fuel smeared density is 80%. The fuel pin is helically wrapped with wire to maintain the pin spacing so that the coolant can flow freely through the pin bundle. The wire-wrap helical pitch is 20.32 cm and the wire diameter is 0.1 cm.

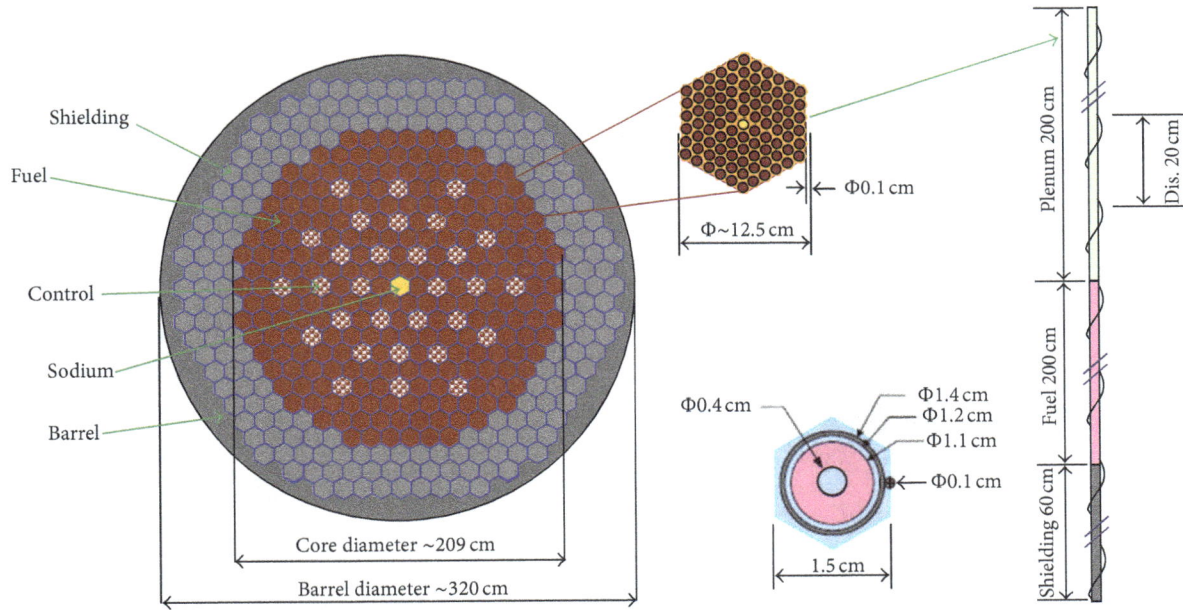

FIGURE 5: Schematic for the reference core of in situ breeding and burning reactor.

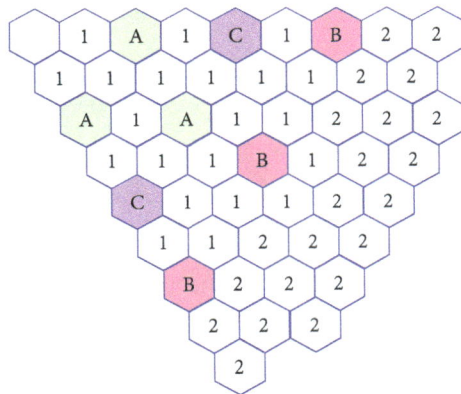

FIGURE 6: Grouping and Layout of Control Assembly.

The control assemblies consist of an absorber bundle contained within a duct. The absorber bundle is a closely packed array of tubes containing compacted boron carbide pellets. The natural boron whose B-10 enrichment is 19.9 a/o is used. Thirty control assemblies are grouped into A, B, and C banks, where bank A is the primary manipulated bank, bank B is the secondary control bank, and bank C is the shutdown bank. The grouping of the control assemblies and its layout are given in Figure 6.

3. The Multiphysics Model Development for ISBBR

Traditionally, nuclear design, thermal-hydraulic analysis, and fuel performance analysis for a reactor core are performed independently. Actually, the neutronics behavior, thermal-hydraulic behavior, fuel thermodynamics behavior, and fuel irradiation behavior in a reactor core are tightly coupled with each other. In PWR core design, thermal-hydraulic feedback has been considered widely in core analysis code due to the significant spectrum effect of coolant density and Doppler effect of fuel temperature. As for ISBBR, besides the thermal-hydraulic feedback, the reactivity effect of thermal expansion and irradiation swelling also have significant influence on the core reactivity and the cycle life.

Figure 7 illustrates the main physics phenomena in the core using metallic fuel. Basically, the neutronics process (Cross Section Parameter → Diffusion/Transport Solution → Neutron Flux → Reaction Rate → Burnup → Cross section Parameter, see Figure 7) is the main driving force for the multiphysics phenomena in the core. Generally, the interactions of neutronics with other disciplines can be grouped into instant effect and historical effect. As for the instant coupling effect, power density derived from the reaction rate determines the fuel rod surface heat flux and the coolant flow field (coolant density and temperature), and also, the power density dominates the fuel temperature distribution in pellet. On the other hand, the coolant density and temperature and fuel temperature and the fuel thermal expansion have direct influence on the cross section parameter, neutron flux, and power density. As for the historical accumulation effect, the accumulated neutron irradiation connects to the fuel burnup, deformation (creep and swell) of cladding and pellet, and fission gas release; all these effects will have influence on the local reaction cross section; in addition, the deformation of cladding and pellet affects the free volume of the fuel rod plenum, which has inversely proportional relations with the rod internal pressure.

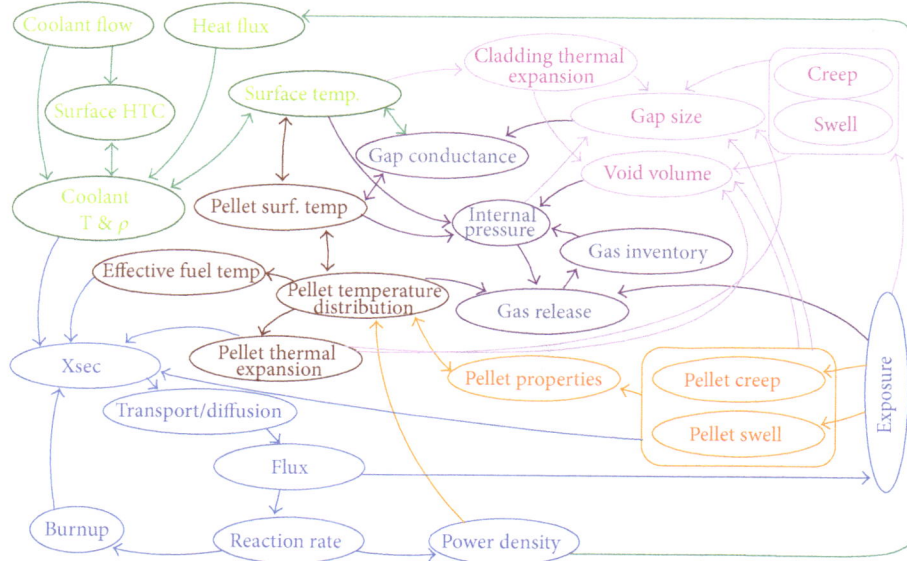

FIGURE 7: Multiphysics behaviors for the core with metallic fuel.

Based on the understanding to the above multiphysics phenomena in the core using metallic fuel and the specific ISBBR fuel design of metallic pellet and ceramic cladding, we developed the specific multiphysics model and the core simulation code TANG for ISBBR. The following sections briefly describe the technical characteristics of the multiphysics model and computer code TANG.

3.1. Neutronics Model.

Neutronics model is the kernel of multiphysics model and also the driving force for heat conduction, heat transfer, and deformation of fuel rod. Neutronics model involves depletion model, parameterized cross section model, and multidimension/multigroup neutron diffusion model.

The depletion model solves nonlinear depletion chains by using Matrix Exponential Algorithm and tracks the evolution of nodal-wise number density for major actinides based on the 3D neutron flux and reaction rate during full cycle life. Figure 8 presents the simplified depletion chain for actinides tracked in TANG code.

Parameterized cross section model captures the instant effect and historical effect of various factors on homogenized assembly cross sections. The instant effect is caused by the deviation of instant local conditions from the reference conditions, such as local burnup (bu), local coolant temperature/density (D_c/T_c), local fuel temperature (T_f), and local axial deformation factor (α), and the historical effect is the accumulated effect caused by the long-term deviation of local conditions from the reference conditions since the deviation of local conditions causes the different local neutron spectrum from the reference case, it results in the different depletion rate or production rate for important actinides from the reference case. The local axial deformation factor (α) models the effect of number densities variation caused by axial deformation on homogenized assembly cross

sections. The formula (1) gives the fundamental expression of the parameterized cross section model:

$$\sum \left(\text{bu}, T_f, D_c, \alpha \right)$$
$$= \frac{\sum_{\text{ref}} (\text{bu}) + \Delta \sum \left(\text{bu}, T_f, D_c \right) + \sum_n^N \sigma_n \left(T_f, D_c \right) \Delta N_n}{\alpha}.$$
(1)

Multidimension/multigroup neutron diffusion model adopts multidimension and multigroup nodal expansion method code MGNEM [4], developed by author previously, to solve multidimension and multigroup neutron diffusion equations for rectangular and hexagonal fuel assembly; also, the universal algorithm of stiffness confinement method (UASCM) [5], which was also developed by author earlier, is coupled with MGNEM to solve multidimension and multigroup time-space kinetic problem. Additional, in order to effectively model the growth of nodal axial mesh caused by thermal expansion and/or irradiation swelling, MGNEM code is using floating mesh in axial direction, so that MGNEM is able to automatically model the axial mesh variation of each node during the iteration process of thermal-hydraulic feedback and finally captures the reactivity effect of axial deformation.

The integration of the above methodologies and technologies endows TANG code abundant calculation functions and flexible simulation ability.

3.2. Fuel Rod Deformation Model.

The thermal expansion coefficient of metallic fuel is large (approximately 2 times of ceramic fuel), and the irradiation swelling effect of metallic fuel is also significant (around 30% at high burnup). As a result, the geometrical change of metallic fuel pellet during heating process and/or irradiation process will not only give penalty on the extra reactivity and cycle life of the core, but also will directly affect the transient behavior of the

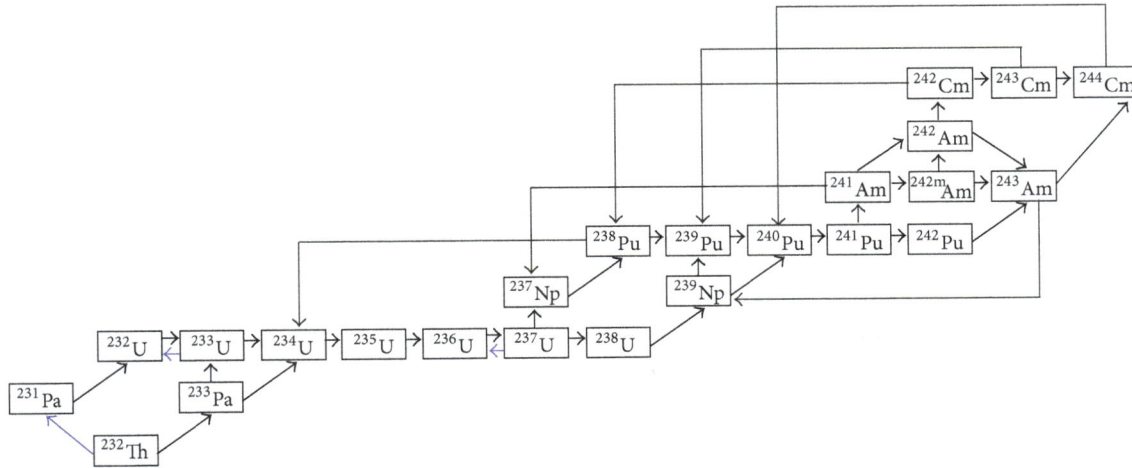

FIGURE 8: Depletion chain for actinides tracked in TANG code.

core, as the negative feedback effect of axial growth can effectively restrain the increase of the reactor core power and automatically bring the reactor to a safe lower power level. The experiments on EBR-II had demonstrated this inherent safety characteristic of the core with metallic fuel [6].

Relative to the metallic pellet, the SiC$_f$/SiC cladding has very good thermal stability, irradiation stability, and mechanics stability. Firstly, the thermal expansion coefficient of SiC$_f$/SiC is only about half of the zircaloy; secondly, the existing irradiation experiment (about 43 DPA) shows that the swelling and creeping phenomena are very weak [3]; in addition, since the strength and hardness of SiC$_f$/SiC are very outstanding, the strain caused by stress for SiC$_f$/SiC material is also very limited. Table 2 presents the properties of relative materials.

According to the above analysis on the metallic pellet and SiC$_f$/SiC ceramic cladding, we proposed a "rigid cladding model" to describe the fuel rod deformation behavior in ISBBR, which assumes the following.

(i) The deformation of SiC$_f$/SiC ceramic cladding is only due to thermal expansion, irradiation swelling, creeping, and stress/strain are ignored.

(ii) The deformation of metallic pellet may be caused by thermal expansion, irradiation swelling, and creeping.

(iii) After the metallic pellet contacts with ceramic cladding, the contact stress shall not cause any strain to ceramic cladding.

(iv) After the metallic pellet contacts with ceramic cladding, the metallic pellet shall become yielded consequently due to the contact stress, and according to the Prandtl-Reuss flow rule, the expansion shall develop to the inner hole of the annular pellet; once the inner hole is closed, the expansion shall switch to axial direction.

Thermal expansion is recoverable and the irradiation deformation (swelling and creeping) is irrecoverable. Therefore,

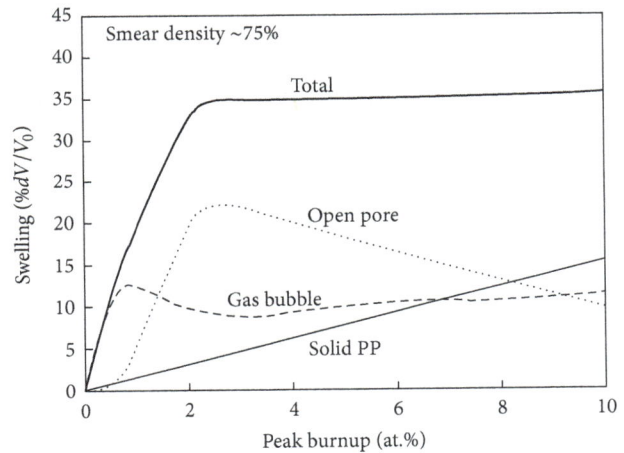

FIGURE 9: Radial swelling versus burnup for metallic fuel.

with the assumption of "rigid cladding model," the cladding deformation is recoverable, and the pellet deformation involves the recoverable and irrecoverable compositions. The recoverable deformation of metallic pellet can affect the core dynamic parameters and the transient behavior of the core, and the irrecoverable deformation of the metallic pellet has direct penalty on the core extra reactivity and the cycle life.

Metallic fuel had significant irrecoverable deformation (swelling and creeping). Figures 9 and 10 present the radial deformation and axial deformation behavior of metallic fuel versus the fuel burnup, respectively, which are based on the experimental results of EBR-II. It can be seen from Figures 9 and 10 that the irradiation deformation of metallic fuel is anisotropic in radial and axial direction; the radial deformation is much more significant than the axial deformation, and the radial deformation grows quickly before 2 at% burnup, and thereafter it maintains around 35% of volume change rate. The irradiation deformation models used in TANG code are just taken from Figures 9 and 10 [7].

TABLE 2: Properties of several typical fuel materials.

	Thermal expansion coefficient ppm/°C	Brinell hardness kg/mm^2	Young's modular (E) GPa	Poisson's ratio (ν)
UO2 (ceramic)	~9	~2000	~96	~0.3
UPuZr (alloy)	~18	~260	~85	~0.3
Zircaloy	~10	~120	~100	~0.3
Stainless steel	~17	~100	~196	~0.3
SiC$_f$/SiC (ceramic) [3]	~4	~2800	~300	~0.14

3.3. Fuel Rods Heat Conduction/Transfer Model. Each fuel assembly in ISBBR is contained within a duct, which directs the coolant flow to fuel rods of the fuel assembly, and there is no exchange of coolant mass and momentum among the assemblies. Therefore, it is reasonable for TANG code to adopt the "single channel model" to simulate the heat conduction within fuel rod and heat transfer between rod surface and coolant; TANG code has a "single channel model" for each fuel assembly modeled in the core to calculate the averaged effect of coolant density/temperature, cladding temperature, and fuel temperature in each elevation of the fuel assembly, and then, the 3D nodal-wise coolant density/temperature, cladding temperature, and fuel temperature are passed to 3D neutronics model, rod deformation model, and other models, so that all models are tightly coupled into a multiphysics model.

The "single channel model" in TANG code uses finite difference method to solve time-dependent heat conduction equation in cylindrical R-Z geometry for steady-state and transient solution, which shall be coupled with steady-state or transient 3D neutronics model. The heat conduction along z-direction is considered due to the high heat conductivity of metallic pellet.

The discrete of heat conduction equation is based on the nominal rod sizes so as to maintain the stability during equation solution, but the deformed rod sizes are used for the gap conductance calculation.

3.4. Fuel Rod Internal Pressure Model. The fuel rod in ISBBR might endure extra high internal pressure and even endanger the fuel rod integrity due to the fission gas release and accumulation under ultralong cycle and ultrahigh burnup. Therefore, the fuel rod internal pressure in ISBBR is an important design constraint. Based on the "single channel model" and 3D burnup distribution of the core, TANG code tracks the fission gas release fraction for 3D nodes and calculates assembly-averaged fuel rod internal pressure at each burnup step.

The main components of fission gas are Xe and Kr, and the total fission yield of Xe and Kr is about 0.25 (totally 2.0). Fission gas is gathered in the grain boundary in earlier stage; with the burnup accumulation, the fission gas gradually gathers into bubble; bubbles grow along with the increased inner pressure and connected mutually; finally, it forms a coherent tunnel. Eventually, fission gas is driven by the temperature and the pressure and release to the gap between pellet and cladding and then the plenum of the rod. The coherent tunnel in metallic fuel is formed at around 2~3 at%

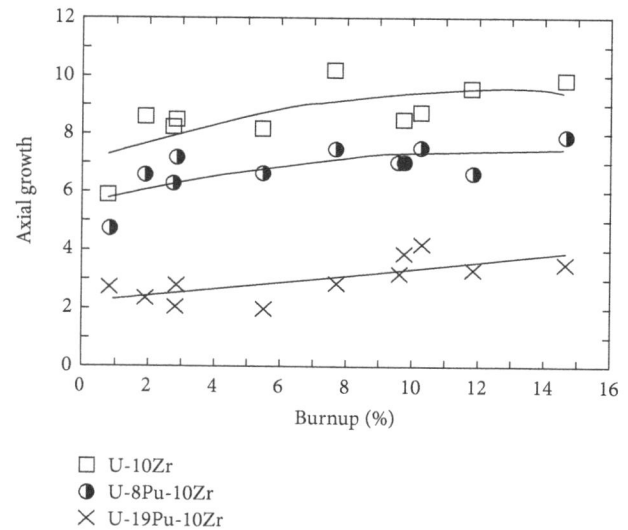

FIGURE 10: Axial swelling versus burnup for metallic fuel.

□ U-10Zr
◐ U-8Pu-10Zr
✕ U-19Pu-10Zr

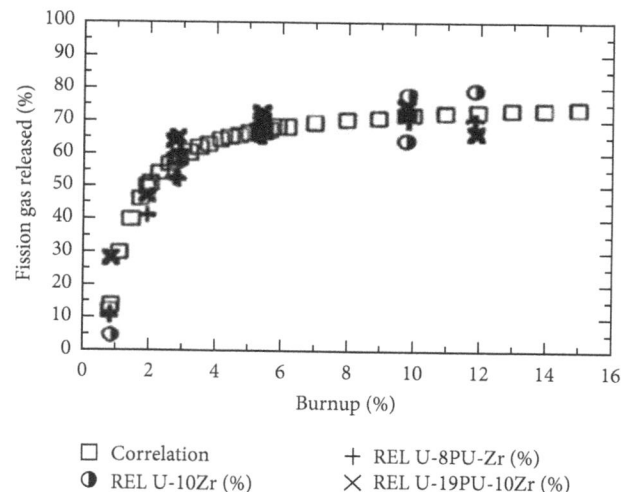

FIGURE 11: Fission gas releases fraction for metallic fuel.

□ Correlation   + REL U-8PU-Zr (%)
◐ REL U-10Zr (%)   ✕ REL U-19PU-10Zr (%)

of burnup, the release fraction of fission gas increases quickly prior to 3 at% of burnup; thereafter, the release fraction maintains at around 70%. Figure 11 presents the fission gas release fraction of metallic fuel versus the fuel burnup, which is based on the experimental results of EBR-II. The fission gas release model used in TANG code is just taken from Figure 11 [7].

The fuel rods internal pressure is calculated by using the free gas state equation $PV = MRT$, in which, P is the fuel rods internal pressure; V is the volume of free space in fuel rods, involving the comprehensive effect of the ceramic cladding deformation, metallic pellet thermal expansion, metallic pellet swelling, and the liquid sodium surface rise squeezed by the deformed cladding and pellet; M is the fuel rods internal gas mole number, including the fission gas and the initial fill gas; R is a universal gas constant ($R = 8.34$ J/Mole-K); T is the average temperature of the fuel plenum.

3.5. Fuel Rod Exposure Dose Model. The structure material, especially the cladding material, in ISBBR might endure severe radiation damage due to ultralong time exposure of fast neutron spectrum. The radiation damage to material usually is measured by displacement per atom (DPA), which means the accumulative displacement number of each medium atom. TANG code equips a fuel rod exposure dose model, so that the maximum cladding dose can be monitored during core design process.

The multigroup DPA cross sections (σ_g^{DPA}) are stored in the parameterized cross section library as the same as other cross sections expressed in formula (1); TANG code shall evaluate the accumulated cladding dose for each node of core based on the 3D flux solution (ϕ_g) and the depletion time increment (ΔT, Sec) with the following formula:

$$\mathrm{DPA}^{n+1} = \mathrm{DPA}^n + \Delta T \sum_g \sigma_g^{\mathrm{DPA}} \phi_g. \tag{2}$$

4. Calculation Results

The reference ISBBR core was analyzed with TANG code, and the calculation results are introduced in following sections.

4.1. Steady-State Results Analysis. Figure 12 shows the relation of atom burnup and energy burnup for reference ISBBR core during 25 effective full power years (EFPYs); it can be seen that 1 at% burnup is equivalent to about 10 GWD/tHM for the reference core; the slight deviation at end of life is due to the fission energy of minor actinide (MA) which is a little higher than plutonium.

Figure 13 presents the critical position of control assembly and the evolution of the peaking power factors during the cycle life. It can be seen from Figure 13 that ISBBR can achieve stable and flat power distribution during the 25 EFPYs of cycle life; Figure 14 further shows the radial power distribution of reference core at different burnup steps and demonstrates that the radial power distribution of reference core is well maintained without shuffling and refueling within an ultralong cycle life. Therefore, ISBBR can avoid the significant distortion and variation of the power distribution existing in other conceptual breeding and burning reactors [8–11]. It is beneficial to core design and power distribution control in engineering sense.

Figure 15 presents the assembly-wise distribution for key design parameters at EOL (25 EFPYs) tracked by the multiphysics models in TANG code. Maximum dose is the

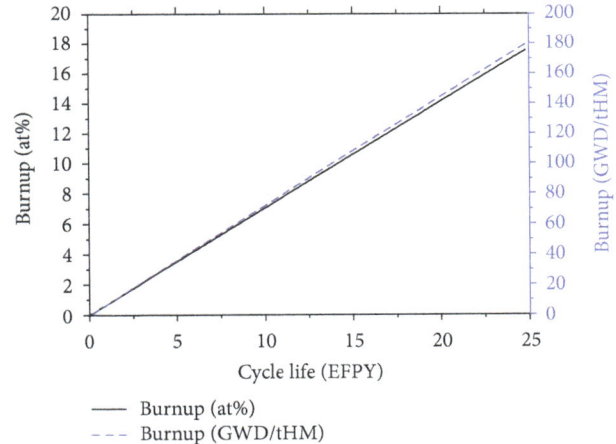

FIGURE 12: Burnup versus cycle length for reference core.

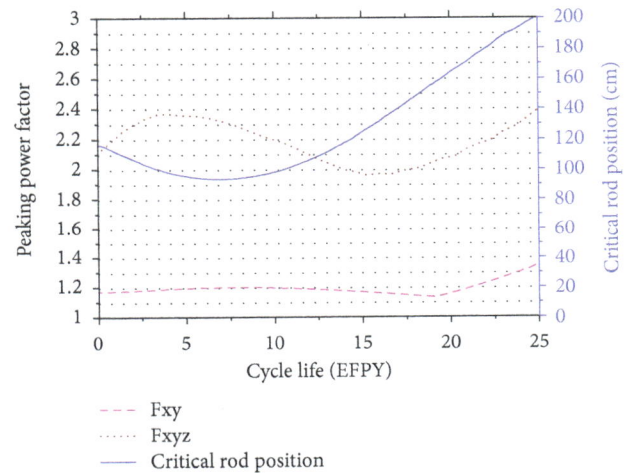

FIGURE 13: Critical rod position and power peaking factor versus cycle lifetime.

peaking irradiation dose for each assembly, and the results show that the maximum irradiation dose at EOL for the SiC$_f$/SiC ceramic cladding in ISBBR core is about 167.8 DPA, which is far smaller than that using stainless steel cladding (e.g., for TWR, the peaking irradiation dose may be as high as 500 DPA); internal pressure is an assembly-averaged rod internal pressure, and the tracked results demonstrate that the maximum internal pressure of fuel rod at EOL in reference core is about 13 MPa, which is thought to be safe to the rod integrity according to the experience of PWR rod performance analysis; the axial deformation factor is an assembly-averaged axial deformation factor, and the calculation results show that the axial deformation of the reference core is more balanced among the full core, which means that the annular pellet has accommodated the radial pellet swelling and the axial deformation is mainly composed of thermal expansion and axial swelling. Figure 16 illustrates the influence of axial thermal expansion and swelling on the core reactivity.

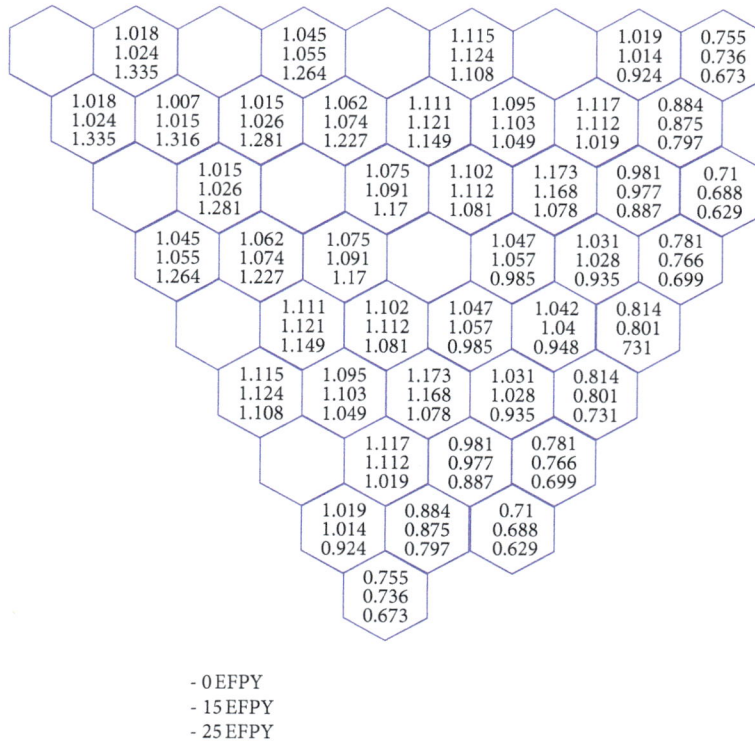

FIGURE 14: Assembly-wise power distribution at 0 EFPY/15 EFPY/25 EFPY.

- 0 EFPY
- 15 EFPY
- 25 EFPY

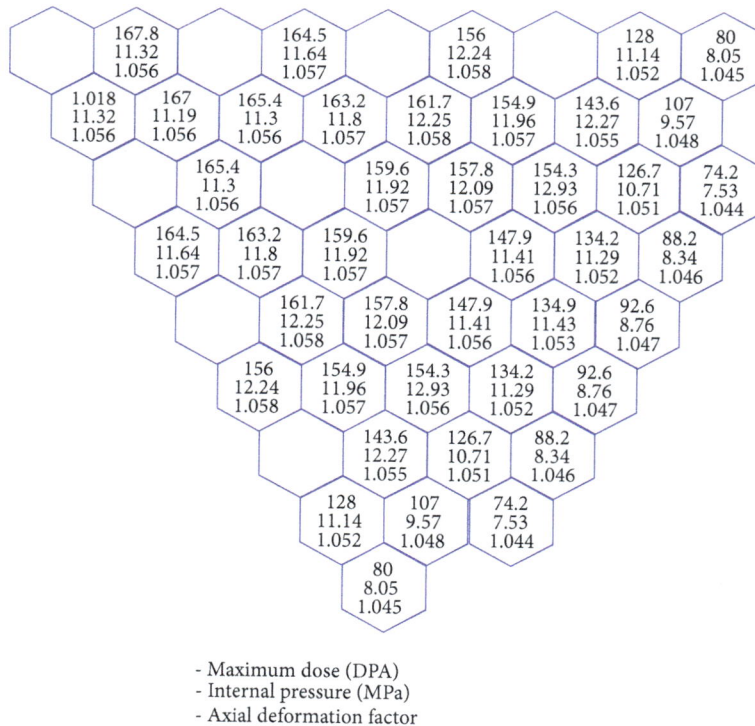

- Maximum dose (DPA)
- Internal pressure (MPa)
- Axial deformation factor

FIGURE 15: Assembly-wise maximum cladding dose, rod internal pressure, and average axial deformation factor at EOL.

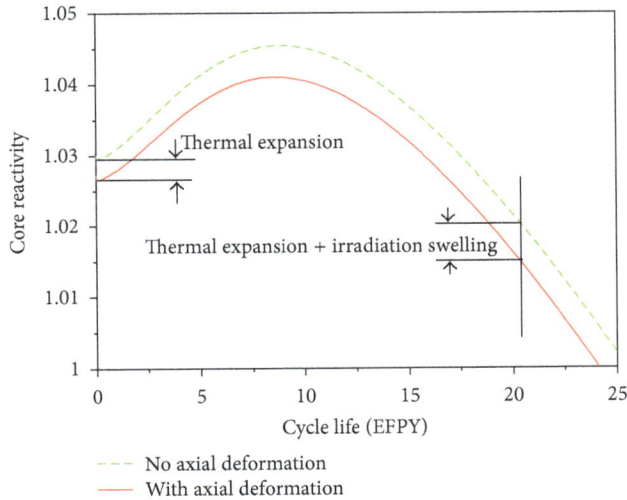

FIGURE 16: Influence of axial expansion on cycle lifetime.

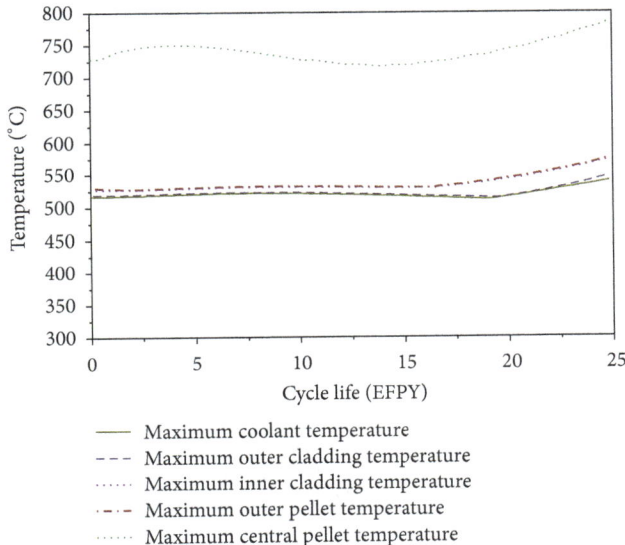

FIGURE 17: Rod peaking temperatures versus cycle lifetime.

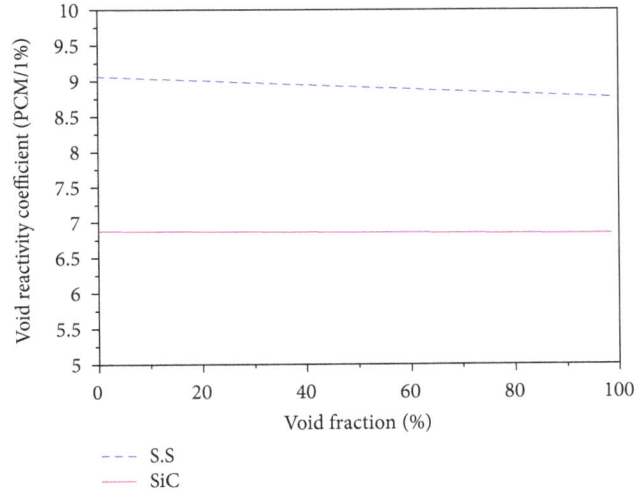

FIGURE 18: Influence of SiC$_f$/SiC on void reactivity.

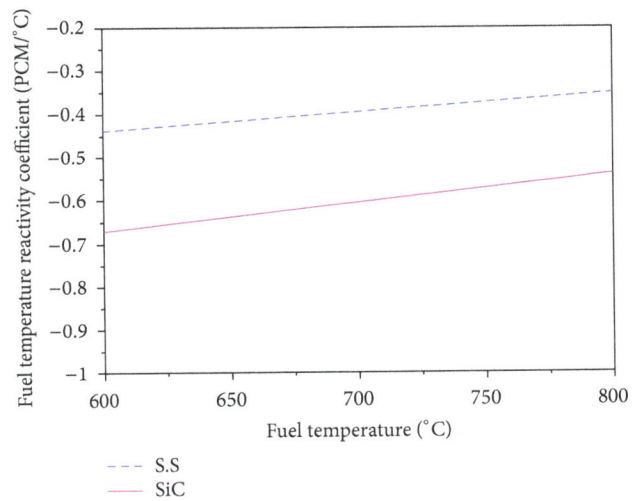

FIGURE 19: Influence of SiC$_f$/SiC on Doppler effect.

Figure 17 shows evolution of the peaking coolant temperature and the peaking fuel rod temperatures versus the cycle life of the reference core. It can be seen from Figure 17 that each indicator has undergone an even history and maintains enough margin to their corresponding design limit, for example, the boiling point of sodium coolant is around 892°C, while the peaking coolant temperature within the reference core cycle life is around 550°C; the maximum operating temperature for SiC$_f$/SiC ceramic cladding is 1600°C, while the peaking cladding temperature during cycle life of the reference core is only around 550°C, and the molten point of metallic pellet (ternary alloy of Uranium-Plutonium-Zirconium) is about 1200°C, but the peaking fuel central temperature during the cycle life is lower than 800°C. Therefore, the reference core of ISBBR has enough thermal margins during the whole cycle life for steady-state operation.

4.2. Transient Results Analysis. The positive Void Reactivity Coefficient and smaller Fuel Doppler Temperature Coefficient are the main characteristics of the sodium-cooled fast reactor and also the main concerns of people to the safety of the sodium-cooled fast reactor. Some studies in [12] show that the positive Void Reactivity Coefficient and smaller Fuel Doppler Temperature Coefficient may be improved if some moderator is introduced into the sodium-cooled fast reactor. Actually, SiC is also a kind of good moderator. Therefore, the Void Reactivity Coefficient and Fuel Doppler Temperature Coefficient of the reference ISBBR core can have significant improvement compared with the traditional SFR using stainless steel cladding, see Figures 18 and 19.

The early demonstration experiments on EBR-II had proved that the SFR using metallic fuel could safely approach lower power level during the anticipated transient without

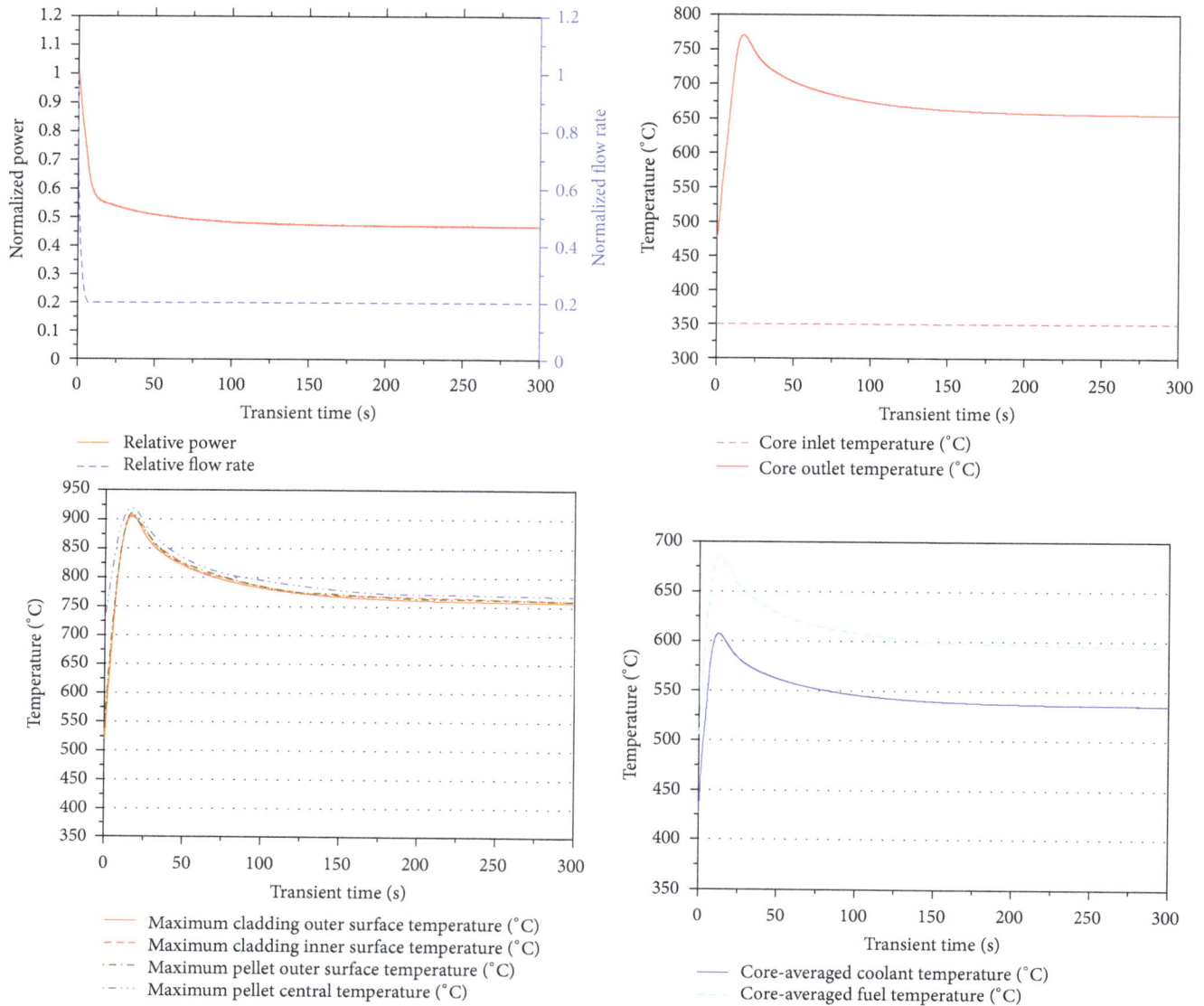

FIGURE 20: Core key parameters response during ULOFA.

scram (ATWS) [6]. The demonstration experiments on EBR-II included unprotected loss of flow (ULOFA) and unprotected loss of heat sink (ULOHSA). In these two events, the initial state of the reactor was full power operation; after the transients were triggered, the reactor power was gradually decreased to lower level due to the inherent safety characteristics of neutronics and thermal-hydraulic and thermal dynamics of metallic fuel, and no damaged consequence happened, such as coolant boiling, cladding defect, or pellet molten.

TANG code is used to simulate the transient response for ULOFA and ULOHSA of reference ISBBR core. The calculated results are given in Figures 20 and 21 for ULOFA and ULOHSA, respectively. It can be seen from Figure 20 that the reactor power shall automatically decrease to 50% or so without any control assembly inserted after the flow rate rapidly reduced to 20% of the nominal value, and the peaking

coolant temperature, the peaking cladding temperature, and the peaking pellet central temperature are far lower than their corresponding operation limits; similarly in Figure 21, after loss of heat sink, the core inlet temperature is rapidly increased from 350°C to 520°C, and at the same time, the power level is gradually decreased to 45% or so without any protective action, and all crucial design parameters are in safe state. These analysis results demonstrate that the reference ISBBR core has inherent passive safety during typical transient processes.

4.3. Fuel Cycle Analysis. ISBBR makes use of the outstanding breeding capability of metallic fuel, produces more fissile material as consuming existing fissile material, and finally achieves ultralong cycle life and ultrahigh burnup.

Figure 22 shows the core-averaged breeding ratio of the reference ISBBR versus cycle life, and it is clear that the

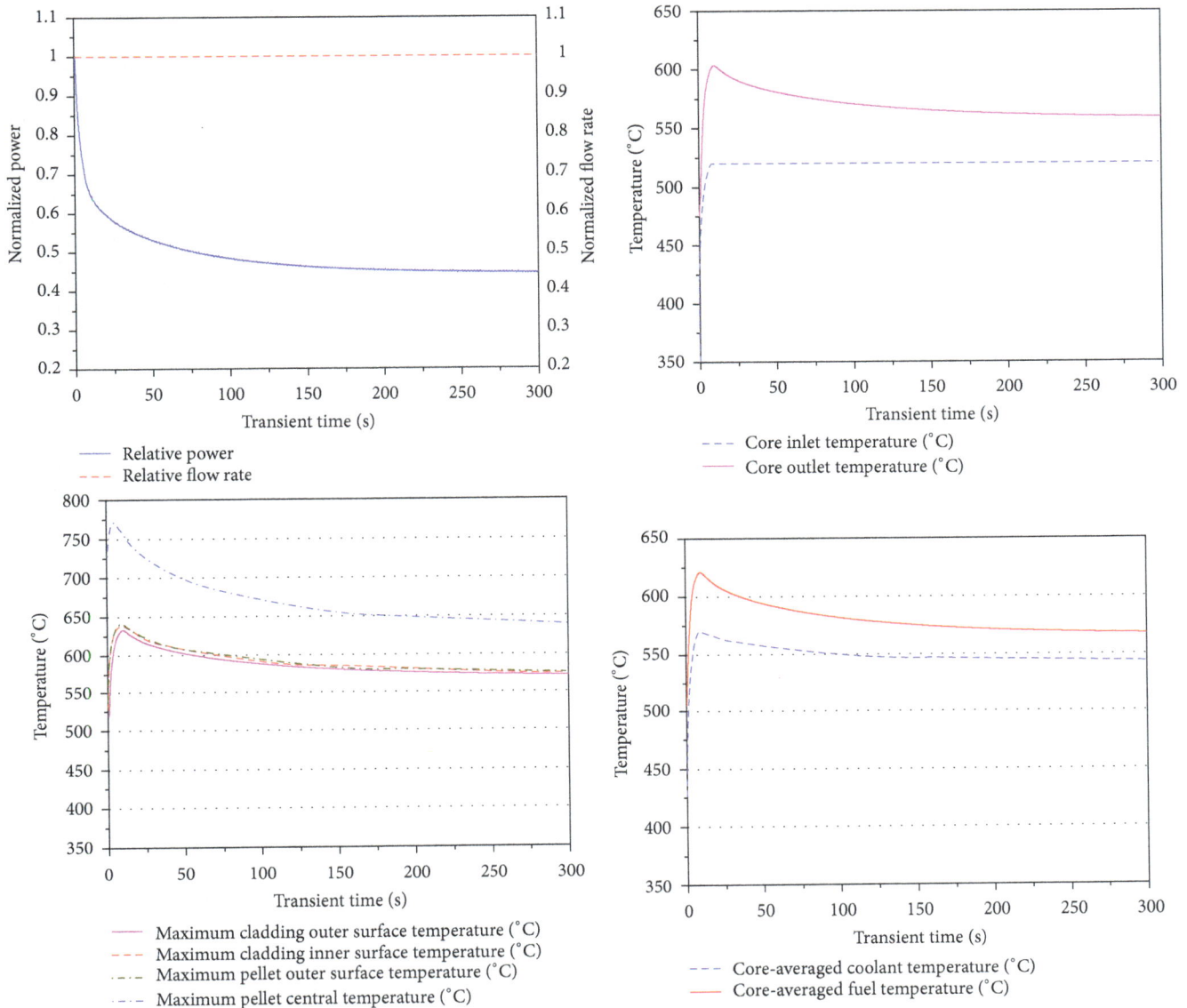

FIGURE 21: Core key parameters response during ULOHSA.

core is always breeding; the heavy metal inventory of the reference core versus cycle life is also given in Figure 22. The initial inventory of heavy metal is about 40.4 tons (including 35.8 tons of depleted uranium and 4.6 tons of Reactor Grade Plutonium), and the final inventory of heavy metal at EOL is around 33 tons, so totally there are about 7.4 tons of heavy metal burnt out, and the fuel utilization rate is about 18 at%. Therefore, the fuel cycle economy of ISBBR is significantly improved compared with current water-cooled reactor, where the typical utilization is only 0.5~0.6 at%.

The motivation to propose ISBBR is not only to pursue safe and economical energy, but also for the following strategic prospects:

(1) continuously consume the huge volume of the depleted uranium and spent fuel accumulated by the development and deployment of water-cooled reactor, and finally achieve the minimization of the waste volume;

(2) Support sustainable development and deployment of fission energy and provide abundant and reliable energy for the peace and development of human society.

The precondition to achieve the above strategic goals is that ISBBR should implement closed fuel cycle. Fortunately, the low molten point (around 1200°C) property of metallic fuel has provided a very favorable condition for economical reprocessing of spent fuel. And then, the prerequisite to achieve sustainable closed fuel cycle for ISBBR is the quantity and quality of the fission material in the spent fuel which has no degradation compared with the initial inventory. Figure 23 illustrates the evolution of plutonium inventory versus the cycle life; it can be seen that the plutonium inventory is

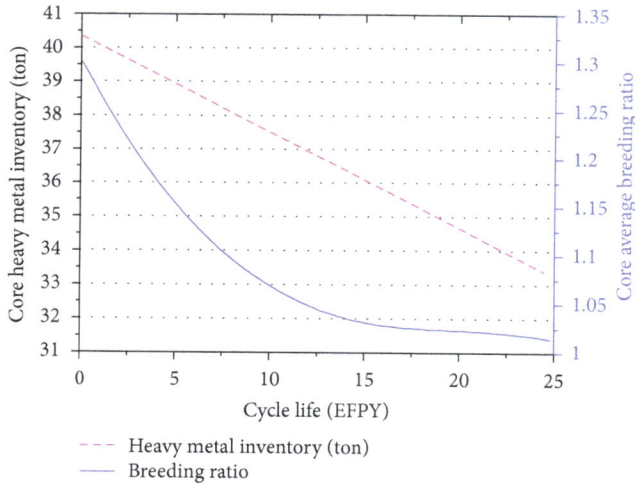

FIGURE 22: Core heavy metal inventory and breeding ratio versus cycle lifetime.

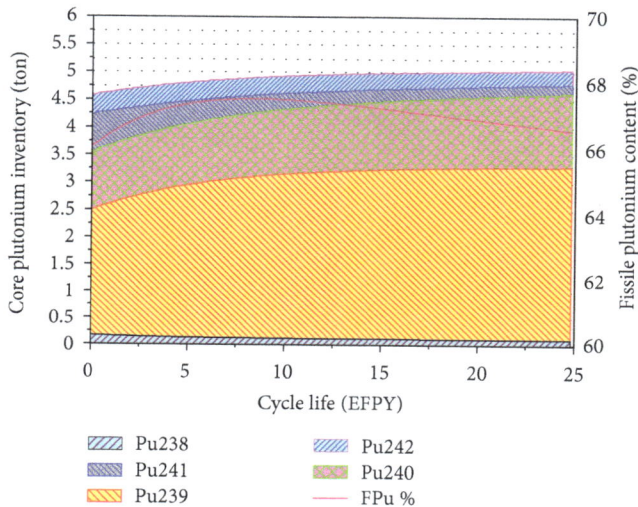

FIGURE 23: Core plutonium inventory and fissile plutonium fraction versus cycle lifetime.

increased from initial 4.3 tons to final 5.1 tons during the cycle life, the incremental is about 10%, and the content of fissile plutonium (^{239}Pu and ^{241}Pu) in total plutonium also has been increased compared with the initial 65%. Therefore, both the quantity and quality of the fission material in ISBBR spent fuel satisfy the prerequisite to maintain sustainable closed fuel cycle.

ISBBR does not pursue the accumulation or doubling of the extra plutonium, but the synchronized breeding and burning of fissile isotopes within an ultralong cycle life; therefore, it naturally satisfies the requirement of nonproliferation.

And also, only relying on the huge amount of depleted uranium and spent fuel accumulated by water-cooled reactor, ISBBR can achieve sustainable fission energy supply for long term and finally achieve the minimization of the waste volume. Let us assume that all the natural Uranium resource

on the earth shall be utilized by PWR, it means that there will be about 6 million tons of depleted uranium and about 0.6 million tons of spent fuel accumulated finally; usually, the content of Reactor Grade Plutonium in the PWR spent fuel is nearly 1%, so the accumulated Reactor Grade Plutonium shall be 6000 tons; taking the reference ISBBR core as an example, where the initial plutonium inventory is 4.6 tons, then, the accumulated Reactor Grade Plutonium is enough to equip $6000/4.6 \approx 1300$ units of reference ISBBR and provides approximate 300 MWe × 1300 = 390 GWe of power supply (a little higher than the current nuclear power installed capacity 377 GWe in the world); now, the 1300 units of ISBBR shall be loaded with 1300 × 35.8 tons ≈ 47,000 tons of depleted uranium and 1300 × 4.6 tons ≈ 6,000 tons of Reactor Grade Plutonium for a fuel cycle, in which 1300 × 7.4 ton ≈ 9600 tons of heavy metal shall be burnt out; taking another assumption, let us ignore the 10% increment of the Reactor Grade Plutonium in spent fuel and think it is used to compensate for the loss during spent fuel reprocessing and the new fuel fabrication, then, the time for these 1300 units of ISBBR to consume the accumulated 6 million tons of depleted uranium can be approximately estimated as follows:

$$\frac{6000000 + (600000 - 6000) - (47000 + 6000)}{9600} \times 25 \text{ year}$$

$$\approx 17000 \text{ years.}$$

(3)

Humans, of course, may not need 17,000 years of fission energy supply; in turn, they may want to have much more installed capacity. In this case, people can have some breeder reactors in earlier stage and produce Reactor Grade Plutonium, and then, they can get their wanted installed capacity by building enough units of ISBBR. Even if the number of ISBBR unit was increased to 13,000 units, the accumulated 6 million tons of Depleted Uranium still can support over 1000 years of nuclear power supply.

5. Conclusion

Based on the platform of traditional sodium-cooled fast reactor, making use of the innovative fuel design and core design, ISBBR can achieve ultralong cycle and ultrahigh burnup and maintain stable radial power distribution during the cycle life without refueling and shuffling.

Primary calculation results, provided by specifically developed computer code TANG, demonstrate that the ISBBR core has enough thermal margins during steady-state operation and inherent passive safety during anticipated transients.

The fuel cycle analysis indicates that the fuel utilization rate of ISBBR can approach 18% and have significant fuel cycle economy compared with the current water-cooled reactors, where fuel utilization is about 0.5~0.6%.

In addition, the features of ultralong cycle, no extra plutonium accumulation, and sustainable closed fuel cycle demonstrate that ISBBR can realize the minimization of waste volume, nonproliferation, and sustainable nuclear power supply.

In conclusion, ISBBR can well satisfy the requirements of Gen-IV nuclear energy system, such as sustainability, economy, safety, and nonproliferation and has a great development potential.

References

[1] IAEA/NEA/OECD, *Uranium2009: Resources, Production and Demand*, IAEA/NEA/OECD, 2010.

[2] World Nuclear Association, *The Global Nuclear Fuel Market: Supply and Demand 2011–2030*, World Nuclear Association, London, UK, 2011.

[3] R. H. Jones, D. Steiner, H. L. Heinisch, G. A. Newsome, and H. M. Kerch, "Radiation resistant ceramic matrix composites," *Journal of Nuclear Materials*, vol. 245, no. 2-3, pp. 87–107, 1997.

[4] S. Si, "3D coarse mesh NEM embedded with 2D fine mesh NDOM for PWR core analysis," in *Proceedings of the Advances in Reactor Physics to Power the Nuclear Renaissance Conference (PHYSOR '10)*, pp. 257–268, Pittsburgh, Penn, USA, May 2010.

[5] S. Si, "Algorithm development and verification of UASCM for multi-dimension and multi-group neutron kinetics model," in *Proceedings of the Advances in Reactor Physics Conference—Linking Research, Industry, and Education (PHYSOR '12)*, Knoxville, Tenn, USA, April 2012.

[6] S. H. Fistedis, Ed., *The Experimental Breeder Reactor II Inherent Safety Demonstration*, vol. 101 of *Nuclear Engineering and Design No. 1*, North-Holland, New York, NY, USA, 1987.

[7] R. G. Pahl, D. L. Porter, D. C. Crawford, and L. C. Walters, "Irradiation behavior of metallic fast reactor fuels," *Journal of Nuclear Materials*, vol. 188, pp. 3–9, 1992.

[8] T. Ellis, R. Petroski, P. Hejzlar et al., "Traveling-wave reactors: a truly sustainable and full-scale resource for global energy needs," in *Proceedings of the International Congress on Advances in Nuclear Power Plants (ICAPP '10)*, Paper 10189, pp. 13–17, San Diego, Calif, USA, June 2010.

[9] G. J. van Tuyle, G. C. Slovik, B. C. Chan, K. J. Kennett et al., "Summary of advanced LMR evaluation-PRISM and SAFR," Tech. Rep. NUREG/CR-5464, 1989.

[10] Y. I. Chang, P. J. Finck, C. Grandy et al., "Advanced burner test reactor preconceptual design report," Tech. Rep. ANL-ABR-1, 2006.

[11] T. K. Kim and T. A. Taiwo, "Feasibility study of ultra-long life fast reactor core concept," in *Proceedings of the Advances in Reactor Physics to Power the Nuclear Renaissance Conference (PHYSOR '10)*, pp. 1756–1766, Pittsburgh, Penn, USA, May 2010.

[12] B. Merk, "On the effect of different placing ZrH moderator material on the performance of a SFR core," *Annual of Nuclear Energy*, vol. 38, pp. 2374–2385, 2011.

8

Perturbation-Theory-Based Sensitivity and Uncertainty Analysis with CASMO-4

Maria Pusa

VTT Technical Research Centre of Finland, P.O. Box 1000, VTT 02044, Finland

Correspondence should be addressed to Maria Pusa, maria.pusa@vtt.fi

Academic Editor: Kostadin Ivanov

The topic of this paper is the development of sensitivity and uncertainty analysis capability to the reactor physics code CASMO-4 in the context of the UAM (Uncertainty Analysis in Best-Estimate Modelling for Design, Operation, and Safety Analysis of LWRs) benchmark. The sensitivity analysis implementation is based on generalized perturbation theory, which enables computing the sensitivity profiles of reaction rate ratios efficiently by solving one generalized adjoint system for each response. Both the theoretical background and the practical guidelines for modifying a deterministic transport code to compute the generalized adjoint solutions and sensitivity coefficients are reviewed. The implementation to CASMO-4 is described in detail. The developed uncertainty analysis methodology is deterministic, meaning that the uncertainties are computed based on the sensitivity profiles and covariance matrices for the uncertain nuclear data parameters. The main conclusions related to the approach used for creating a covariance library compatible with the cross-section libraries of CASMO-4 are presented. Numerical results are given for a lattice physics test problem representing a BWR, and the results are compared to the TSUNAMI-2D sequence in SCALE 6.1.

1. Introduction

The topic of this paper is the development of sensitivity and uncertainty analysis capability to the reactor physics code CASMO-4 [1] in the context of the UAM (Uncertainty Analysis in Best-Estimate Modelling for Design, Operation and Safety Analysis of LWRs) benchmark [2]. At VTT, CASMO-4 is the standard tool for lattice physics calculations, and therefore it was a natural choice as the development platform for a sensitivity and uncertainty calculation system for the pin cell and fuel assembly exercises in the benchmark.

Sensitivities with respect to uncertain parameters can be computed efficiently by utilizing the adjoint system of the criticality equation. The propagated parameter uncertainty can then be calculated deterministically by the Sandwich rule by combining the sensitivity profiles with the covariance matrices of the parameters. As a first step, classical perturbation theory (CPT) was implemented to CASMO-4 to enable the computation of critical eigenvalue sensitivities with respect to nuclear data parameters. In this context, a methodology was devised for processing the covariance matrices from SCALE 6 [3] to become compatible with the

cross-section libraries of CASMO-4 to enable uncertainty analysis. This work has been reported in detail in [4]. Recently, generalized perturbation theory (GPT) has been added to the code as a new feature. This enables performing sensitivity analysis for responses that can be presented as reaction rate ratios. In this framework, one generalized adjoint system needs to be solved for each response, after which the response sensitivity profiles for all parameters of interest can be computed in an efficient manner.

This paper is organized as follows. Section 2 reviews the theoretical background for sensitivity and uncertainty analysis based on generalized perturbation theory, and Section 3 focuses on the implementation to CASMO-4. In Section 3.1, the computation of generalized adjoint solutions is considered and practical guidelines are presented for modifying a deterministic transport code to solve the adjoint problems needed in sensitivity analysis. Section 3.2 concerns the computation of sensitivity and uncertainty profiles. Finally, in Section 4, numerical results are presented for a lattice physics test problem representing a BWR, and they are compared to the TSUNAMI-2D sequence in SCALE 6.1.

2. Theoretical Background

The purpose of sensitivity analysis is to study how sensitive a mathematical model is to perturbations in its uncertain parameters. The target of uncertainty analysis is to estimate how the uncertainty in these parameters is propagated to a response dependent on the mathematical model under consideration. In this work the mathematical model is the neutron transport eigenvalue problem, which can be written in operator form as

$$\mathbf{A}\Phi = \frac{1}{k}\mathbf{B}\Phi, \qquad (1)$$

where $\Phi \in H_\Phi$ is the neutron flux, H_Φ is a Hilbert space, and k is the multiplication factor. The uncertain parameters consist of nuclear data parameters and they are denoted by the vector $\sigma \in E_\sigma$. It should be noted that both the continuous-energy criticality equation and the various systems derived from it in numerical computations can be written in the form of (1).

2.1. Sensitivity Analysis. The object of local sensitivity analysis is to determine how the response R depends on the uncertain parameters in the vicinity of their best-estimate values. In this work, the responses under consideration include homogenized assembly parameters and the multiplication factor, whereas the uncertain parameters are neutron cross-sections. When considering the continuous-energy eigenvalue problem, the cross-sections are functions of energy and location, and the appropriate derivative is the functional directional derivative called the Gâteaux variation [5]. It follows that the sensitivity of R with respect to the perturbation $\mathbf{h} = [\delta\Phi, \delta\sigma] \in D = H_\Phi \times E_\sigma$ at the point $\hat{\mathbf{e}} = [\hat{\Phi}, \hat{\sigma}] \in D$ may be defined as

$$\delta R(\hat{\mathbf{e}}; \mathbf{h}) = \lim_{t \to 0} \frac{R(\hat{\mathbf{e}} + t\mathbf{h}) - R(\hat{\mathbf{e}})}{t}. \qquad (2)$$

When the parameters σ are perturbed, also the solution Φ is affected and therefore the computation of the sensitivity $\delta R(\hat{\mathbf{e}}; \mathbf{h})$ requires that the perturbation $\delta\Phi$ is known. In principle, $\delta\Phi$ can be computed to first order from the following *forward sensitivity system*:

$$\delta\mathbf{A}(\hat{\mathbf{e}}; \mathbf{h}) = -\frac{1}{k^2}\delta k(\hat{\mathbf{e}}; \mathbf{h})\mathbf{B}\Phi + \frac{1}{k}\delta\mathbf{B}(\hat{\mathbf{e}}; \mathbf{h})$$

$$\iff \mathbf{A}'_\sigma(\hat{\mathbf{e}})\delta\sigma + \mathbf{A}(\hat{\mathbf{e}})\delta\Phi = -\frac{1}{k^2}\delta k(\hat{\mathbf{e}}; \mathbf{h})\mathbf{B}\Phi$$

$$+ \frac{1}{k}\mathbf{B}'_\sigma(\hat{\mathbf{e}})\delta\sigma + \frac{1}{k}\mathbf{B}(\hat{\mathbf{e}})\delta\Phi, \qquad (3)$$

which can be derived by taking the Gâteaux variation of system (1) with respect to a perturbation $\mathbf{h}$ on both sides. However, when computing several sensitivities, this approach would require the repetitive solving of (3).

Fortunately, the sensitivities can be computed more efficiently by exploiting the adjoint of (1), which is defined as the system that satisfies the following relation: (In some cases the adjoint relation needs to be written in the form $\langle \mathbf{A}\Phi + (1/k)\mathbf{B}\Phi, \Psi \rangle = \langle \Phi, \mathbf{A}^*\Psi + (1/k)\mathbf{B}^*\Psi \rangle + [\mathbf{P}(\Psi, \Phi)]_{\mathbf{x} \in \partial\Omega}$, where $[\mathbf{P}(\Psi, \Phi)]_{\mathbf{x} \in \partial\Omega}$ is a bilinear form associated with the system. We will only consider cases where it is straightforward to force this term to vanish.)

$$\left\langle \mathbf{A}\Phi - \frac{1}{k}\mathbf{B}\Phi, \Psi \right\rangle = \left\langle \Phi, \mathbf{A}^*\Psi - \frac{1}{k}\mathbf{B}^*\Psi \right\rangle, \qquad (4)$$

where the brackets $\langle \cdot, \cdot \rangle$ denote an inner product. When considering the continuous-energy criticality equation, it is customary to employ the L^2 inner product [6, 7]. The solution to the adjoint problem

$$\left(\mathbf{A}^* - \frac{1}{k}\mathbf{B}^* \right)\Psi = 0 \qquad (5)$$

is called the *fundamental adjoint*. Physically, the solution to this system can be interpreted to represent the average contribution, that is, importance of a neutron to the multiplication factor. Interestingly, the adjoint system of (5) can be derived solely based on this physical interpretation [8]. Like the neutron flux, the fundamental adjoint has an arbitrary normalization, and the concept of importance should be understood in relative terms. Therefore, the value $\Psi(\mathbf{r}, \Omega, E)$ represents the importance of a neutron located at the point $[\mathbf{r}, \Omega, E]$ compared to the importance of neutrons elsewhere in the phase space [9]. Based on this physical reasoning, it can be deduced that the fundamental adjoint must always be nonnegative.

By utilizing (4) and (5), it is straightforward to obtain the following expression for the relative sensitivity of the multiplication factor with respect to a perturbation $\delta\sigma$ (For derivation, see for example, [4, 10]):

$$\frac{\delta k(\hat{\mathbf{e}}; \mathbf{h})}{k} = -\frac{\langle (\mathbf{A}'_\sigma(\hat{\mathbf{e}}) - (1/k)\mathbf{B}'_\sigma(\hat{\mathbf{e}}))\delta\sigma, \Psi \rangle}{\langle (1/k)\mathbf{B}\Phi, \Psi \rangle}. \qquad (6)$$

This equation is known in reactor physics as *classical perturbation theory*. In addition, the adjoint system can be utilized in the sensitivity analysis of the eigenvalue problem for other responses fulfilling the following properties. Firstly, the response R must be Fréchet-differentiable with respect to Φ, in which case we can write

$$\delta R(\hat{\mathbf{e}}; \mathbf{h}) = R'_\sigma(\hat{\mathbf{e}})\delta\sigma + \langle \nabla_\Phi R(\hat{\mathbf{e}}), \delta\Phi \rangle_\Phi. \qquad (7)$$

In addition, the response's Fréchet derivative $\nabla_\Phi R$ (also called gradient) must be orthogonal to the forward solution

$$\langle \nabla_\Phi R, \Phi \rangle = 0. \qquad (8)$$

When these assumptions are fulfilled, the *generalized adjoint* corresponding to the response R can be defined as the solution to the following inhomogeneous system:

$$\left(\mathbf{A}^* + \frac{1}{k}\mathbf{B}^* \right)\Gamma = \frac{\nabla_\Phi R}{R}. \qquad (9)$$

Notice that in the previous equation the eigenvalue k is fixed to correspond to the solution of (1) and therefore the operator $\mathbf{A}^* + (1/k)\mathbf{B}^*$ is singular, which necessitates (8) in order for the solution $\mathbf{\Gamma}$ to exist. Also, when a solution $\mathbf{\Gamma}_0$ to (9) exists, there exists an infinite amount of solutions of the form

$$\mathbf{\Gamma} = \mathbf{\Gamma}_0 + a\mathbf{\Psi}, \quad a \in \mathbb{R}. \tag{10}$$

In this case, it is possible to choose a solution that is orthogonal to the (forward) fission source. This particular solution can be written as

$$\begin{aligned}
\mathbf{\Gamma}_p &= \mathbf{\Gamma}_0 - \frac{\langle \mathbf{\Gamma}_0, \mathbf{B}\mathbf{\Phi} \rangle}{\langle \mathbf{\Psi}, \mathbf{B}\mathbf{\Phi} \rangle}\mathbf{\Psi} \\
&= \mathbf{\Gamma}_0 - \frac{\langle \mathbf{B}^*\mathbf{\Gamma}_0, \mathbf{\Phi} \rangle}{\langle \mathbf{B}^*\mathbf{\Psi}, \mathbf{\Phi} \rangle}\mathbf{\Psi}.
\end{aligned} \tag{11}$$

We can now derive a practical expression for the response sensitivity with respect to a perturbation $\delta\boldsymbol{\sigma}$:

$$\begin{aligned}
\frac{\delta R(\hat{\mathbf{e}}, \mathbf{h})}{R} &= \frac{R'_\sigma(\hat{\mathbf{e}})\delta\boldsymbol{\sigma}}{R} + \left\langle \frac{\nabla_\Phi R(\hat{\mathbf{e}})}{R}, \delta\mathbf{\Phi} \right\rangle_\Phi \\
&\overset{(9)}{=} \frac{R'_\sigma(\hat{\mathbf{e}})\delta\boldsymbol{\sigma}}{R} + \left\langle (\mathbf{A}^* + \tfrac{1}{k}\mathbf{B}^*)\mathbf{\Gamma}, \delta\mathbf{\Phi} \right\rangle_\Phi \\
&\overset{(4)}{=} \frac{R'_\sigma(\hat{\mathbf{e}})\delta\boldsymbol{\sigma}}{R} + \left\langle \mathbf{\Gamma}, \left(\mathbf{A} + \tfrac{1}{k}\mathbf{B}\right)\delta\mathbf{\Phi} \right\rangle_\Phi \\
&\overset{(3)}{=} \frac{R'_\sigma(\hat{\mathbf{e}})\delta\boldsymbol{\sigma}}{R} - \left\langle \mathbf{\Gamma}, \left(\mathbf{A}'_\sigma(\hat{\mathbf{e}}) - \tfrac{1}{k}\mathbf{B}'_\sigma(\hat{\mathbf{e}})\right)\delta\boldsymbol{\sigma} \right\rangle_\Phi \\
&\quad - \frac{\delta k(\hat{\mathbf{e}}; \mathbf{h})}{k^2}\langle \mathbf{\Gamma}, \mathbf{B}\mathbf{\Phi} \rangle_\Phi \\
&= \frac{R'_\sigma(\hat{\mathbf{e}})\delta\boldsymbol{\sigma}}{R} - \left\langle \mathbf{\Gamma}_p, \left(\mathbf{A}'_\sigma(\hat{\mathbf{e}}) - \tfrac{1}{k}\mathbf{B}'_\sigma(\hat{\mathbf{e}})\right)\delta\boldsymbol{\sigma} \right\rangle_\Phi.
\end{aligned} \tag{12}$$

This framework is often referred to as *generalized perturbation theory* when the response R is of the form:

$$R(\mathbf{e}) = \frac{\langle \mathbf{\Phi}, \mathbf{\Sigma}_1 \rangle}{\langle \mathbf{\Phi}, \mathbf{\Sigma}_2 \rangle}. \tag{13}$$

In this case, it is straightforward to show that (8) is satisfied and that R is Fréchet-differentiable, the relative gradient being

$$\frac{\nabla_\Phi R}{R} = \frac{\mathbf{\Sigma}_1}{\langle \mathbf{\Phi}, \mathbf{\Sigma}_1 \rangle} - \frac{\mathbf{\Sigma}_2}{\langle \mathbf{\Phi}, \mathbf{\Sigma}_2 \rangle}. \tag{14}$$

The generalized adjoint $\mathbf{\Gamma}(\mathbf{r}, \mathbf{\Omega}, E)$ can be physically interpreted as the average contribution of an additional neutron at the phase space point $[\mathbf{r}, \mathbf{\Omega}, E]$ to the response under consideration. The generalized adjoint is normalized according to the value of the response. It should also be noticed that since an additional neutron may also reduce the value of the response, generalized adjoints can also have negative values. The gradient of the response may also be negative in some parts of the phase space.

In practice, the eigenvalue problem and the corresponding adjoint equations are solved numerically, which gives rise to some complications in the perturbation theory formalism. Ideally, the discretizations should be performed in a consistent manner so that the respective adjoint relations are satisfied at all stages of the computation [5]. However, this is usually infeasible in reactor physics calculations, and therefore it is customary to take the eigenvalue problem discretized with respect to energy and direction as the starting point for sensitivity analysis. This issue is discussed in more detail in [4].

2.2. Uncertainty Analysis. The uncertainty of the uncertain parameters $\boldsymbol{\sigma}$ should be understood in terms of the Bayesian probability interpretation. In this framework, probability is defined as a subjective measure that characterizes the plausibility of various hypotheses. When estimating parameters, all knowledge about a parameter σ_j is assumed to be incorporated into its marginal probability distribution $p(\sigma_j)$. This distribution is defined so that the integral $\int_a^b p(\sigma_j)\, d\sigma_j$ corresponds to the (Bayesian) probability that the value of σ_j belongs to the interval $[a, b]$. The distribution $p(\boldsymbol{\sigma})$ can then be used to form an estimate for the parameters and their associated uncertainties. Usually, the mean value or the mode is chosen as the estimate for the parameters, whereas the covariance matrix of the distribution is chosen as the descriptive statistic for the uncertainty.

In the Bayesian formalism, the outcome of the uncertainty analysis should ideally be the full posterior distribution $p(\mathbf{R})$. However, determining $p(\mathbf{R})$ analytically is usually extremely challenging and the distribution can only be estimated pointwise based on a simulation. In deterministic uncertainty analysis, the objective is not to form the entire distribution $p(\mathbf{R})$, but to compute an estimate for the covariance matrix $\mathrm{Cov}[\mathbf{R}]$ by linearizing the response $\mathbf{R} \approx \mathbf{S}\boldsymbol{\sigma}$. Here $\mathbf{S} \in \mathbb{R}^{J \times K}$ is the response vector sensitivity matrix, J is the number of responses, and K is the number of uncertain parameters. After linearizing the response, the covariance matrix can be computed simply using the identity

$$\mathrm{Cov}[\mathbf{R}] \approx \mathrm{Cov}[\mathbf{S}\boldsymbol{\sigma}] = \mathbf{S}\,\mathrm{Cov}[\boldsymbol{\sigma}]\mathbf{S}^T, \tag{15}$$

known as the first-order uncertainty propagation formula or the *Sandwich rule*.

3. Implementation

3.1. Computation of Generalized Adjoint Fluxes. This section reviews the guidelines for modifying a deterministic transport solver to compute the adjoint solutions needed in generalized perturbation theory and describes the methodology used in the implementation to CASMO-4. As mentioned previously, the description on the implementation of classical perturbation theory to CASMO-4 has been recently published in [4], and therefore, in this paper, the emphasis is placed on the GPT-specific features.

As explained in Section 2.1, it is customary to take the energy- and direction-discretized system as the starting point for perturbation theory. In CASMO-4, the multigroup criticality equation is solved by the method of characteristics assuming isotropic scattering. Therefore, the following system of equations may be taken as the forward problem:

$$\Omega_m \cdot \nabla \Phi^g(\mathbf{r}, \Omega_m) + \Sigma^g \Phi^g(\mathbf{r}, \Omega_m)$$

$$= \frac{1}{4\pi} \sum_{h=1}^{G} \Sigma_s^{h \to g} \phi^h(\mathbf{r}) + \frac{\chi_g}{4\pi k} \sum_{h=1}^{G} \bar{\nu} \Sigma_f^h \phi^h(\mathbf{r}), \qquad (16)$$

$$g = 1, \ldots, G.$$

In (16) the scalar flux is approximated by the quadrature formula

$$\phi^h(\mathbf{r}) = \sum_{m=1}^{M} \omega_m \Phi^h(\mathbf{r}, \Omega_m). \qquad (17)$$

In order to simulate an infinite lattice, the boundary conditions are often assumed to be reflective, that is,

$$\Phi(\mathbf{r}, \Omega_m, E) = \Phi(\mathbf{r}, \Omega'_m, E), \quad \mathbf{r} \in \Gamma, \; \Omega_m \cdot \mathbf{n} < 0, \qquad (18)$$

where $\Omega_m = \Omega'_m - 2(\mathbf{n} \cdot \Omega'_m)\mathbf{n}$ is the reflection direction. The inner product corresponding to this discretization can be defined in a consistent manner as

$$\langle \Phi, \Psi \rangle = \sum_{g=1}^{G} \sum_{m=1}^{M} \omega_m \int_D d^3\mathbf{r} \Phi^g(\mathbf{r}, \Omega_m) \Psi^g(\mathbf{r}, \Omega_m). \qquad (19)$$

The adjoint system can now be written

$$-\Omega_m \cdot \nabla \Psi^g(\mathbf{r}, \Omega_m) + \Sigma^g \Psi^g(\mathbf{r}, \Omega_m)$$

$$= \frac{1}{4\pi} \sum_{h=1}^{G} \Sigma_s^{g \to h} \psi^h(\mathbf{r}) + \frac{\bar{\nu} \Sigma_f^g}{4\pi k} \sum_{h=1}^{G} \chi_h \psi^h(\mathbf{r}), \quad g = 1, \ldots, G, \qquad (20)$$

with the boundary conditions

$$\Psi(\mathbf{r}, \Omega_m, E) = \Psi(\mathbf{r}, \Omega'_m, E), \quad \mathbf{r} \in \Gamma, \; \Omega_m \cdot \mathbf{n} > 0. \qquad (21)$$

It is straightforward to check that the systems (16) and (20) with their respective boundary conditions satisfy (4) with respect to the inner product defined by (19).

The generalized adjoint problem for a response of the form of (13) can now be written

$$-\Omega_m \cdot \nabla \Gamma^g(\mathbf{r}, \Omega_m) + \Sigma^g \Gamma^g(\mathbf{r}, \Omega_m)$$

$$= \frac{1}{4\pi} \sum_{h=1}^{G} \Sigma_s^{g \to h} \gamma^h(\mathbf{r}) + \frac{\bar{\nu} \Sigma_f^g}{4\pi k} \sum_{h=1}^{G} \chi_h \gamma^h(\mathbf{r}) \qquad (22)$$

$$+ \frac{\Sigma_1^g(\mathbf{r})}{\langle \Phi, \Sigma_1 \rangle} - \frac{\Sigma_2^g(\mathbf{r})}{\langle \Phi, \Sigma_2 \rangle}, \quad g = 1, \ldots, G,$$

where the generalized adjoint of the scalar flux has been denoted by $\gamma^h(\mathbf{r})$. As explained in Section 2.1, this system may have an infinite number of solutions, of which we wish to solve the one that satisfies

$$\langle \mathbf{B}^* \Gamma_p, \Phi \rangle = 0. \qquad (23)$$

In deterministic transport solvers, the iteration for fixed source calculations is generally of the form

$$\mathbf{A}\Phi^{n+1} = \mathbf{B}\Phi^n + \mathbf{S}, \qquad (24)$$

where $\mathbf{S}$ is an external source. This iteration scheme with a fixed eigenvalue is also well suited for solving the generalized adjoint problem of (22), in which case the iteration takes the form

$$\mathbf{A}^* \Gamma^{n+1} = \frac{1}{k} \mathbf{B}^* \Gamma^n + \frac{\nabla_\Phi R}{R}. \qquad (25)$$

During the iteration, however, the convergence to the particular solution that is orthogonal to the fission source must be ensured. It is straightforward to show that if the initial guess for the generalized adjoint flux satisfies (23), this orthogonality property is preserved during the iteration. Firstly,

$$\langle \mathbf{A}^* \Gamma^{n+1}, \Phi \rangle \overset{(25)}{=} \frac{1}{k} \langle \mathbf{B}^* \Gamma^n, \Phi \rangle + \left\langle \frac{\nabla_\Phi R}{R}, \Phi \right\rangle$$

$$\overset{(8)}{=} \frac{1}{k} \langle \mathbf{B}^* \Gamma^n, \Phi \rangle. \qquad (26)$$

On the other hand,

$$\langle \mathbf{A}^* \Gamma^{n+1}, \Phi \rangle = \langle \Gamma^{n+1}, \mathbf{A}\Phi \rangle \overset{(1)}{=} \frac{1}{k} \langle \Gamma^{n+1}, \mathbf{B}\Phi \rangle$$

$$= \frac{1}{k} \langle \mathbf{B}^* \Gamma^{n+1}, \Phi \rangle. \qquad (27)$$

Therefore, for each iteration n,

$$\langle \mathbf{B}^* \Gamma^{n+1}, \Phi \rangle = \langle \mathbf{B}^* \Gamma^n, \Phi \rangle, \qquad (28)$$

from which the result follows. In practice, however, due to round-off errors and the unavoidable inconsistencies in formulating the discretizations and adjoint relations, a refinement of the iteration scheme is necessary to guarantee that (23) remains satisfied [11]. A suitable procedure is to force the orthogonality of the solution with each outer iteration. In this case, in accordance with (11), the iteration takes the form

$$\mathbf{A}^* \Gamma^{n+1} = \frac{1}{k} \mathbf{B}^* \left(\Gamma^n - \frac{\langle \mathbf{B}^* \Gamma^n, \Phi \rangle}{\langle \mathbf{B}^* \Psi, \Phi \rangle} \Psi \right) + \frac{\nabla_\Phi R}{R}. \qquad (29)$$

Notice that this iteration scheme requires that the forward solution and the fundamental adjoint solution have been previously computed and that they are accessible during the iteration.

By comparing (29) with the forward problem of (16), it can be seen that if the forward system had an external source, the systems would be of the same form with the exception that the adjoint system is solved in the opposite direction. Therefore, if the transport solver does not rely on the assumption of the nonnegativity of the flux or the sources, relatively few modifications are needed to transform the solver to also compute the generalized adjoint functions. For example, the method of characteristics, used in CASMO-4, does not require that the solution or the sources are nonnegative. In this case, the following operations need to be performed *before* the adjoint calculation [10].

(1) Transpose the scattering matrix.

(2) Interchange the vectors $\overline{\nu}\sigma_{\mathrm{f}}$ and χ.

(3) Invert the group indices for all variables as follows: $G \leftrightarrow 1, (G-1) \leftrightarrow 2, \ldots$.

After these operations, the transport solver can be used to compute the fundamental adjoint solution. Notice also that these operations automatically convert the forward boundary conditions to the adjoint boundary conditions. When solving a generalized adjoint problem, the following changes need to be additionally implemented *within* the (forward) transport solver.

(1) Add the response gradient $\nabla_\Phi R/R$ to the variable for an external source.

(2) Modify the fission source F_g to the form

$$F_g = \frac{\chi_g}{4\pi k} \sum_{h=1}^{G} \overline{\nu}\, \Sigma_{\mathrm{f}}^h \left(\phi^h(\mathbf{r}) - \frac{\langle \mathbf{B}\Phi, \Phi_{\mathrm{F}} \rangle}{\langle \mathbf{B}\Psi, \Phi_{\mathrm{F}} \rangle} \Psi^g \right), \quad (30)$$

where Φ_{F} denotes the forward solution of (16) and Ψ the adjoint solution of (20).

The multigroup solution Φ given by the solver must then be interpreted so that, for example, $\Phi^g(\mathbf{r}, \Omega)$ corresponds to $\Gamma^{G+1-g}(\mathbf{r}, -\Omega)$. Notice that if the transport solver is based on a numerical scheme that relies on the nonnegativity of the flux or the sources, some additional modifications are necessary in addition to the ones described above. For further details, see for example, [11].

3.2. Computation of Sensitivity and Uncertainty Profiles. After obtaining the adjoint solutions, the sensitivities with respect to the multigroup nuclear data parameters can be computed according to (6) and (12). Notice that even after the multigroup approximation, these parameters are still spatial functions and therefore the derivatives in the equations refer to functional derivatives. The inner product in the sensitivity expressions can be discretized as

$$\langle \Phi, \Psi \rangle \approx \sum_{i=1}^{I} \sum_{g=1}^{G} \sum_{m=1}^{M} \omega_m V_i\, \overline{\Phi}^{g,i,m}\, \overline{\Psi}^{g,i,m}, \quad (31)$$

where i denotes the mesh index and $\overline{\Phi}^{g,i,m}$ and $\overline{\Psi}^{g,i,m}$ denote the average fluxes.

In order to compute the uncertainties using the Sandwich rule, the sensitivities and covariance matrices need to be formed with respect to the same parameters using the same energy group structure. In the SCALE 6 covariance library [3], the available covariance matrices are given in a 40-group structure for the parameters listed in Table 1. Most of these covariance matrices are nuclide specific. It should be emphasized that there is no covariance data for the group-to-group transfer cross-sections.

Multigroup covariance matrices can in principle be transformed to another multigroup structure by simple mathematical techniques. The applicability of this approach depends on the differences between the group structures.

TABLE 1: Parameters for which there exists covariance data in the SCALE library.

Parameter	MT number
σ_{t}	1
σ_{e}	2
σ_{i}	4
$\sigma_{\mathrm{n,2n}}$	16
σ_{f}	18
σ_{γ}	102
$\sigma_{\mathrm{n,p}}$	103
$\sigma_{\mathrm{n,d}}$	104
$\sigma_{\mathrm{n,t}}$	105
$\sigma_{\mathrm{n,He}}$	106
$\sigma_{\mathrm{n,\alpha}}$	107
$\overline{\nu}$	456
χ	1018

In particular, the widths of the energy groups should not dramatically change. In this work, the code Angelo 2.3 [12] was used to transform the matrices to the energy group structure used in the sensitivity calculations with CASMO-4. The transformation procedure used in the code is based on flat-flux approximation, where the resampled values on the new grid are computed as lethargy overlap weighted averages. For further details, see [13]. When modifying the energy group structure of fission spectrum covariance matrices, further correction procedures are necessary in order to guarantee that the covariance matrices are in accordance with the normalization condition $\sum_g \chi^g = 1$ [14]. The correction can also be applied to the fission spectrum sensitivities in which case the sensitivities are called *constrained* [14]. This was the approach chosen in this work.

In order to utilize the covariance data given for the parameters in Table 1, sensitivity profiles should be computed with respect to the same parameters. However, many lattice physics codes such as CASMO, HELIOS [15], WIMS [16], and DRAGON [17] employ nuclear data libraries that do not contain cross-section data for the individual capture and scattering reactions, but only for the total capture and scattering cross-section. There are generally three different approaches to overcome this difficulty. The most natural approach is perhaps to add the missing cross-sections to the code, either by creating a new cross-section library or by modifying the cross-sections inside the code [18]. Another option, suitable for deterministic analysis, is not to use problem-dependent cross-sections in the sensitivity analysis. In this case, the sensitivity coefficients can be computed outside the code based on the forward and adjoint fluxes and any set of cross-sections. This was the idea, for example, behind connecting DRAGON with the sensitivity and uncertainty analysis code SUSD3D after a generalized adjoint mode was implemented to DRAGON [19]. The third option is to form the covariance matrices corresponding to the total capture and scattering cross-sections [4]. This is the approach that was chosen in this work.

Since the relationships between the total and individual capture and scattering reactions are linear, the covariance matrices corresponding to the total capture and scattering reactions can be computed with the Sandwich rule without introducing any approximation. The method used for combining the covariance matrices has been recently described in detail in [4]. Therefore, only the most important conclusions related to the methodology are repeated here.

Firstly, in the context of the capture reactions, the results are expected to be fully consistent with the case where the sensitivities are computed with respect to the individual capture reactions. In the case of the scattering reactions, however, the sensitivity profiles with respect to the individual and the total scattering cross-sections cannot be defined in a consistent manner and this affects the uncertainty results. In this context, it should be emphasized that the treatment of the covariance matrices involves no approximations and the inconsistency is solely related to the computation of the sensitivities. As mentioned previously, there is no cross-section data for the transfer cross-sections $\sigma_x^{h \to g, j}$ but only for $\sigma_x^{g, j} = \sum_{h=1}^{G} \sigma_x^{g \to h, j}$, where x refers to a scattering reaction (e.g., elastic, inelastic) and j is the nuclide index. Therefore, in order to use the scattering covariance data, the sensitivity profiles should be computed with respect to $\sigma_x^{g, j}$. Because of the scattering source term in (16), however, the derivative with respect to $\sigma_x^{g, j}$ is not mathematically well defined without additional constraints. Typically it is assumed that the probabilities of transfers to various groups are fixed, that is,

$$\sigma_x^{g \to h, j} = \sigma_x^{g, j} p_x^{g \to h, j}, \tag{32}$$

where $p_x^{g \to h}$ is the proportion of neutrons scattered from energy group g to energy group h, which is assumed to remain fixed even if the scattering cross-section $\sigma_x^{g, j}$ is perturbed [20]. Based on this assumption, the scattering source in (16) can be written as

$$S^g = \frac{1}{4\pi} \sum_{h=1}^{G} \Sigma_s^{h \to g} \phi^h = \frac{1}{4\pi} \sum_{x} \sum_{j} N^j \sum_{h=1}^{G} \sigma_x^{h, j} p_x^{h \to g} \phi^h, \tag{33}$$

where the summations over x include all scattering reactions. After this assumption, the derivative with respect to $\sigma_x^{g, j}$ is well defined and can be computed as usual. It is straightforward to show that this approach corresponds to computing the sensitivity coefficients with respect to the transfer cross-sections $\sigma_x^{g \to h, j}$ and summing them over h.

However, the sensitivity with respect to the total scattering cross-section $\sigma_s^j = \sum_x \sigma_x^j$ is not well defined if the constraint (32) is enforced. In order to define this sensitivity, fixed transfer rates must be assumed for the total scattering cross-section. Also, computing the total scattering sensitivity as the sum of the individual scattering sensitivities implicitly enforces this constraint. Since the two assumptions required to compute the individual and total scattering sensitivities are inconsistent, the chain rule of derivation does not apply to them, and, for example, although $\sigma_s^{g, j} = \sigma_e^{g, j} + \sigma_i^{g, j}$ holds, $dR/d\sigma_e^{g, j} \neq (dR/d\sigma_s^{g, j})(d\sigma_s^{g, j}/d\sigma_e^{g, j})$.

4. Numerical Results for PB-2 Lattice Physics Exercise

The calculation framework was applied to the BWR test case from the UAM benchmark lattice physics Exercise 1.2 considering a single fuel assembly with reflective boundary conditions [2], and the results were compared against the TSUNAMI-2D sequence in SCALE 6.1 [21]. The test problem represents Peach Bottom 2 (PB-2) under hot zero power conditions. Two-group homogenized cross-sections have been considered as responses in the GPT framework.

The outline of the CASMO-4 calculations is presented in Figure 1. The calculations were carried out using the cross-section library E60200 that contains 70 energy groups and is based on ENDF/B-VI data [22]. The covariance data were taken from the SCALE 6 library ZZ-SCALE6.0/COVA-44G [3] according to the guidelines of the benchmark. The library is based on evaluations from various sources (including ENDF/B-VII, ENDF/B-VI, JENDL-3.1) and approximate covariance data. The covariances in the library are given in relative terms, and therefore the library is intended to be used with all cross-section libraries including the ones that are inconsistent with the evaluations. While this is not strictly correct, it is considered to be acceptable due to the scarcity of comprehensive covariance data among other reasons [23].

The list of the nuclides present in these test cases can be found in the benchmark specification [2]. Apart from the isotopes of chromium and iron, all available covariance data in the library was included in the uncertainty computations. The reason for excluding these isotopes is that the employed cross-section library E60200 does not contain isotope-specific cross-sections for these materials but only cross-sections for natural chromium and iron.

The covariance matrices from ZZ-SCALE6.0/COVA-44G were processed for compatibility with CASMO-4. The sensitivity profiles in CASMO-4 were computed using the 40-group structure option that was the closest match to the amount of groups in the covariance data and, as mentioned in Section 3.2, the code Angelo 2.3 [12] was used to process the covariance matrices to this energy group structure. Next, the nuclear data processing code NJOY [24] was used to transform the 40-group covariance files to the BOXR format. Auxiliary FORTRAN programs were written for combining the covariance matrices according to the principles described in Section 3.2.

The TSUNAMI-2D calculations were performed using the ENDF/B-VI-based cross-section library V6-238 containing 238 energy groups. The module CENTRM was used for self-shielding. Implicit sensitivity analysis [9] was omitted in the TSUNAMI calculations in order to facilitate the comparison of the results given by CASMO-4 and TSUNAMI-2D.

4.1. Results Based on Classical Perturbation Theory. A summary of the results based on classical perturbation theory for the multiplication factor is presented in Table 2. The relative difference between the multiplication factors computed with CASMO-4 and TSUNAMI-2D is 52 pcm in both forward and adjoint cases. For the total uncertainty, the

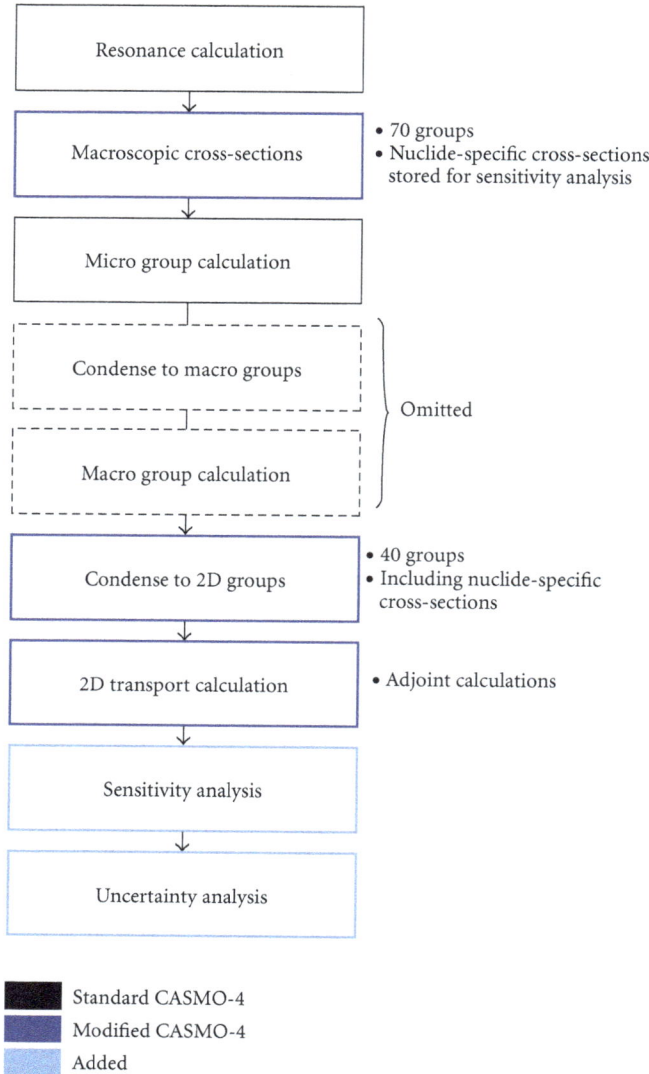

FIGURE 1: Outline of the CASMO-4 calculations.

TABLE 2: Summary of the results for the multiplication factor.

Code	Forward k	Adjoint k	Rel. uncertainty, $\Delta k/k$ (%)
CASMO-4	1.10548	1.10546	0.508
TSUNAMI-2d	1.10490	1.10490	0.506

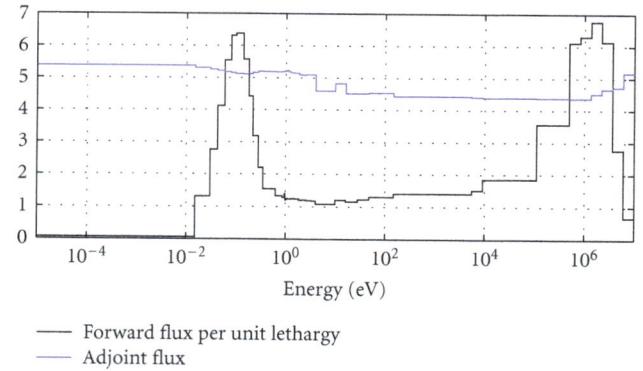

FIGURE 2: Volume-averaged forward flux and fundamental adjoint flux.

values given by CASMO-4 and TSUNAMI-2D are also very consistent. Table 3 shows the five most significant sources of uncertainty together with the corresponding energy- and region-integrated sensitivity coefficients. As can be seen from this table, both the sensitivity and the uncertainty results are in good accordance. The greatest difference occurs for the capture cross-section of ^{238}U, for which CASMO-4 yields a greater sensitivity. This appears to originate from the differences in the cross-section libraries. In particular, the cross-section library E60200 used in the CASMO-4 calculation has not been reduced in terms of the ^{238}U resonance integral, which is known to be overestimated in the ENDF/B-VI data [22].

Figure 2 shows the volume-averaged forward flux and the volume-averaged fundamental adjoint $\overline{\Psi}$ corresponding to this test case. As explained in Section 2.1, the value $\overline{\Psi}^g$ represents the average importance of neutrons in the energy group g to the multiplication factor in comparison to neutrons

in other energy groups. The plot can be interpreted from this point of view. For example, it can easily be seen from the figure how the ^{238}U capture cross-section resonances reduce the importance of neutrons in the corresponding energy groups. This phenomenon is particularly clear in the energy group E_{14} = [4.00 eV, 9.88 eV], where the multigroup capture cross-section attains its maximum value. It can also be clearly distinguished from the plot how the adjoint function has a higher value in the energy groups corresponding to the peaks in the fission cross-section of ^{235}U. To further demonstrate this, Figure 3 shows a plot of the problem-dependent ^{235}U fission and ^{238}U capture cross-sections in the same 40-group structure. The increase in the adjoint values in the highest energy groups corresponds mainly to the increase in the value of $\overline{\nu}$ at these energies.

Figure 4 shows the multiplication factor sensitivity profiles for the parameters, whose integrated sensitivity coefficients have the greatest absolute values, excluding the sensitivity profile with respect to the fission spectrum of ^{235}U, which was constrained in the computation. As can be seen from the figure, the multiplication factor is the most sensitive to the fission parameters of ^{235}U, the capture cross-section of ^{238}U, and the scattering cross-section of ^{1}H. The positive sensitivity to the capture of ^{238}U in the highest energy group follows from the fact that in CASMO-4 the (n, 2n) reaction cross-section has been included in the capture cross-section with a negative sign in this group. It is instructive to compare the sensitivity profiles with the forward and adjoint fluxes plotted in Figure 2. Notice that the peaks in the sensitivity profiles of ^{235}U coincide with the thermal peak of the neutron flux, where most of the fissions occur. In general, perturbing a nuclear parameter has a greater impact on the results in the energy groups, where the flux is higher. On the contrary, the values of the fundamental adjoint represent the *average* importance

TABLE 3: The five most significant sources of uncertainty for the multiplication factor and the corresponding energy- and region-integrated relative sensitivity coefficients.

Nuclide	Parameter pair	Sensitivity		Contribution to $\Delta k/k$ (%)	
		CASMO	TSUNAMI	CASMO	TSUNAMI
^{238}U	σ_c, σ_c	-2.434×10^{-1}	-2.143×10^{-1}	3.198×10^{-1}	2.902×10^{-1}
^{235}U	$\bar{\nu}, \bar{\nu}$	9.160×10^{-1}	9.370×10^{-1}	2.720×10^{-1}	2.773×10^{-1}
^{235}U	σ_c, σ_c	-1.027×10^{-1}	-1.025×10^{-1}	1.454×10^{-1}	1.422×10^{-1}
^{235}U	σ_f, σ_f	4.038×10^{-1}	4.212×10^{-1}	1.372×10^{-1}	1.409×10^{-1}
^{235}U	σ_f, σ_c	4.038×10^{-1}	4.212×10^{-1}	1.238×10^{-1}	1.245×10^{-1}
		-1.027×10^{-1}	-1.025×10^{-1}		

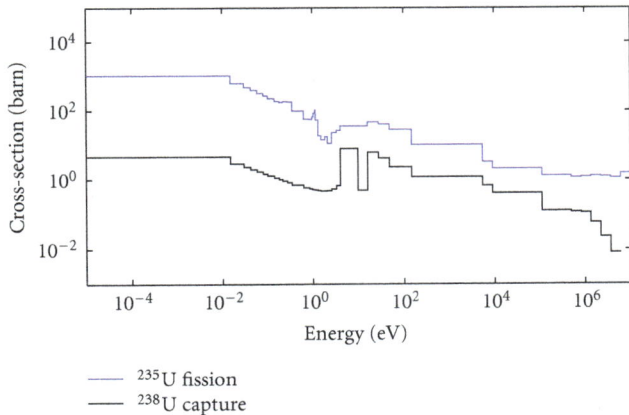

FIGURE 3: Self-shielded multigroup cross-sections corresponding to the test problem.

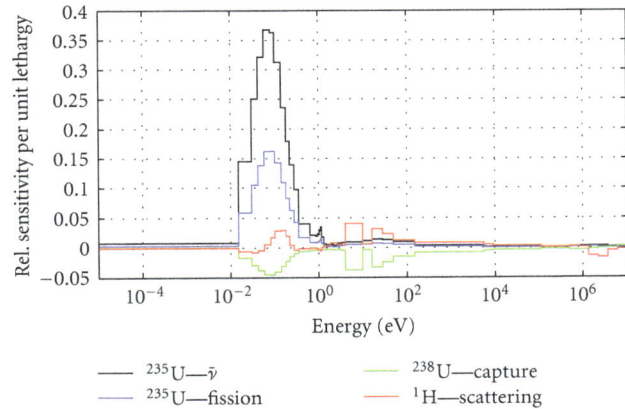

FIGURE 4: Multiplication factor sensitivity profiles.

of neutrons in different energy groups. In particular, the lowest energy group has the highest importance, but this is not manifested in the sensitivity profiles, as the flux is very close to zero in this group. The negative sensitivities to the scattering reaction of ^{1}H in the four lowest energy groups can be attributed to the fact that in these groups upscattering is more likely than downscattering. Therefore, neutrons are scattered to energy groups with a lower importance. The same reasoning applies to the scattering sensitivity of ^{1}H in the highest energy groups, where neutrons are scattered downwards and the values of the adjoint function decrease rapidly with energy.

4.2. Results Based on Generalized Perturbation Theory. Table 4 presents the values and the total uncertainties of the homogenized two-group cross-sections that were considered as responses in the GPT-based sensitivity and uncertainty analysis. In computing the responses, the thermal cut-off was set at 0.625 eV. It can be seen from the table that all total uncertainty values are in good agreement with the thermal responses, whereas for the fast responses the uncertainties given by TSUNAMI-2D are consistently greater.

Tables 5 and 6 show more detailed sensitivity and uncertainty results for the two-group homogenized production cross-sections $\nu\Sigma_{f,1}$ and $\nu\Sigma_{f,2}$. As can be seen from Table 6, in the case of $\nu\Sigma_{f,1}$, the difference in the total uncertainty

values given by CASMO-4 and TSUNAMI-2D is attributable to the scattering of ^{238}U, for which TSUNAMI-2D yields a significantly greater uncertainty value, although the total scattering sensitivity coefficients given by both codes are very close. As explained in Section 3.2, the sensitivity with respect to the total scattering cross-section can only be defined if the group-to-group transfer probabilities are assumed to be fixed for the total scattering. Also, defining the total scattering sensitivity as the sum of the individual scattering sensitivities implicitly enforces this assumption. However, in the TSUNAMI-2D computation, the total scattering uncertainty is computed based on the individual scattering sensitivities, which rely on the assumption of fixed transfer rates for each scattering reaction. The difference in the total scattering uncertainties is hence explained by incompatible constraints in the two uncertainty calculations. This phenomenon is more evident for the fast group responses since they are more sensitive to the inelastic scattering of ^{238}U.

Figure 5 shows the volume-averaged generalized adjoint solutions for the responses $\nu\Sigma_{f,1}$ and $\nu\Sigma_{f,2}$, denoted by $\bar{\Gamma}_{f,1}$ and $\bar{\Gamma}_{f,2}$, respectively. As previously explained, the adjoint values in each energy group can be interpreted to represent the average importance of neutrons in that group to the response under consideration. Therefore, it is not surprising that thermal neutrons are more important to the response $\nu\Sigma_{f,2}$, whereas fast neutrons are more important to the response $\nu\Sigma_{f,1}$. The positive values of $\bar{\Gamma}_{f,2}$ in the fast groups result

TABLE 4: Values and uncertainties of the responses considered in the GPT framework.

Response R	Value		Relative uncertainty $\Delta R/R$ (%)	
	CASMO	TSUNAMI	CASMO	TSUNAMI
$\nu\Sigma_{f,1}$	4.976×10^{-3}	4.951×10^{-3}	8.399×10^{-1}	9.754×10^{-1}
$\nu\Sigma_{f,2}$	6.922×10^{-2}	6.938×10^{-2}	4.490×10^{-1}	4.478×10^{-1}
$\Sigma_{c,1}$	5.348×10^{-3}	5.380×10^{-3}	1.098×10^{0}	1.168×10^{0}
$\Sigma_{c,2}$	2.653×10^{-2}	2.672×10^{-2}	5.066×10^{-1}	5.040×10^{-1}
$\Sigma_{f,1}$	1.935×10^{-3}	1.927×10^{-3}	5.563×10^{-1}	6.820×10^{-1}
$\Sigma_{f,2}$	2.841×10^{-2}	2.847×10^{-2}	3.244×10^{-1}	3.226×10^{-1}

TABLE 5: The five most significant sources of uncertainty for the response $\nu\Sigma_{f,2}$ and the corresponding energy- and region-integrated relative sensitivity coefficients.

Nuclide	Param. pair	Sensitivity		Contribution to $\Delta R/R$ (%)	
		CASMO	TSUNAMI	CASMO	TSUNAMI
^{235}U	$\bar{\nu}, \bar{\nu}$	9.996×10^{-1}	9.998×10^{-1}	3.105×10^{-1}	3.106×10^{-1}
^{235}U	σ_f, σ_f	7.985×10^{-1}	7.941×10^{-1}	2.893×10^{-1}	2.869×10^{-1}
^{235}U	σ_f, σ_c	7.985×10^{-1}	7.941×10^{-1}	1.134×10^{-1}	1.139×10^{-1}
		-3.599×10^{-2}	-3.667×10^{-2}		
^{238}U	σ_c, σ_c	-4.406×10^{-2}	-4.255×10^{-2}	7.257×10^{-2}	7.222×10^{-2}
^{235}U	σ_c, σ_c	-3.599×10^{-2}	-3.667×10^{-2}	5.613×10^{-2}	5.672×10^{-2}

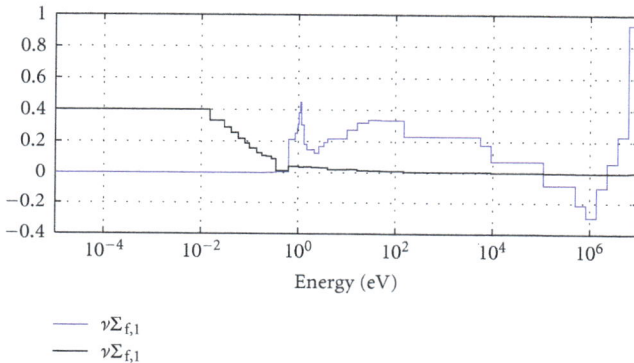

FIGURE 5: Volume-averaged generalized adjoints corresponding to the responses $\nu\Sigma_{f,1}$ and $\nu\Sigma_{f,2}$.

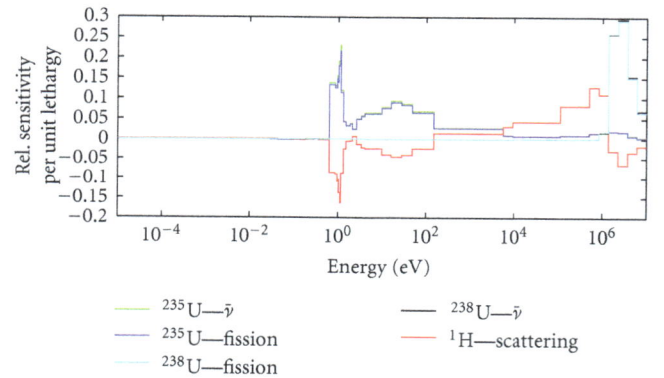

FIGURE 6: Sensitivity profiles of the response $\nu\Sigma_{f,1}$.

from the downscattering of neutrons. Notice that $\overline{\Gamma}_{f,1}$ also has a small positive value in the first thermal group, which corresponds to the possibility of upscattering. For the most part, both adjoint fluxes qualitatively follow the fission cross-section of ^{235}U plotted in Figure 3. In the highest energy groups, the values of $\overline{\Gamma}_{f,1}$ increase rapidly due to the increase in the values of $\bar{\nu}$. The negative values of $\overline{\Gamma}_{f,1}$ between 0.111 MeV and 2.231 MeV signify that additional neutrons in those energy groups would on average contribute more to the denominator $\langle\Phi, 1\rangle_1$ than to the numerator $\langle\Phi, \nu\Sigma_f\rangle_1$. This in accordance with the fact that fission is unlikely to occur in this energy region.

Figure 6 shows the sensitivity profiles of $\nu\Sigma_{f,1}$ with respect to the parameters, whose integrated sensitivity coefficients have the greatest absolute values. As can be anticipated, the response is the most sensitive to the fission parameters of ^{235}U and ^{238}U and in addition to the scattering of ^{1}H. It

is interesting to compare these profiles with the plot of the generalized adjoint $\overline{\Gamma}_{f,1}$ in Figure 5. The sensitivity to the scattering of ^{1}H has the smallest values in the groups with the highest importance, as this reaction transfers neutrons to energy groups with a lower importance. Since fast neutrons mostly scatter downwards, the scattering sensitivity has positive values in the groups between 149 eV and 1.35 MeV, where the importance decreases with increasing energy. This trend is reversed at 1.35 MeV, where the importance of the energy groups begins to increase with energy, mainly due to the increase in the values of $\bar{\nu}$ at these energies.

The sensitivity profiles corresponding to the response $\nu\Sigma_{f,2}$ are plotted in Figure 7. It is noteworthy that the profiles qualitatively resemble the respective profiles of the multiplication factor in the thermal region, whereas they quickly fall to nearly zero in the fast region. From the perspective of the GPT framework, it is again enlightening

TABLE 6: The five most significant sources of uncertainty for the response $\nu\Sigma_{f,1}$ and the corresponding energy- and region-integrated relative sensitivity coefficients. The sensitivity coefficients with respect to the parameter χ have been constrained.

Nuclide	Param. pair	Sensitivity		Contribution to $\Delta R/R$ (%)	
		CASMO	TSUNAMI	CASMO	TSUNAMI
^{235}U	χ, χ	4.657×10^{-9}	-2.757×10^{-10}	5.934×10^{-1}	6.150×10^{-1}
^{238}U	$\bar{\nu}, \bar{\nu}$	3.975×10^{-1}	3.879×10^{-1}	4.623×10^{-1}	4.544×10^{-1}
^{238}U	σ_f, σ_f	3.931×10^{-1}	3.834×10^{-1}	2.084×10^{-1}	1.994×10^{-1}
^{238}U	σ_s, σ_s	-2.743×10^{-2}	-2.718×10^{-2}	2.015×10^{-1}	5.148×10^{-1}
^{235}U	σ_f, σ_f	5.826×10^{-1}	5.866×10^{-1}	1.588×10^{-1}	1.466×10^{-1}

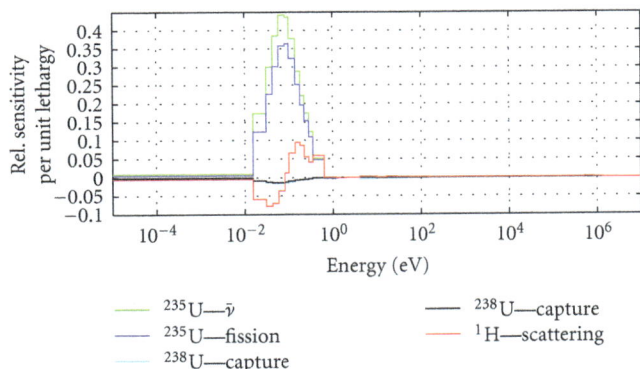

FIGURE 7: Sensitivity profiles of the response $\nu\Sigma_{f,2}$.

to compare the sensitivity plots with the adjoint function $\overline{\Gamma}_{f,2}$ plotted in Figure 5 and the flux $\overline{\Phi}$ shown in Figure 2. In the case of this response, the average importance of neutrons increases steadily with decreasing energy. Therefore, it is reasonable that the scattering sensitivities are again negative in the groups where upscattering is more likely than downscattering. Also, the sensitivities peak in the energy region coinciding with the thermal peak of the forward flux.

5. Summary and Conclusions

Sensitivity and uncertainty analysis capability has been developed to the reactor physics code CASMO-4 in the context of the UAM benchmark. Sensitivities with respect to nuclear data parameters can be computed efficiently by utilizing the adjoint system of the criticality equation. The propagated nuclear data uncertainty can then be calculated deterministically by the Sandwich rule.

Initially, classical perturbation theory was implemented to the code, which enabled sensitivity analysis of the critical eigenvalue. In this context, covariance matrices from scale 6 were transformed to become compatible with CASMO-4, and the resulting covariance library was connected with the code. Since the cross-section libraries of CASMO-4 do not contain data for the individual capture and scattering reactions, the covariance matrices of the individual subreactions were combined in the covariance library. This work has been reported in detail in [4], and the main conclusions related to the methodology were summarized in this paper. In particular, the sensitivities with respect to total scattering

and individual scattering cross-sections cannot be defined in a consistent manner, which leads to some systematic differences in the uncertainty results.

Recently, generalized perturbation theory was added to the code as a new feature, which enables performing sensitivity analysis for responses that can be represented as reaction rate ratios. For each response, the computation of sensitivity profiles with respect to all parameters of interest requires solving one generalized adjoint system. The mathematical background as well as the physical interpretation of the generalized adjoint solutions were reviewed, and practical guidelines were given for modifying a deterministic transport code to solve the generalized adjoint systems needed in sensitivity analysis. The theory for computing the sensitivity profiles was presented both from the perspective of function space analysis and numerical computations.

Numerical results were presented for a lattice physics test problem representing a BWR in hot zero power conditions, and they were compared to the results given by the TSUNAMI-2D sequence in SCALE 6.1. Two-group homogenized cross-sections were considered as responses in the generalized perturbation theory framework. The results were in very good agreement with the thermal responses, whereas in the case of fast responses, the uncertainties given by TSUNAMI-2D were consistently greater. Detailed sensitivity and uncertainty results were presented and analyzed for the homogenized fast and thermal production cross-sections. The differences in the uncertainty results for the fast responses were explained by the incompatible constraints used in computing the scattering uncertainties.

In the future, the work will continue by extending the GPT framework to other responses in addition to two-group homogenized cross-sections with the eventual goal of modifying CASMO-4 to provide uncertainty estimates for all homogenized assembly data, which can then be propagated to coupled neutronics/thermal hydraulics calculations.

Acknowledgment

This work was funded through the Finnish National Research Programme on Nuclear Power Plant Safety 2011–2014, SAFIR2014.

References

[1] J. Rhodes and M. Edenius, *CASMO-4, A Fuel Assembly Burnup Program, Users Manual*, 2001.

[2] K. Ivanov, M. Avramova, I. Kodeli, and E. Sartori, "Benchmark for uncertainty analysis in modeling (UAM) for design, operation, and safety analysis of LWRs," NEA/NSC/DOC(2007) 23, 2007.

[3] ZZ-SCALE6.0/COVA44G, "A 44-group cross section covariance matrix library retrieved from the scale-6.0 package," NEA Data Bank Code Package USCD1236/03, 2011.

[4] M. Pusa, "Incorporating sensitivity and uncertainty analysis to a lattice physics code with application to CASMO-4," *Annals of Nuclear Energy*, vol. 40, no. 1, pp. 153–162, 2012.

[5] D. G. Cacuci, *Sensitivity and Uncertainty Analysis*, vol. 1, Chapman & Hall/CRC, Boca Raton, Fla, USA, 2003.

[6] B. G. Carlson and K. D. Lathrop, "Transport theory-the method of discrete ordinates," in *Computing Methods in Reactor Physics*, H. Greenspan, C. N. Kelber, and D. Okrent, Eds., Gordon and Breach Science Publishers, New York, NY, USA, 1968.

[7] E. E. Lewis and J. W. F. Miller, *Computational Methods of Neutron Transport*, John Wiley & Sons, New York, NY, USA, 1984.

[8] J. Lewins, *Importance: The Adjoint Function*, Pergamon Press, Oxford, UK, 1965.

[9] M. L. Williams, B. L. Broadhead, and C. V. Parks, "Eigenvalue sensitivity theory for resonance-shielded cross sections," *Nuclear Science and Engineering*, vol. 138, no. 2, pp. 177–191, 2001.

[10] M. L. Williams, "Perturbation theory for nuclear reactor analysis," in *CRC Handbook of Nuclear Reactors Calculations*, Y. Ronen, Ed., vol. 3, CRC Press, Boca Raton, Fla, USA, 1986.

[11] R. L. Childs, "Generalized perturbation theory using two-dimensional, discrete ordinates transport theory," Tech. Rep. ORNL/CSD/TM-127, Oak Ridge National Laboratory, Oak Ridge, Tenn, USA, 1980.

[12] I. Kodeli, "Manual for ANGELO2 and LAMBDA codes," NEA-1798/03 Package, 2010.

[13] I. Kodeli and E. Sartori, "Neutron cross-section covariance data in multigroup form and procedure for interpolation to users' group structures for uncertainty analysis applications," in *Proceedings of the PHYSOR International Conference on the Physics of Reactors: Operation, Design and Computation*, Marseille, France, 1990.

[14] I. Kodeli, M. Ishikawa, and G. Aliberti, "Evaluation of fission spectra uncertainty and their propagation," in *OECD/NEA WPEC Subgroup 26 Final Report: Uncertainty and Target Accuracy Assessment for Innovative Systems Using Recent Covariance Data Evaluations*, C. Appendix, Ed., OECD, Paris, France, 2008.

[15] "HELIOS Methods," Studsvik Scanpower, 2000.

[16] WIMS9A, "NEW FEATURES, A Guide to the New Features of WIMS Version 9A," Serco Assurance, http://www.sercoassurance.com/answers/, 2005.

[17] G. Marleau, A. Hébert, and R. Roy, "A User Guide For Dragon Version 4," IGE294, http://www.polymtl.ca/nucleaire/DRAGON/, 2009.

[18] W. Wieselquist, A. Vasiliev, and H. Ferroukhi, *Nuclear Data Uncertainty Propagation in a Lattice Physics Code Using Stochastic Sampling*, ANS Physics of Reactors (PHYSOR 2012): Advances of Reactor Physics, Knoxville, Tenn, USA, 2012.

[19] A. Bidaud, G. Marleau, and E. Noblat, "Nuclear data uncertainty analysis using the coupling of DRAGON with SUSD3D," in *Proceedings of the International Conference on Mathematics, Computational Methods & Reactor Physics (M&C '09)*, May 2009.

[20] C. R. Weisbin, J. H. Marable, J. L. Lucius et al., "Application of FORSS sensitivity and uncertainty methodology to fast reactor benchmark analysis," Tech. Rep. ORNL/TM-5563, 1976.

[21] "SCALE: a modular code system for performing standardized computer analyses for licensing evaluation," Tech. Rep. ORNL/TM-2005/39, Radiation Safety Information Computational Center at Oak Ridge National Laboratory as CCC-725, Oak Ridge, Tenn, USA, 2009, Version 6, Vols. I–III.

[22] J. Rhodes, *JEF 2.2 and ENDF/B-VI 70 Group Neutron Data Libraries*, Studsvik, Nykoping, Sweden, 2005.

[23] M. L. Williams, D. Wiarda, G. Arbanas, and B. L. Broadhead, "Scale nuclear data covariance library," in *SCALE: A Modular Code System for Performing Standardized Computer Analyses for Licencing Evaluation, Version 5*, ORNL/TM-2005/39, Oak Ridge National Library/U.S. Nuclear Regulatory Commission, Oak Ridge, Tenn, USA, 20052009.

[24] R. E. MacFarlane and D. W. Muir, "The NJOY Nuclear Data Processing System, Version 91," Manual LA-12740-M, Los Alamos National Laboratory, Los Alamos, NM, USA, 1994.

9

Remarks on Consistent Development of Plant Nodalizations: An Example of Application to the ROSA Integral Test Facility

J. Freixa and A. Manera

Laboratory for Reactor Physics and Systems Behaviour (LRS), Paul Scherrer Institut (PSI), 5232 Villigen, Switzerland

Correspondence should be addressed to J. Freixa, jordi.freixa@psi.ch

Academic Editor: Klaus Umminger

Experimental results obtained at integral test facilities (ITFs) are used in the validation process of system codes for the transient analyses of light water reactors (LWRs). The expertise and guidelines derived from this work are later applied to transient analyses of nuclear power plants (NPPs). However, the boundary conditions at the NPPs will always differ from those at the ITF, and hence, the soundness of the ITF model needs to be maximized. An unaltered ITF nodalization should prove to be able to simulate as many tests as possible, before any conclusion is derived to NPP analyses. The STARS group at the Paul Scherrer Institut (PSI) actively participates in several international programs, where ITFs are being used (e.g., ROSA, PKL). Several tests carried out at the ROSA large-scale test facility operated by the Japan Atomic Energy Agency (JAEA) have been simulated in recent years by using the United States Nuclear Regulatory Commission (US-NRC) system code TRACE. In this paper, 5 different posttest analyses are presented, along with the evolution of the employed TRACE nodalization and the process followed to track the consistency of the nodalization modifications. The ROSA TRACE nodalization provided results in a reasonable agreement with all 5 experiments.

1. Introduction

In the last decades, integral test facilities (ITFs) have been used to validate thermal-hydraulic codes with the final objective of performing safety analyses for nuclear power plants (NPPs). The experiments carried out at facilities like PKL, LOFT, LOBI, BETHSY, or LSTF have been employed to build up expertise in the usage of system codes. User guidelines have been derived, models and correlations have been corrected or further developed, and limitations on the use of system codes have been identified. Such expertise is later on applied to the modelling of NPPs. One of the lessons learned with the use of ITFs for system codes validation is that even though a given code with a given nodalization is able to capture correctly the phenomena occurring in one test, it might not be the case in a successive test with different boundary conditions. This can be true even for single phenomena such as the critical flow at a break location, so that one should not assume that the phenomenon will be well simulated in an NPP model. As a matter of fact, it has to be considered that the conditions to be simulated in

the NPPs differ from those simulated with the ITF models, so that confidence in the system code and the associated nodalization can be built only after successful simulations of a wide range of tests.

The large number of available experiments in test facilities like the ROSA large-scale test facility (LSTF) operated by the Japanese Atomic Energy Agency (JAEA) offers the possibility to validate various physical models under different conditions, at the same time developing guidelines and strategies for building up the system nodalization. When developing an ITF nodalization, it is essential to simulate as many tests as possible by using exactly the same nodalization. The extend of expertise and guidelines obtained by a model that is able to reproduce several tests at the same time is far larger than what one can obtain with few tests, because the performance of the models and correlations is tested in different conditions. Users must not resort to simulate the different tests independently, because if done so, the robustness and the soundness of the model will be lost. Generally speaking, each time a new test is simulated and modifications are introduced to the main model other than initial or boundary conditions, all previous tests should be recalculated to assure the consistency of the modifications. In principle, any improvement undertaken in the nodalization must be valid for all previous tests. Obviously, this supposes a considerable effort, especially when various full models of facilities or power plants are to be maintained. Hence, it is recommendable to establish a clear and easy methodology, aimed at reducing the efforts of the analyst. Examples of this consistent quality assurance can be found in [1, 2].

The STARS group [3] of the Paul Scherrer Institut (PSI) actively participates in several international programs where ITFs are employed (e.g., ROSA, PKL). Several tests of the ROSA/LSTF have been simulated at PSI in the recent years, within the OECD/NEA ROSA 1 and 2 projects. These projects aim at addressing thermal-hydraulic safety issues relevant for light water reactors through experiments making use of the ROSA/LSTF facility. The experiments are used for the development and validation of simulation methodologies of the complex phenomena occurring during design basis accidents (DBAs) and beyond DBAs. ROSA/LSTF, operated by the Japan Atomic Energy Agency (JAEA), is a full-height and 1/48 volumetrically scaled test facility of a 1100 MWe-class pressurized water reactor (PWR). The facility allows to perform system integral experiments simulating the thermal-hydraulic responses at full-pressure conditions during small break loss-of-coolant accidents (LOCAs) and other transients. The reference plant is Unit-2 of Turuga NPP of the Japan Atomic Power Company.

A nodalization of the ROSA facility has been built in the PSI STARS group using the United States Nuclear Regulatory Commission (US-NRC) thermal-hydraulic code TRACE. The nodalization has been employed to calculate 5 different ROSA experiments, focusing mainly on small and intermediate break LOCAs. From test to test, sensitivity studies have been performed which have led to nodalization modifications and corrections. This paper focuses on the evolution of the nodalization, and the work carried out in order to maintain a sole nodalization able to satisfactorily simulate all the transients.

2. Model Description

The ROSA/LSTF nodalization has been produced by using the US-NRC thermal-hydraulic code TRACE (version 5.0). The nodalization had been derived combining the information of existing TRAC-p and RELAP input decks provided by JAEA together with the original technical drawings of the facility.

The TRACE model of the LSTF, shown in Figure 1, consists of a 3D vessel, two separate loops with two steam generators, and a pressurizer. The primary system is completed with pressurizer control systems (spray system, relief and safety valves, and base and proportional heaters) and safety injection systems (accumulators and low- and high-pressure injections). The secondary system is provided with main and auxiliary feedwater systems, a set of relief and safety valves, a main steam line, and a common steam header.

The 3D vessel component is composed by 20 axial levels, 4 radial rings, and 10 azimuthal sectors. The three first rings cover the core region, and the fourth ring represents the downcomer (DC). The heater rods are simulated by means of 30 heat structures, one for each section in the radial plane. The rods are grouped in three categories according to the power regions of the LSTF.

All bypasses are nodalized according to the geometrical specifications, and the friction k-factors were adjusted such that the mass flow at the bypasses during the steady-state calculation matched the experimental values. An additional leakage between the DC and the UP was added, as it was seen to be the only way to represent the evolution of the void fraction in the downcomer. The considerations that lead to the nodalization of this additional leakage will be discussed in Section 4.1.1.

The 8 control rod guide tubes (CRGTs) are simulated by 12 pipes connecting the core outlet with the upper head. The two inner CRGTs are linked geometrically with three different core volumes as shown in Figure 1. In order to keep the same connections from the core exit to the upper head (UH), the two inner CRGTs were split into three pipes. Volumes and sections were divided by three, while the hydraulic diameter was kept. Hence, each of the two inner tubes is simulated by means of three parallel pipes connected with three out of ten azimuthal cells of the inner ring. The other guide tubes are simulated by using a single pipe component for each tube.

The correct simulation of metal structures and heater rods is of main importance to capture the system behavior during small break LOCA (SBLOCA) and intermediate break LOCA (IBLOCA) scenarios. The ROSA nodalization developed at PSI includes all heat structures corresponding to walls, heated rods, unheated rods, internal metal structures, and support plates. All these components are simulated by a total of 53 heat structures.

The ROSA/LSTF has a core protection logic aimed at avoiding damage to the heater rods in the core, by limiting the maximum allowed core temperatures. Basically, the core power is decreased when the rods temperature rises above ≈960 K. The logic of this protection is included in the control system simulated within the TRACE input data set.

The two loops are simulated separately, and a single pipe component is used in the nodalization to represent the U-tube bundle. The pressurizer is nodalized with a 7-cell pipe. The steam generators are represented by separator components, and the heat losses through the outer walls are taken into account as well.

The countercurrent flow limitation (CCFL) model was activated in correspondence of two locations:

(i) the top plate at the top of the core region. The Wallis correlation [4] was used with the same coefficients specified in the OECD BEMUSE project ($m = 1.0$ and $c = 0.8625$) [5];

(ii) the steam generator (SG) inlet plenums and U-tube section. Again the Wallis correlation was used although in this case the coefficients suggested by Yonomoto [6] were used ($m = 1.0$, $c = 0.75$).

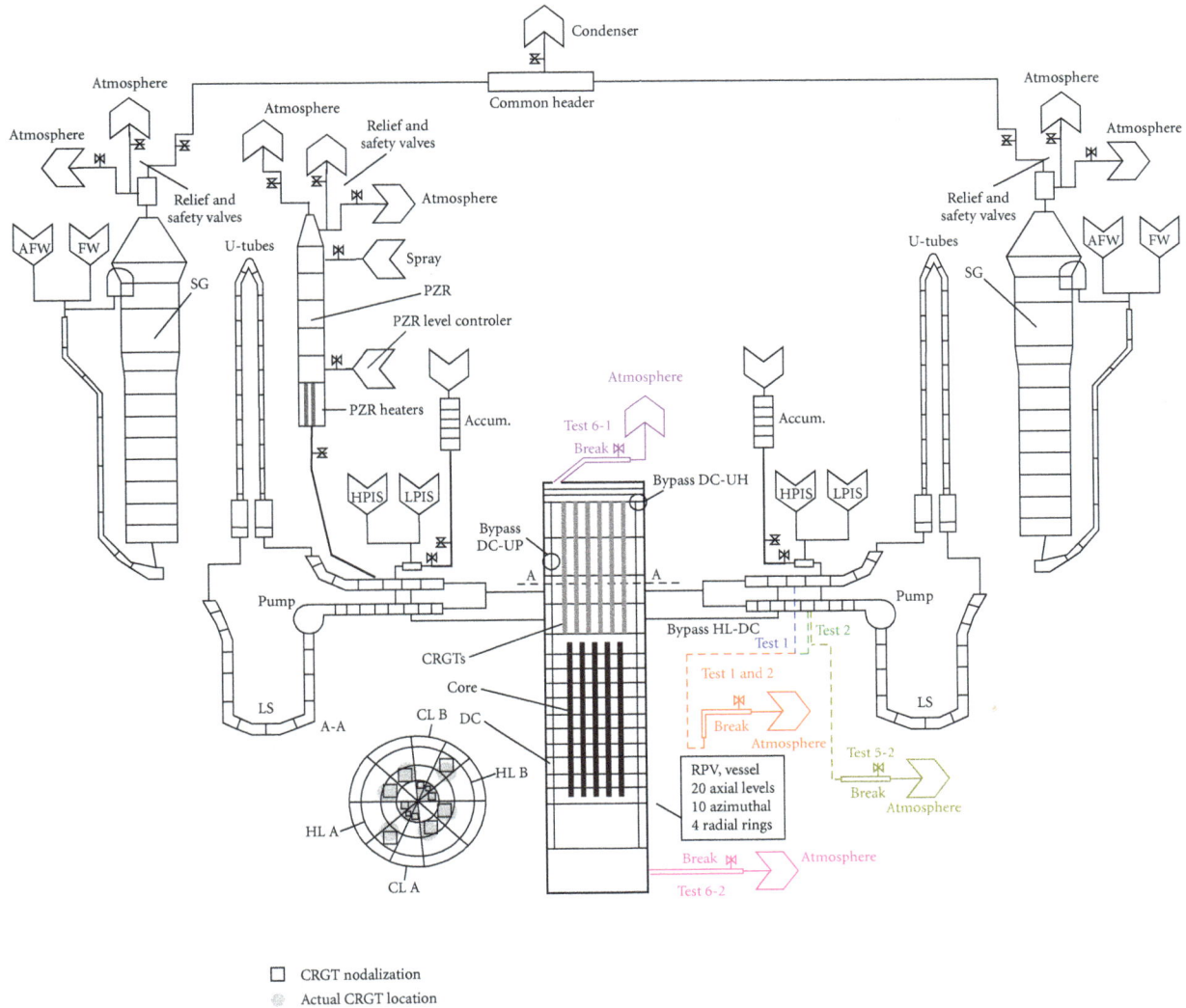

FIGURE 1: TRACE nodalization of the ROSA/LSTF.

2.1. *Steady-State and Initial Conditions.* A steady-state calculation was carried out in order to reach the experimental initial conditions. The initial conditions obtained by the TRACE model (Table 1) after the steady-state calculation are in quite close agreement with the experimental values. All steady-state parameters are within the error bands of the measurements. The most significant discrepancy is a 3% error in the steam flow rate. The high percentual errors in the bypasses flows are within the error bands of the measurements which are quite large.

The use of a three-dimensional vessel was seen to be crucial. Different flows were detected in the CRGTs depending on the position of the pipe relative to the vessel cross-section, that is, depending on the radial and azimuthal cell where the CRGTs were connected. While some CRGTs showed upward flow, others experienced downward flow, allowing natural convection in the upper head. Capturing this phenomenon was found to be crucial in order to correctly reproduce the UH temperature, which would have been otherwise too high.

TABLE 1: Percentual error between the calculated steady state and the experimental values.

	Percentual error (%)
Core power	0.0
DC to UH bypass flow	17.0
Hot leg to DC bypass flow	17.0
Primary pressure	0.3
Hot leg fluid temperature	0.3
Cold leg fluid temperature	0.2
Mass flow rate	0.0
Pressurizer level	2.0
Main steam pressure	0.0
Secondary side liquid level	0.5
Steam flow rate	1.9
Main feedwater flow rate	1.5

TABLE 2: Control logic of Test 6-1 [7].

Event	Condition
Break	Time zero
Generation of scram signal	Primary pressure 84% of initial value
PZR heater off	Generation of scram signal or PZR liquid level below 2.3 m
Initiation of core power decay curve simulation	
Initiation of primary coolant pump coastdown	
Turbine trip (closure of stop valve)	Generation of scram signal
Closure of main steam isolation valve	
Termination of main feedwater	
Opening and closing of the SG relief valves	SG pressures 110%/103%
Generation of SI signal	Primary pressure below 79% of initial value
Initiation of auxiliary feedwater	Generation of SI signal
Initiation of SG secondary-side depressurization as accident management (AM) action by fully opening relief valves	Core exit temperature reaches 623 K
Initiation of accumulator system	Primary pressure below 29% of initial value
Initiation of LPSI system	LP pressure below 8% of initial value

3. Tests

This section describes the final results of 5 different ROSA/LSTF tests that were simulated with the ROSA TRACE nodalization. The results shown in this section were obtained by the last integrated version of the model after simulating test 5-2 of the ROSA-1 project. All the relevant modifications undertaken during the process are described in Section 4. All tests start with the same initial conditions.

3.1. Test 6-1. Test 6-1 is an SBLOCA case with the break located in the UH of the reactor pressure vessel (RPV). The break was realized in the facility by using a sharp-edge orifice mounted downstream of a horizontal pipe that was connected to the upper head (the orifice flow area corresponded to 1.9% of the volumetrically scaled cross-sectional area of the reference PWR cold leg). The test started by opening the break valve, and by increasing at the same time the rotation speed of the primary coolant pumps up to 1500 rpm, for a better simulation of the expected pressure and flow values that would prevail in a similar transient in the reference PWR. Proportional heaters were used in the pressurizer (PZR) to control the primary pressure, while backup heaters compensated for system heat losses. The proportional heater power was turned off simultaneously with the activation of the scram signal, while the backup power was reduced and only completely shut down immediately after the PZR liquid level became lower than 2.3 m. This particular operation of the reactor coolant pumps and the PZR heaters was performed in all the tests analyzed in the work presented here. The description of the control logic of Test 6-1 is detailed in Table 2.

The scram signal was set to be dependent on the primary pressure by a set point of 84% of the initial pressure and initiated the core power decay, the primary coolant pumps coastdown, the termination of the feedwater system, and the closure of the main steam isolation valve. The safety injection

(SI) signal was generated when the primary pressure fell below 79%. When the core exit temperature reached 623 K, depressurization of the steam generator secondary side as accident management action by fully opening the relief valves was initiated. The auxiliary feedwater was also triggered by the SI signal. The ECC system was directed towards the two cold legs and consisted of two accumulators (Accs) injecting at a primary pressure of 29% of the initial primary pressure and two low-pressure safety injection (LPSI) pumps that were started at a lower plenum (LP) pressure of 8%. The high-pressure safety injection (HPSI) system was not actuated in this test. Further details on the experimental procedures and results can be found in [7].

3.1.1. Results. Table 3 shows the chronology of the main events that occurred in Test 6-1, comparing the experimental values with the calculated ones. All the events in the TRACE simulation occur similarly as in the experiment. Discrepancies are found to be within a reasonable range of less than 80 seconds. The calculation was ended at 3000 seconds as all main events had occurred by that time.

The most important parameters for Test 6-1 are shown in Figure 2. Since the break was located at the top of the RPV, the transient can be easily divided into three different phases according to the break flow conditions: blowdown phase, break discharge in two-phase choked regime, and break discharge in single-phase vapor.

The first two phases, blowdown and two-phase flow phases, were simulated very well by the TRACE model. The distribution of coolant was correctly reproduced, as shown in the bottom graph of Figure 2. The early drop of the DC level and cold leg level (see [8]) (at about 250 s and 200 s, resp.,) could only be simulated assuming a hypothetical bypass between the UP and the DC (this assumption which will be explained and justified in Section 4.1.1). During this phase, the primary and secondary pressures were steady around the SG relief valves set points (middle graph of Figure 2). After

TABLE 3: Chronology of the main events in Test 6-1 [7].

Event	Exp.	TRACE
Break valve opened	0	0
Scram signal	26	20
SI signal	27	21
Break flow from subcooled to two-phase flow	50	46
Primary coolant pumps stopped	277	270
Break flow to single-phase vapor	700	741
Primary pressure lower than SG pressure, core level decreases	784	761
Full opening of SG relief valves	1090	1071
Core protection activated	1205	1271
Initiation of Acc system	1305	1225
Initiation of LPSI	2893	—
End of test	3266	3000

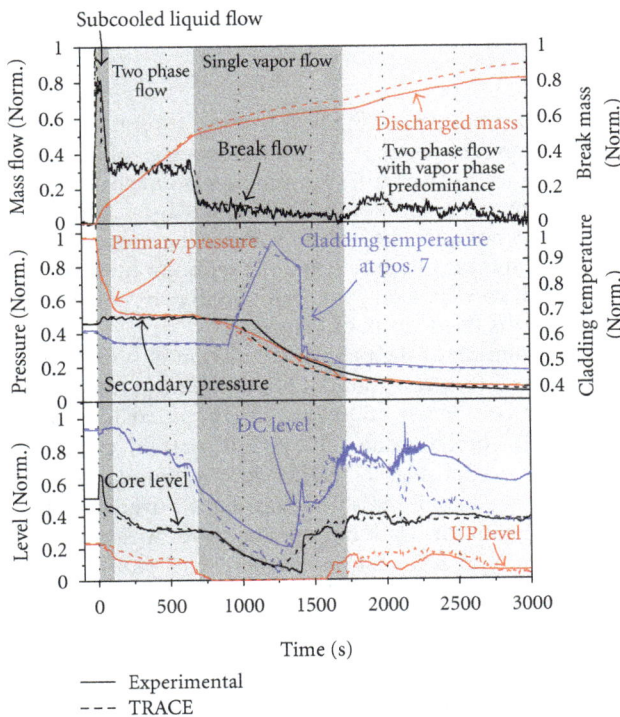

FIGURE 2: TRACE results for Test 6-1. From top to bottom: (1) break flow and integrated discharged mass, (2) primary and secondary pressure along with the maximum cladding temperature, and (3) RPV water levels.

this first reduction of mass inventory, the UP level dropped and the penetration holes at the bottom of the CRGTs started to void, and thus steam flew to the upper head and was expelled through the break. The transition from two-phase flow to single-vapor flow at the break was well matched by the TRACE nodalization although a delay in the transition of 50 seconds was detected. Due to this delay, the integrated discharged mass at the end of the two-phase flow part was higher in the simulation. Afterwards, the single-phase vapor

choked flow at the break reported slightly higher values (see top graph in Figure 2). As a consequence, the primary pressure fell at a higher rate and the UH mass experienced a larger reduction. The drop of core level occurring during this phase was satisfactorily simulated (Figure 2). The maximum PCT, one of the most interesting signals of this test, was also correctly simulated by the model as shown in the middle graph of Figure 2. The PCT reached the core protection set points similarly in both the test and the calculation, and the power was reduced to 10% after the PCT reached 970 K. Afterwards, the PCT started to decrease. The core level started to rise again right after the opening of the accumulator valves (at $t = 1225$ s). Since the primary pressure decreased faster in the calculation, the valves opened earlier than in the experiment, so the core level rose earlier as well.

Further details on the simulation of Test 6-1 can be found in [8].

3.2. Test 6-2. Test 6-2 was started by opening a 0.1% break in the LP. The break was realized by using an inner-diameter sharp-edge orifice mounted downstream a horizontal pipe connected to the LP. The scram signal was triggered when the primary pressure was lower than 84% of the initial value [9]. The PZR heaters were operated as explained in Section 3.1 (page 4). The control logic of Test 6-2 is detailed in Table 4.

The SI signal was generated when the primary pressure decreased below 79%, and it was ensued by initiation of the auxiliary feedwater (AFW). Thirty minutes after the generation of the SI signal, asymmetrical steam generator secondary-side depressurization as accident management action was activated in order to achieve a depressurization rate of 55 K/h in the primary system. A loss of offsite power was considered, so that the HPSI pumps were not available. The LPSI and accumulator systems were fully operable and were actuated as shown in Table 4. Further details on the experimental procedures and results can be found in [9].

3.2.1. Results. Table 5 shows the chronology of the main events that occurred in Test 6-2, comparing the experimental values with the calculated ones.

The most relevant results of Test 6-2 are shown in Figure 3, which displays the cladding temperature, RPV collapsed water levels, primary and secondary pressures, and the loop mass flow rates. Overall good agreement between simulation and experimental results was obtained for the whole transient.

The primary and secondary pressures are shown in the third graph of Figure 3. During the first part of the transient, the primary pressure decreases slightly faster in the calculation, and therefore, some of the associated events occurred earlier (see the chronology, Table 5). Once the depressurization is started in loop B, the primary pressure remains close to the loop A secondary pressure until the reflux-condenser mode in this loop is interrupted (around 4000 s). Afterwards, the primary pressure follows the loop B depressurization. As shown in Figure 3, the primary pressure stabilizes around 15 bars during the latter period

TABLE 4: Control logic of Test 6-2 [9].

Event	Condition
Break	Time zero
Generation of scram signal	Primary pressure 84% of initial value
PZR heater off	Generation of scram signal or PZR liquid level below 2.3 m
Initiation of core power decay curve simulation	
Initiation of primary coolant pump coastdown	
Turbine trip (closure of stop valve)	Generation of scram signal
Closure of main steam isolation valve	
Termination of main feedwater	
Generation of SI signal	Primary pressure below 79% of initial value
Opening and closing of the SG relief valves	SG pressures 110%/103%
Initiation of auxiliary feedwater	Generation of SI signal
Initiation of asymmetrical SG secondary-side depressurization as AM action to achieve a depressurization rate of 55 K/h in the primary system	30 minutes after generation of SI signal
Initiation of accumulator system	Primary pressure below 29% of initial value
Initiation of LPSI	LP pressure below 8% of initial value

TABLE 5: Chronology of the main events in Test 6-2 [7].

Event	Exp.	TRACE
Break valve opened	0	0
Scram signal	569	453
SI signal	736	545
Primary coolant pumps stopped	819	708
Asymmetrical SG secondary-side depressurization	2548	2345
Closure of SG RV in loop with PZR	2679	2345
Initiation of accumulator system	≈5150	5450
Inflow of nitrogen gas from Acc, loop with PZR	≈10030	10995
Inflow of nitrogen gas from Acc, loop w/o PZR	≈11070	11013
Core uncovery	≈20400	22370
Initiation of LPSI	≈21940	—
Core protection activated	≈23270	23510
Second actuation of LPSI system	≈23320	—
End of transient	24034	24000

of the test (12000–25000 seconds), which overestimates the experimental results. On the one hand, the secondary-side depressurization stops at a slightly higher pressure. This depressurization depends on the capacity of the relief valve, and thus on its area and friction losses along the line. The k-factors along this line were adjusted according to the correct flows expected at high pressures. On the other hand, the higher pressure can also be caused by an overestimation of the injected nitrogen, or by an inaccurate treatment of the heat transfer in the U-tubes under the presence of nitrogen. The nitrogen concentrates in the U-tubes and induces a degradation of the heat transfer from the primary to the secondary system. As a consequence of having a slightly higher primary pressure, the set point for the LPSI intervention was never reached in the calculation, and hence, the core was not quenched as in the experimental test. In any case, it is important to notice that the set point in the facility was reached very late and with a very small margin (see small window in the third graph of Figure 3).

The evolution of the natural circulation in both loops is well predicted (bottom graph of Figure 3). After the primary pumps are completely stopped, the computed mass flows drop to the correct value. The end of natural circulation in loop A occurs slightly earlier in the simulation, and afterwards, the loop B mass flow was slightly overpredicted.

The maximum cladding temperature was measured in position 7 on rod (4,4) of element B17 (top graph of Figure 3). Two peaks were observed in the experiment, the first one was quenched by a temporary injection of the LPSI system, while the second peak reached the core protection set point, triggering a reduction of the core power. Afterwards, the LPSI system started again and quenched the core rapidly. In the calculation, the first peak did not appear, and the second one reached the first protection set point, and as the core power was reduced, the cladding temperature started to decrease slowly. However, quenching did not occur since the LPSI did not intervene in view of the too high primary pressure. Further calculations showed that the temperature would have kept rising when disabling the LSTF core protection system, leading eventually to core damage.

The DC, core, and upper plenum levels are displayed in the second graph of Figure 3. The overall behavior as simulated by means of the PSI TRACE nodalization is in accordance with the experiment. Reasonably, large discrepancies are observed when the LPSI system starts in the experiment and the primary system is refilled. There is also a time delay on the plunging of the core level at 2000 seconds.

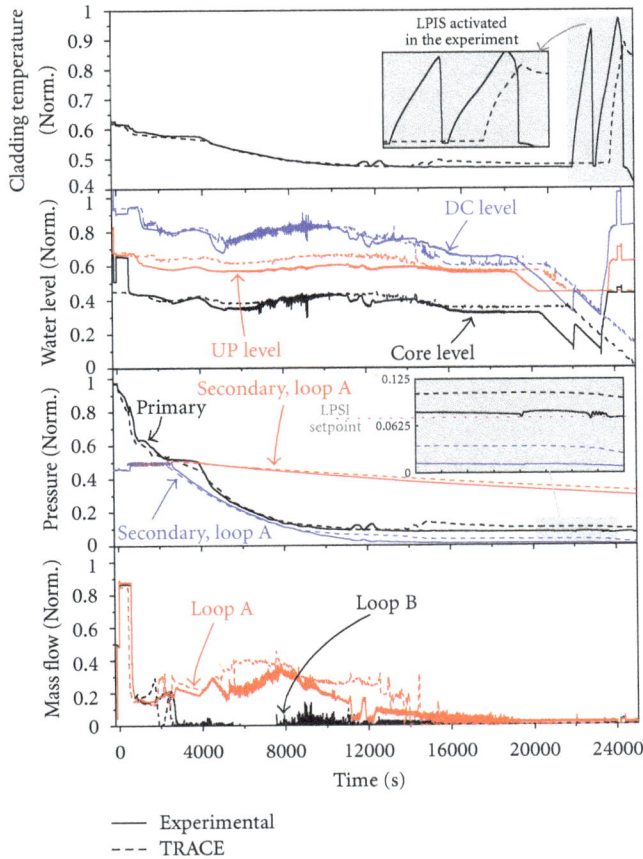

FIGURE 3: TRACE results for Test 6-2. From top to bottom: (1) cladding temperature in position 7 of rod (4,4), (2) RPV water levels, (3) primary and secondary pressures, and (4) loop mass flows.

3.3. Test 1. Test 1 is an intermediate break LOCA transient, in which a complete rupture of the surge line and the concurrent unavailability of the HPSI system are assumed. The control logic of the test is displayed in Table 6. The test is started by isolating the pressurizer and by opening the break valve located in the hot leg. Since a large amount of coolant is released through the break short after its opening, the primary pressure and the RPV level plunge. At the time when the hot leg of the broken loop becomes empty and the break flow switches from two-phase flow to single-phase vapor flow, the coolant that remains in the core starts to flash, and the core level is further reduced. An increase in the cladding temperature occurs as the heater rods become uncovered. Almost at the same time, the accumulator's valves open due to the drop of the primary pressure below the accumulators (Accs) set point. When the core level starts to drop, a large amount of steam is produced in the core, so that the pressure difference between the UP and the DC increases. The pressure drop is further enhanced by steam condensation in the cold legs due to the Acc's injection. When this pressure drop is large enough to drag the mass of water in the loop seal (LS), an LS clearance occurs. The core is refilled by means of the Acc's injection and the LS clearance. The timing between these three phenomena (Acc's injection,

LS clearance, and break flow conditions) will confine the maximum peak cladding temperature.

The pressure set points for the opening and closure of the SG relief valves are 110% and 103%, respectively. Due to the assumed loss of offsite power, the HPSI is disabled for this test. The LPSI and Acc systems are available for this test and are actuated as shown in Table 6. Further details on this test and its simulation can be found in [10, 11].

3.3.1. Results. The chronology of the main events of Test 1 is shown in Table 7. All the events as simulated by TRACE occurred at very similar times as for the experiment (within 5 seconds difference), except for the initiation of the LPSI which was delayed in the experiment due to an increase of the primary pressure after the core was quenched. This increase in the primary pressure could not be explained by the experimentalists at the moment of writing.

The most important results of Test 1 are shown in Figure 4 where three different graphs display (from top to bottom): the break mass flow and integrated discharged mass, the primary and secondary pressures along with the PCT and the core, and UP and DC collapsed water levels.

The break flow in Test 1 was overpredicted by TRACE even though a two-phase discharge coefficient of 0.85 was used. The primary and secondary pressures were very well predicted once the discharge coefficient was reduced. The calculated maximum PCT was also found to be in accordance with the experimental results; however, the maximum PCT took place at a different elevation (around 1 meter difference). The maximum temperature in the TRACE calculation was detected at position 7 (see [12] for details), which is the place where the maximum PCT is usually detected. In the experiment, the maximum temperature was measured at position 5 instead (see [12] for details), which is about 1 meter below. This might be due to a slightly different core level evolution in the experiment. The water levels in the vessel are shown in the bottom graph of Figure 4 presenting a good agreement with the experiment.

3.4. Test 2. Test 2 is an intermediate break LOCA transient for which a complete rupture of the ECC line connected to the cold leg is assumed. The 17% break, corresponding to the volumetrically scaled cross-sectional area of the reference PWR cold leg, was realized by using a nozzle mounted on the top of the cold leg (upward direction). Due to the ECC line rupture, the ECC is only available in the intact loop. The failure of a diesel generator was assumed as well. Considering that the test intends to simulate a 4-loop Westinghouse design transient, the HPSI and LPSI flows injected into the intact loop correspond to 3/8 of the total nominal flow because, from the nominal 8 available pumps (2 per each cold leg), four are lost due to the failure of the diesel generator, and an extra one is lost because of the ECC line rupture. The Acc's volume to be injected in the intact loop corresponds to the volumetrically scaled volume of three Accs of the reference plant.

The control logic of the test is reported in Table 8. The test was started by opening the break; after the scram signal

TABLE 6: Control logic of Test 1 [10].

Event	Condition
Close of PZR spray line valves and PZR heaters	30 minutes before break
Isolation of PZR by closing surge line valve	1 minute before break
Break opening	Time zero
Generation of scram signal	4 seconds
Initiation of primary coolant pump coastdown	
Initiation of core power decay simulation	
Turbine trip (closure of stop valve)	Generation of scram signal
Closure of main steam isolation valve	
Termination of main feedwater	
Opening and closing of the SG relief valves	SG pressures 110% and 103%
Initiation of accumulator system	Primary pressure below 29% of initial value
Initiation of low-pressure injection system	LP pressure below 8% of initial value

TABLE 7: Chronology of the main events in Test 1 [10].

Event	Exp.	TRACE
PZR isolation	−60 s	−60 s
Onset of break	0 s	0 s
Scram signal	1 s	4 s
Turbine trip	1 s	4 s
Primary coolant pumps stopped	4 s	4 s
Main FW isolated	8 s	4 s
Start of decay heat	20 s	22 s
Actuation of ACC	154 s	155 s
PCT increases	164 s	163 s
Maximum PCT	182 s	185 s
Start of LPSI	504 s	349 s
End of test	1533 s	500 s

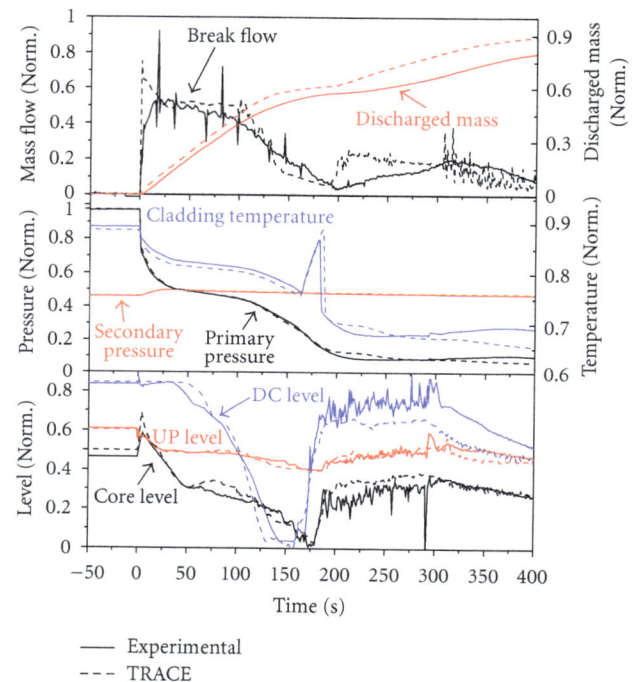

FIGURE 4: TRACE results for Test 1. From top to bottom: (1) break flow and integrated discharged mass, (2) primary and secondary pressure along with the maximum cladding temperature, and (3) RPV water levels.

was reached, the typical sequence of action is realized: initiation of core power decay, initiation of primary coolant pumps coastdown, turbine trip, closure of main steam isolation valve, and termination of main feedwater flow.

Further details on this test can be found in [13].

3.4.1. Results. The chronology of the events taking place during Test 2 is shown in Table 9. Most of the events took place similarly in the simulation and in the experiment. The main difference is that in the simulation the core protection limit was not reached, and therefore the power was not reduced. As a consequence, the primary pressure remained slightly higher, and the LPSI setpoint was also not reached before the first 500 s of the transient. Some of the other discrepancies in the chronology are due to inconsistencies between the actuated experimental procedures and the ones described in the protocols since some experimental timings differ from the control logic shown in Table 8. For example, the initiation of the HPSI system was defined to be 12 seconds after the SI signal, but in the experiment, the HPSI started 25 seconds after the SI signal.

The most relevant results of Test 2 are displayed in Figure 5: primary and secondary pressures, maximum PCT, break flow and integrated discharged mass, and the RPV collapsed water levels.

The top graph is showing the primary and secondary pressures along with the PCT at position 7. The primary and secondary pressures were found to be in good agreement with the experiment; however, the PCT was underestimated by about 20 degrees although the experimental temperature reached the LSTF protection system, and thus, the hypothetical maximum value that the PCT would have reached cannot

TABLE 8: Control logic of Test 2 [13].

Event	Condition
Break	Time zero
Generation of scram signal	Primary pressure 84% of initial value
PZR heater off	Generation of scram signal or PZR liquid level below 2.3 m
Initiation of core power decay curve simulation	
Initiation of primary coolant pump coastdown	
Turbine trip (closure of stop valve)	Generation of scram signal
Closure of main steam isolation valve	
Termination of main feedwater	
Generation of SI signal	Primary pressure below 79% of initial value
Initiation of HPIS in the loop with PZR only	12 s after SI signal
Initiation of accumulator system	Primary pressure below 29% of initial value
Initiation of low-pressure injection system	LP pressure below 8% of initial value
Opening and closing of the SG relief valves	SG pressures 110%/103%

TABLE 9: Chronology of the main events in Test 2 [13].

Event	Exp.	TRACE
Onset of break	0	0
Scram signal	5	6
SI signal	7	9
Turbine trip and MSIVs closed	10	9
Primary pumps trip	11	6
Termination of SG FW	13	6
SG relief valves open	≈27–57	≈25–55
Initiation of core power decay	29	24
Initiation of HPSI	≈35	21
Loop seal clearing	≈40	≈60
Primary pressure lower than SG pressure	≈55	74
Initiation of Acc system	≈110	126
Core protection activated	≈140	—
Max. PCT	≈150	141
Whole core was quenched	≈180	212
Primary coolant pumps stopped	260	256
Termination of Acc injection	≈280	≈285
Initiation of LPSI	≈290	—
End of test	1212	500

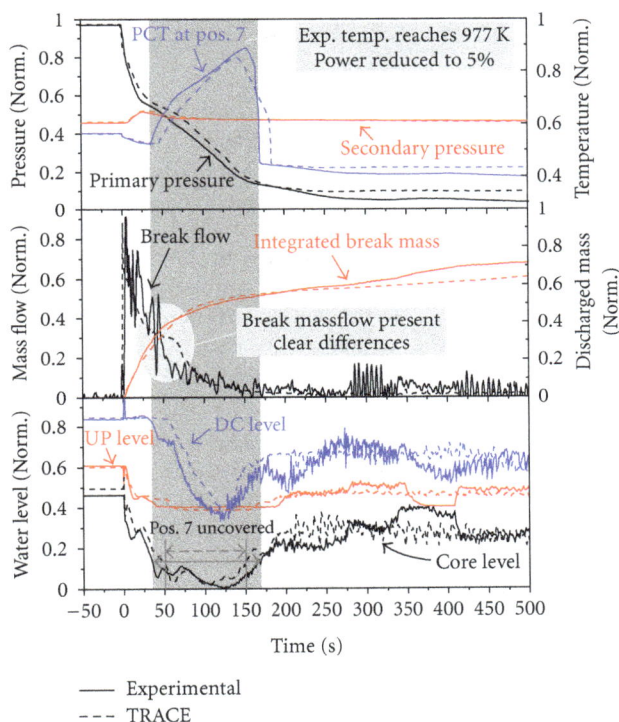

FIGURE 5: TRACE results for Test 2. From top to bottom: (1) primary and secondary pressure along with the cladding temperature at pos. 7 rod (4,4), (2) break flow and integrated discharged mass, and (3) RPV water levels.

be known. In this sense, it is better to examine the core water level shown in the bottom graph of Figure 5. Overall, the water levels in the RPV were well predicted by TRACE. However, as indicated in Figure 5 the window of time where the heated rods were uncovered was slightly longer in the experiment, and therefore, the PCT was underpredicted. The graph in the middle focus on the break mass flow and the integrated discharged mass. The time evolution of the break flow was not correctly simulated by TRACE. Even though the break device was identical to the one used in Test 1, in this case, the break mass flow under subcooled conditions was underpredicted (first 30 seconds, see Figure 16 for details). Figure 5 shows results with a two-phase choked coefficient of 1.1. It was found that the variations on the subcooled choked

coefficients had no effect on the results, probably due to a limitation of the flow within the TRACE code (which could be derived from an internal bug), and the two-phase choked coefficient was increased to compensate the total discharged mass rather than to better predict the flow during this phase. Even though small modifications of the discharge coefficients are in general accepted by the international community, these changes must be consistent, and the user must not use

TABLE 10: Control logic of Test 5-2 [14].

Event	Condition
Break	Time zero
Generation of scram signal	Primary pressure 84% of initial value
PZR heater off	Generation of scram signal or PZR liquid level below 2.3 m
Initiation of core power decay curve simulation	
Initiation of primary coolant pump coastdown	
Turbine trip (closure of stop valve)	Generation of scram signal
Closure of main steam isolation valve	
Termination of main feedwater	
Generation of SI signal	Primary pressure below 79% of initial value
Opening and closing of the SG relief valves	SG pressures 110%/103%
Initiation of SG secondary-side depressurization by fully opening of the relief valves with auxiliary feedwater	10 minutes after generation of SI signal
Initiation of enhanced SG secondary-side depressurization by fully opening of safety valves	Primary pressure below 13% of initial value
Initiation of accumulator system	Primary pressure below 29% of initial value

different values depending on the case (unless the break conditions and geometry are very different). The differences observed between Tests 1 and 2 indicate that a revision of the choked flow of the TRACE code is needed. As a matter of fact, the choked flow in the latest version of TRACE (TRACE 5.0 patch 2) has been improved, and the problems experienced in Test 2 might have been solved, nonetheless additional problems may arise with the new code version (see Section 5 on page 15 for further details on this issue).

3.5. Test 5-2. Test 5-2 is a small break LOCA located in the cold leg of the loop without PZR. The test was started by opening a 0.5% break which was realized in the facility by using an inner-diameter sharp-edge orifice. The boundary conditions and control logic of this test is shown in Table 10. As for previous tests, the scram signal was actuated when the primary pressure fell below 84% of the nominal value and triggered the typical events for an SBLOCA case, as described in Table 10.

The SI signal was generated when the primary pressure decreased below 79%. Ten minutes after the generation of the SI signal, full symmetrical SG secondary-side depressurization as accident management action was activated by fully opening the relief valves. Auxiliary feedwater was also activated concurrently. When the primary pressure fell below 13%, enhanced SG secondary-system depressurization was started by fully opening the safety valves.

A total failure of the HPSI system was assumed in the test. The accumulators started to discharge when the primary pressure was reduced below 29%. The LPSI was disabled for this test, in order to observe the phenomena occurring in the SG under the influences of gas inflow when the primary pressure is below the LPSI actuation pressure. The logic of the ROSA/LSTF core protection system for Test 5-2 was the same as the one used for Test 6-1. Further details on this test can be found in [14].

TABLE 11: Chronology of the main events in Test 5-2 [14].

Event	Exp. (s)	TRACE (s)
Break valve opened	0	0
Scram signal	95	65
SI signal	145	112
Primary coolant pumps stopped	349	319
full opening of SG relief valves	756	696
AFW starts	756	696
Initiation of Acc system	≈1320	1293
Full opening of SG safety valves	≈2420	2586
End of test	13376	12000

3.5.1. Results. The chronology of the main events of Test 5-2 are shown in Table 11, where the experimental results are compared with the calculation. Due to a slightly overprediction of the depressurization rate of the primary side at the beginning of the transient, all the events in the simulation occurred about 30 seconds earlier than in the experiment.

The most relevant results obtained for Test 5-2 are displayed in Figure 6 (break flow and discharged mass, primary and secondary pressures, and RPV collapsed water levels).

The simulation of Test 5-2 is in fair agreement with the experiment. The primary and secondary pressure trends were correctly simulated by the ROSA TRACE nodalization. In Figure 6, the first 1000 seconds are zoomed in for a better view on the comparison between experiment and calculation. In a similar manner as in Test 6-2, the depressurization rate at the beginning of the transient was slightly overpredicted by TRACE. The full depressurization of the system by opening the relief valves and later on the safety valves was steeper at the beginning and slower at the end although the overall evolution in the simulation is acceptable. The initial break flow (first 1000 s of transient) was correctly

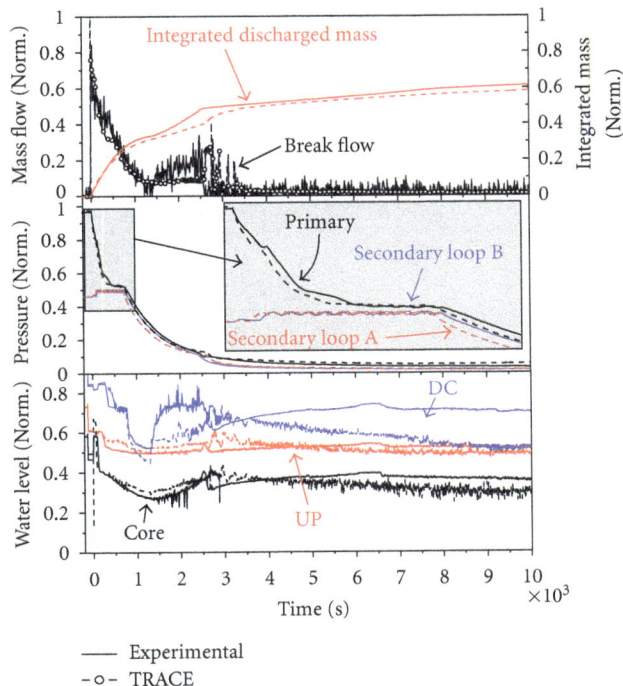

FIGURE 6: TRACE results for Test 5-2. From top to bottom: (1) break flow and integrated discharged mass, (2) primary and secondary pressures, and (3) RPV collapsed water levels.

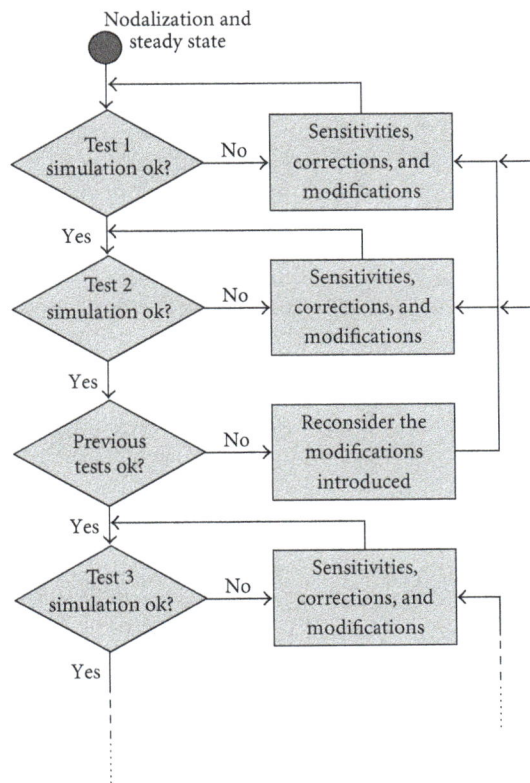

FIGURE 7: Flowchart for the model maintenance.

reproduced. Afterwards, discrepancies on the DC level were observed, as well as for the break flow (from 1000 to 3000 seconds).

All in all, a good agreement with the experimental results was obtained.

4. Model Evolution

The results presented in Section 3 were all obtained with the same TRACE nodalization, and only the boundary conditions specific for each test were modified from case to case. However, when the tests were first simulated, modifications and corrections of the nodalization accompanied by sensitivity studies were needed. The control systems were finally modified with the aim of allowing the user to run all previous cases with a minimum number of changes. In order to ensure the consistency of the modifications introduced, all previous tests were recomputed with the updated model after each posttest. To perform this action, it is important to maintain the model as compact as possible and to keep track of the included modifications. A basic methodology has been drawn for this purpose, which allows the user to recalculate any previous transient at any time in order to assure any important modification. The flowchart followed for the maintenance of the TRACE input deck is shown in Figure 7.

Basically, once the nodalization is developed and a satisfactory steady state is achieved, the nodalization is applied to the simulation of a first experimental test. At this stage, the original nodalization might be modified on the basis of

sensitivity tests, in the case that the comparison with the experimental data is not satisfactory and the deficiencies are identified to be in the nodalization approach and not in the constitutive correlations hardwired in the system code. Then, the same nodalization is used to simulate a second test. Again, changes to the nodalization might be necessary to correctly reproduce the significant phenomena of the given experimental test. This time, however, before proceeding to the simulation of a third test, the nodalization would be applied "as it is" to the simulation of the previous test, in order to ensure that the modifications to the nodalization do not lead to unsatisfactory behaviors (i.e., the results obtained with the new nodalization should remain in close agreement with the previous validation results or result in an even better agreement with the experimental data). Once this step is concluded satisfactorily, the analyst can proceed to a third test, and so on. A compromise might be required between two different nodalizations when the results of previous tests are worsened. In this case, one nodalization must be picked up as the most appropriate.

This very procedure was employed at PSI for the TRACE ROSA nodalization presented in this work.

The ROSA/LSTF TRACE nodalization developed at PSI has been used for more than 5 years and has therefore undergone several modifications. The most relevant of those are described in the following chapters. Since the modifications and corrections to the nodalization were triggered by the discrepancies found when comparing to a given experimental test, they are classified accordingly.

4.1. Corrections Introduced during the Simulation of Test 6-1

4.1.1. Leakage from the Upper Plenum to the Downcomer. The cold leg and DC level evolution during the first part of Test 6-1 (until 700 s) pointed out interesting differences between the TRACE model and the experiment. While the level evolution in the core, upper plenum, upper head, and hot legs were reasonably predicted, the cold leg and DC level could not be correctly simulated. What the TRACE nodalization was not able to simulate was exactly the DC pressure drop, which was reduced in the experiment after the interruption of the natural circulation. The root of the discrepancy was the earlier presence of steam in the upper part of the DC, which modified the pressure drop inducing an earlier drop of the DC and the cold leg water levels. A series of sensitivity calculations was performed to analyse the possible reasons. After investigating all plausible causes and following the recommendations given in [15], it was concluded that the discrepancy could derive from a hypothetical leak from the DC to the upper plenum through the seal of the core support barrel. This seal ring was broken in August 1995, when the UH-UP bypass was plugged and then repaired in November 2001. It was found out that both the cold leg and the DC level evolutions were correctly simulated if a small leakage (1% of steady-state core flow) was assumed to exist in this region. The results of the DC and cold leg levels with and without the leakage are shown in Figure 8. Further details on this analysis can be found in [16]. This modification was then kept for the simulation of all successive tests producing a closer or equal agreement for all tests.

4.1.2. UH Nodalization.

Test 6-1 constituted a perfect experiment to study the performance of the choked flow model, since the location at the top of the RPV allowed a clear differentiation between the phases that a choked flow may experience, namely, subcooled-liquid phase, two-phase, and single-phase vapor flows (Figure 2 on page 5). The TRACE nodalization performed very well for the first two phases; however, when the flow at the break turned into single-phase vapor, the break flow rate was overestimated by TRACE. Sensitivities on the break nodalization and possible modifications on the TRACE model were carried out and constituted the bulk of a previous publication [8]. One of the main modifications to the model consisted in a refinement of the nodalization of the RPV upper head. As a matter of fact, it was found that a finer nodalization of the upper head was essential to correctly predict the void fraction and the saturation temperature at the break location. Initially, only one axial level between the CRGTs exits and the top of the RPV (0.504 m of height) was used. In order to enhance mixing and allow convection in this region, the number of axial levels was finally increased. With this modification, the steam temperature at the break inlet was more accurately predicted (around 2 degrees lower). It is important to notice that the gas velocity under single-phase vapor conditions and choked flow is basically a function of the stagnant temperature and the specific heat ratio; hence, an accurate prediction of the temperature close to the break is crucial. In this simulation, with an increased number of

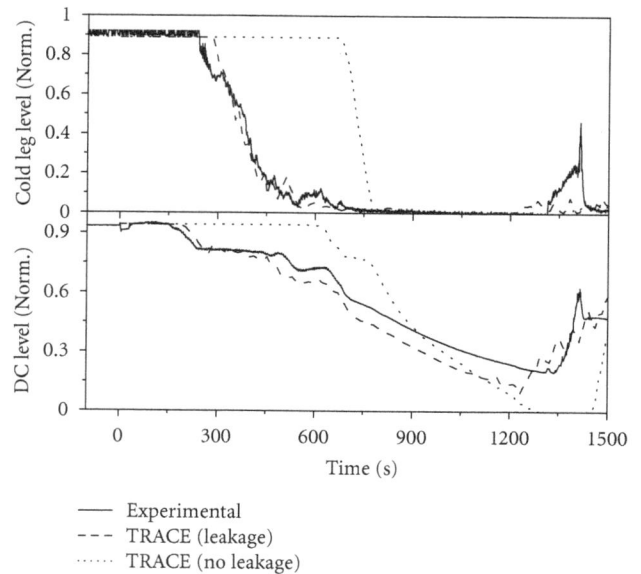

FIGURE 8: Collapsed water levels of the DC and the cold leg for Test 6-1 with and without a leakage between the UP and the DC.

cells, the saturation temperature at the break was reduced and so was the break mass flow (see Figure 9 for details). The initial nodalization with only one axial level was split in three and 6 smaller nodes of 0.168 m and 0.084 m of height, respectively. There was almost no difference between the results obtained with 3 and 6 additional levels, but the computational time was very high for the 6-level calculation. Therefore, 3 levels for the upper part of the UH were used in the final nodalization (i.e., 20 axial levels for the entire RPV). The improvement obtained in the simulation of the break mass flow can be seen with the integrated break flow plotted in Figure 9. Further details can be found in [8].

4.2. Corrections Introduced during the Simulation of Test 6-2.

After a satisfactory simulation of Test 6-1, the same TRACE nodalization was employed for the simulation of Test 6-2. As a first approach, an early interruption of natural circulation was predicted leading to an increase of the primary pressure, as shown in Figure 10. Afterwards, the transient evolution could not be correctly captured by the simulation. The early stop of natural circulation (NC) indicated a wrong estimation of the U-tube levels and heat transfer to the secondary side. The reason could be related to the use of a single U-tube in representation of the whole bundle. However, it was found eventually that the U-tube nodalization (1 pipe) was too coarse and the discrepancies could be fixed by simply refining the axial nodalization of the corresponding pipe component. Figures 10 and 11 show the pressure and the mass flow in loop B, respectively, obtained with the two nodalizations. The old and new nodalization of the SG are shown in Figure 12.

The new nodalization was then used to simulate again Test 6-1. The results were improved with the new nodalization.

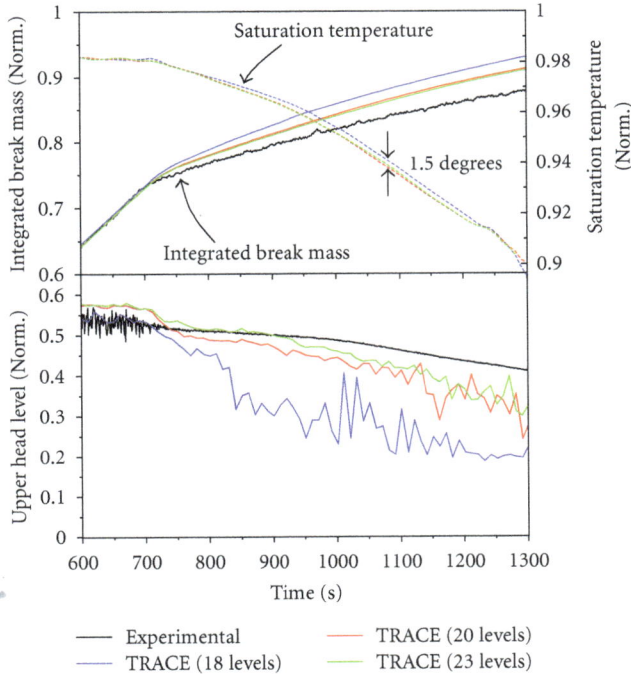

FIGURE 9: Sensitivity on the UH nodalization for Test 6-1 with finer and coarser meshing of the UH. Top: integrated break mass and saturation temperature at the top of the UH dome. Bottom: UH collapsed water level.

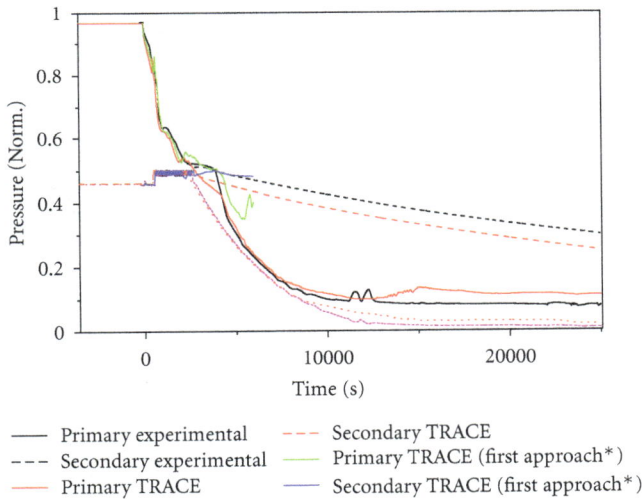

FIGURE 10: Primary and secondary pressures for Test 6-2 using both nodalizations of the SG's primary and secondary sides.

4.3. Corrections Introduced during the Simulation of Test 1.
Test 1 was first carried out as a blind calculation along with the evaluation of its uncertainties [11], performed by applying the GRS SUSA methodology [17, 18]. Even though the maximum PCT temperature in the blind calculation showed lower values than in the experiment, the accompanying uncertainty evaluation helped confining the possible problems. Three modifications were included in the nodalization

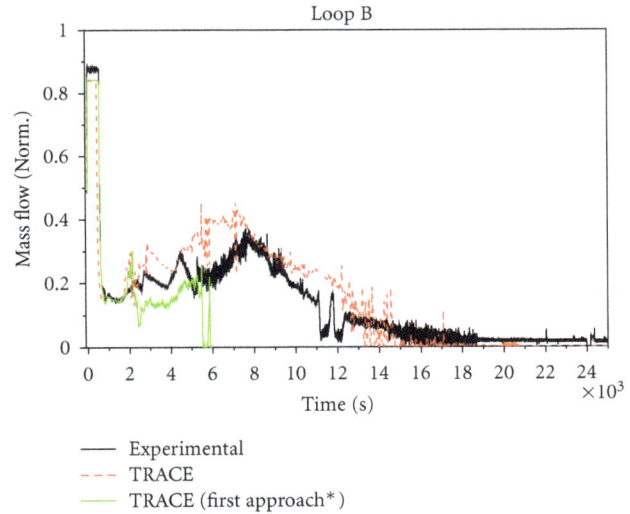

FIGURE 11: Mass flow in loop B using both nodalizations of the SG's primary and secondary sides.

used for the posttest calculation: a correction of the Acc's lines, the reduction of the two-phase choked coefficient, and the inclusion of the CCFL model at the top of the core.

The blind calculation of Test 1 displayed considerably larger accumulator flows than the experiment; hence, this line was revised and finally renodalized from scratch, in order to achieve a better representation of the pressure drops along the line. Figure 13 shows the results of the blind calculation and the modification on the Acc's line. It must be noticed that the rest of modifications included during the posttest process further increased the quality of the Acc's mass flow results (line labeled as "final TRACE result" in Figure 13).

Since the break flow was clearly overestimated during the two-phase flow regime, a parametric study on the two-phase choked flow coefficient was performed. A coefficient of 0.8 provided results in a closer agreement with the experiment, as shown in Figure 14. The final coefficient used for this test was 0.85. This modification was introduced after recalculating the test by following the methodology for a consistent nodalization.

The height of the PCT was confined by two factors. On the one hand, the time difference between the start of the core uncovery and the LS clearance was crucial for this transient. Both the LS clearance and the Acc's injection helped quenching the core. However, it was the LS clearance the phenomenon that triggered the reflood of the core. LS clearance occurs when the pressure difference between the UP and the DC is large enough to drive the LS water column into the DC. This occurred thanks to steam condensation in the cold leg with the injection of cold water from the Accs. The second phenomena that will define the height of the maximum PCT is the speed at which the core level drops during this window of time. The first phenomenon here described was well simulated even in the blind calculation; however, the level decreased at a slower rate, and therefore, the PCT was lower.

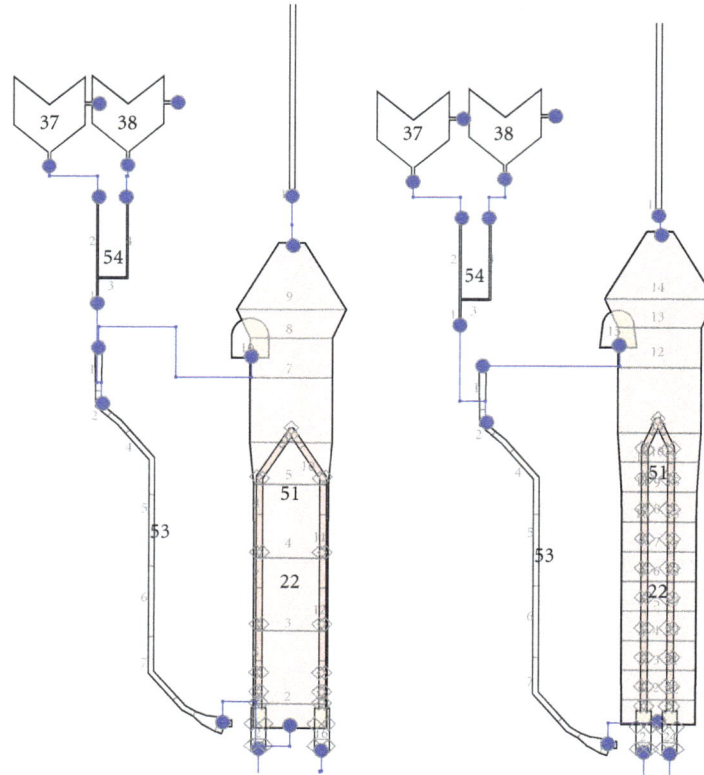

FIGURE 12: Old and new nodalization of the SG and U-tubes. Modification introduced during the analysis of Test 6-2.

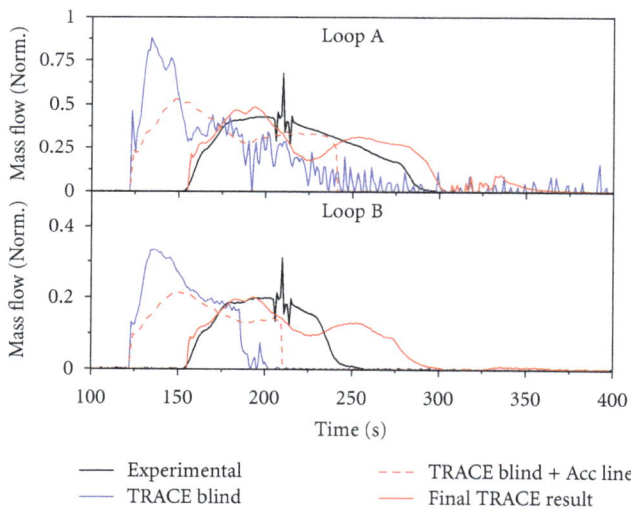

FIGURE 13: Accumulator injection during Test 1 with the new nodalization of the Acc's lines.

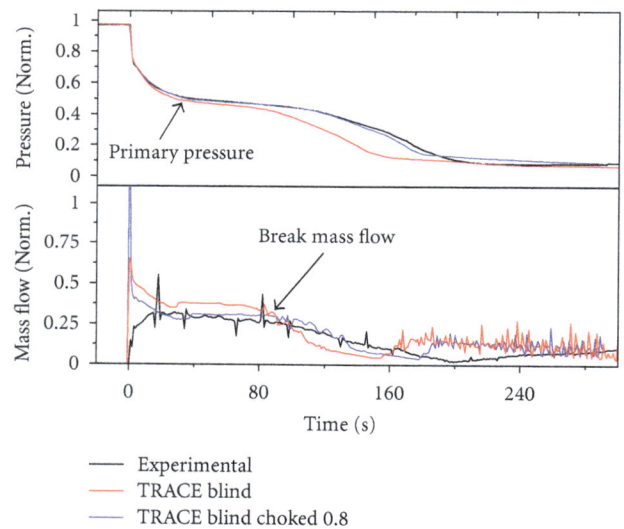

FIGURE 14: Primary pressure and break mass flow for Test 1, pretest results with different two-phase break discharge coefficients.

At this point of the analysis, it became obvious that an additional phenomenon significant for the evolution of this transient was the occurrence of CCFL conditions at the top of the core. Due to the fact that the CCFL was (mistakenly) not activated for the blind calculation, the core level remained higher than in the experiment due to the amount of water falling from the UP. In the experiment, the drop of water falling from the UP was prevented by the steam flowing out of the core. Once the CCFL model at the location of the upper core plate was activated, the core level dropped faster, in better agreement with the experimental data. The effect of using the CCFL model is shown in Figure 15, where the results of the cladding temperature are shown for

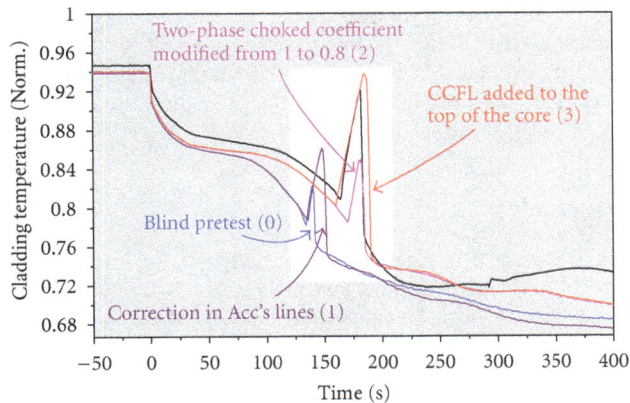

FIGURE 15: Maximum PCT for Test 1, evolution of the results with different modifications.

TABLE 12: Choked flow discharge coefficients used for all break locations, subcooled discharge (C_{sub}) and two-phase flow discharge (C_{2p}) coefficients.

Test	Type	C_{sub}	C_{2p}
Test 6-1	Orifice	1.0	1.0
Test 6-2	Orifice	1.0	1.0
Test 1	Nozzle	1.0	0.85
Test 2	Nozzle	1.0	1.1
Test 5-2	Orifice	1.0	1.0

the different posttest steps. Summarizing, starting from the blind calculation, firstly the accumulator's lines were corrected, then the two-phase choked coefficient was decreased, and finally the CCFL model was used at the top of the core. Further details on these modifications can be found in [11].

After this nodalization modifications, all previous tests were recalculated, and there was no significant difference.

4.4. Corrections Introduced during the Simulation of Test 2. Since the IBLOCA of Test 2 occurs in the cold leg, CCFL phenomena may take place at the entrance of the U-tubes or at the connection between the pipes and the SG plenums. Therefore, the CCFL was activated in this region during the simulation of Test 2, and the simulation results were consequently improved. All cases were recalculated satisfactorily. For Test 1, the choked two-phase flow coefficient was slightly shifted from 0.8 to 0.85.

4.5. Corrections Introduced during the Simulation of Test 5-2. For this test, only corrections on the control systems in order to accommodate the test particularities were introduced. No further modification to the nodalization was necessary. However, all previous tests were recalculated to avoid any inconsistency or possible error introduced during the configuration process.

5. Choked Flow Response

The modeling of choked flows is one of the most important issues in nuclear thermal hydraulics, since it affects the prediction of the flow-rate at break locations, and therefore plays a key role in the evolution of all LOCA transients. The tests performed in LSTF within the ROSA-1 and 2 projects dealt with breaks of different sizes and location and thus provided a good database to evaluate the performance of the TRACE chocked flow models.

The v5.0 version of the TRACE code has been used for all simulations presented in this paper and has provided a good

agreement with the experiments. The discharge coefficients used for each test are described in Table 12.

A coefficient of 1.0 was used in all the orifice breaks providing reasonable enough results. However, for the two nozzle cases, the use of different two-phase flow discharge coefficients (C_{2p}) was needed even though they presented exactly the same geometry.

The disagreement of the choked flow patterns obtained for Test 2 and especially the need to use different discharge coefficients than those used for Test 1 motivated us to test newer versions of the code. Even though small modifications on discharge coefficients are in general accepted by the international community, these must be consistent, and the user must not use different values depending on the case (unless the break conditions and geometry are very different). The differences observed between Tests 1 and 2 indicated that a revision of the choked flow of TRACE was required. Once we tested the same model with the latest official release of the code (TRACE 5.0 patch 2), it was found out that the choked model was providing indeed a much better agreement with the experiment. Further studies indicated that, in Test 2 by using v5.0, the subcooled choked flow (first 33 seconds) was limited within the code subroutines, so that variations of the subcooled discharge coefficient (C_{sub}) would not have an impact on the results. As a matter of fact, a sensitivity on the C_{sub} showed that the results were influenced only when values of C_{sub} under 0.85 were being employed, so all calculations with a higher coefficient were equivalent with those obtained with a coefficient of 0.85 (see Figure 16). This behaviour was not observed by using the latest TRACE release. Since the flow was much lower at the beginning of the transient (with the v5.0 RC3 version), a higher value for the C_{2p} was needed to compensate the total discharged mass (see Table 12). This explains the differences between the C_{2p} used in Tests 1 and 2. It is important to point out that the subcooled break flow in the rest of the tests does not play an important role, and this might be the reason why this problem did not become evident until the simulation of Test 2.

The problems with the subcooled part of the choked flow model have been solved in patch 2; however, other issues have arisen with this version which hinders its usage. The methodology presented in this paper can also be used to quickly test new code versions, and in this case, patch 2 presented several deficiencies indicating that the new version cannot be used consistently.

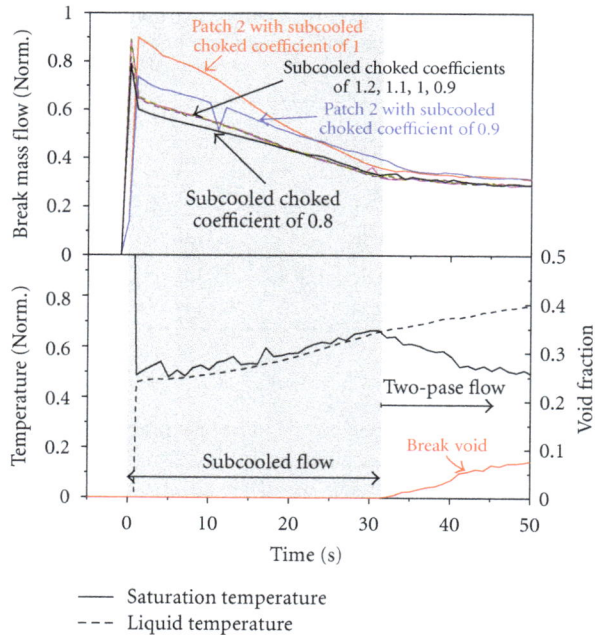

FIGURE 16: Break mass flow during the subcooled phase for Test 2; different subcooled choked coefficients are used for the TRACE v5.0 RC3 and patch 2 versions.

6. Conclusions

A nodalization of the ROSA/LSTF facility has been developed by using the US-NRC system code TRACE v5.0. The nodalization has been used to simulate 5 different tests focused on small and intermediate break LOCA cases. The simulations have been performed during the last 4 years; however, the model evolution has been tracked, and a methodology has been drawn to maintain a single consistent nodalization for all tests. Every time that a new posttest analysis was completed, the previous posttests simulations were carried out again, in order to guarantee that any new modification to the nodalization would only improve the overall performance of the nodalization. Finally, a reasonable agreement with all of the five tests was obtained. However, the calculated subcooled choked flow in Test 2 underestimated the experimental results. It was found out that the subcooled choked flow in the TRACE v5.0 version presented deficiencies, which have been solved in the latest TRACE release (5.0 patch 2). It is important to point out that the subcooled break flow in the previous tests does not play an important role, and thus this problem was not identified earlier. Even though the problems with the subcooled choked flow were solved with patch 2, the methodology presented in this paper allowed a quick assessment of the new version, and different issues were detected; therefore, TRACE5.0 was used in this paper.

Acknowledgments

This work was partly funded by the Swiss Federal Nuclear Safety Inspectorate ENSI (Eidgenössisches Nuklearsicherheitsinspektorat), within the framework of the STARS project (http://stars.web.psi.ch/). The authors are as well grateful to the OECD/NEA ROSA-2 project participants: JAEA for experimental data and the Management Board of the OECD/NEA ROSA-2 project for providing the opportunity to publish the results.

References

[1] F. Reventós, L. Batet, C. Llopis, C. Pretel, M. Salvat, and I. Sol, "Advanced qualification process of ANAV NPP integral dynamic models for supporting plant operation and control," *Nuclear Engineering and Design*, vol. 237, no. 1, pp. 54–63, 2007.

[2] C. Llopis, F. Reventós, L. Batet, C. Pretel, and I. Sol, "Analysis of low load transients for the Vandellòs-II NPP. Application to operation and control support," *Nuclear Engineering and Design*, vol. 237, no. 18, pp. 2014–2023, 2007.

[3] Paul Scherrer Institut, "Steady-state and transient analysis research of the swiss reactors (STARS project)," 2011, http://stars.web.psi.ch/.

[4] G. B. Treybal, *One-Dimensional Two-Phase Flow*, McGraw-Hill, New York, NY, USA, 1969.

[5] NEA-CSNI. Bemuse Phase V Report, "Uncertainty and sensitivity analysis of a LB-LOCA in ZION nuclear power plant," Tech. Rep., Committee on the Safety of Nuclear Installations, OECD, Nuclear Energy Agency, 2011.

[6] T. Yonomoto, "CCFL characteristics of PWR steam generator U-tubes," in *Proceedings of the ANS International Topic Meeting on Safety of Thermal Reactors*, Portland, Ore, USA, 2001.

[7] T. Takeda, M. Suzuki, H. Asaka, and H. Nakamura, "Quick-look data report of OECD/NEA ROSA project test 6-1 (1.9% pressure vessel upper-head small break LOCA experiment)," Tech. Rep. JAEA-Research 2006–9001, Japan Atomic Energy Agency, 2006.

[8] J. Freixa and A. Manera, "Analysis of an RPV upper head SBLOCA at the ROSA facility using TRACE," *Nuclear Engineering and Design*, vol. 240, no. 7, pp. 1779–1788, 2010.

[9] T. Takeda, M. Suzuki, H. Asaka, and H. Nakamura, "Quick-look data report of OECD/NEA ROSA project test 6-2 (0.1% pressure vessel bottom small break LOCA experiment)," Tech. Rep. JAEA-Research 2006–9002, Japan Atomic Energy Agency, 2006.

[10] T. Takeda, M. Suzuki, H. Asaka, and H. Nakamura, "Quick-look data report of test 1, test for hot leg intermediate break loss of coolant accident with break size equivalent to 17Flow area," Tech. Rep. JAEA-Research 2010-, Japan Atomic Energy Agency, 2010.

[11] J. Freixa, T.-W. Kim, and A. Manera, "Thermal-hydraulic analysis of an intermediate LOCA test at the ROSA facility including uncertainty evaluation," in *Proceedings of the 8th International Topical Meeting on Nuclear Thermal-Hydraulics, Operation and Safety (NUTHOS '10)*, Shanghai, China, October 2010.

[12] The ROSA-V Group, "ROSA-V large scale test facility (LSTF) system description for the third and fourth simulated fuel assemblies," Tech. Rep. JAERI-Tech 2003-037, Japan Atomic Energy Agency, 2003.

[13] T. Takeda, M. Suzuki, H. Asaka, and H. Nakamura, "Quick-look data report of ROSA-2/LSTF Test 2 (cold leg intermediate break LOCA IB-CL-03 in JAEA," Tech. Rep. JAEA-Research 2010-, Japan Atomic Energy Agency, 2010.

[14] T. Takeda, M. Suzuki, H. Asaka, and H. Nakamura, "Final data report of ROSA/LSTF Test 5-2 (primary cooling through

steam generator secondary-side depressurization experiment SB-CL-40 in JAEA)," Tech. Rep. JAEA-Research 2009-, Japan Atomic Energy Agency, 2009.

[15] H. Austregesilo and H. Glaeser, "Results of post-test calculation of LSTF Test 6-1 (SB-PV-09) with the code ATHLET," in *OECD/NEA ROSA Project, 4th PRG Meeting*, Tokai-mura, Japan, 2006.

[16] J. Freixa, "Post-test thermal-hydraulic analysis of ROSA Test 6.1," Tech. Rep. TM-41-08-10, Paul Scherrer Institut, 2008.

[17] H. Glaeser, "GRS analysis for CSNI uncertainty methods study (UMS)," Tech. Rep., Committee on the Safety of Nuclear Installations, OECD, Nuclear Energy Agency, 2011.

[18] E. Hofer and M. Kloos, *Relap5/Mod3.3 Code Manual. Volume I: Code Structure, System Models, and Solution Methods*, 2003.

10

Moderating Material to Compensate the Drawback of High Minor Actinide Containing Transmutation Fuel on the Feedback Effects in SFR Cores

Bruno Merk

Department of Reactor Safety, Institute of Resource Ecology, Helmholtz-Zentrum Dresden-Rossendorf, Postfach 51 01 19, 01314 Dresden, Germany

Correspondence should be addressed to Bruno Merk; b.merk@hzdr.de

Academic Editor: Wei Shen

The use of fine distributed moderating material to enhance the feedback effects and to reduce the sodium void effecting SFRs is described. The drawback on the feedback effects due to the introduction of minor actinides into SFR fuel is analyzed. The possibility of compensation of the effect of the minor actinides on the feedback effects by the use of fine distributed moderating material is demonstrated. The consequences of the introduction of fine distributed moderating material into fuel assemblies with fuel configurations foreseen for minor actinide transmutation are analyzed, and the positive effects on the transmutation efficiency are shown. Finally, the possible increase of the Americium content to improve the transmutation efficiency is discussed, the limit value of Americium is determined, and the possibilities given by an increase of the hydrogen content are analyzed.

1. Introduction

The positive coolant density feedback coefficient is inherent to the system in sodium-cooled fast reactors (SFRs). This effect is the basis for the sodium void effect, which is the maximal reduction of the sodium density. The reduction of the positive feedback effects as well as the enhancement of the negative ones is an important point in the design of future sodium-cooled fast reactors. The relevance of the topic has been highlighted in the last year in the IAEA TM on Innovative Fast Reactor Designs with Enhanced Negative Reactivity Feedback Features in Vienna [1].

The nature of feedback effects in fast reactors as well as the sodium void effect itself and the different contributions to the effect are well known since the 1960s. Detailed descriptions have already been given in "Reactivity Coefficients in Large Fast Power Reactors" in 1970 [2]. Already in the 1970s numerical studies were performed with the aim to reduce the sodium void effect [3]. These studies were mostly performed on the basis of full core calculations for the optimization of the core geometry to reduce the sodium void effect.

One important outcome of the full core calculations was the development of high leakage cores with their big core diameter (~5 meters) in combination with a very small core height (≤1 meter). Another method for the enhancement of the leakage in the case of sodium voiding is the replacement of the reflector above the core with a sodium plenum which enhances the leakage significantly in the case of a sodium voiding event in the upper core region. Current publications mostly concentrate on the design of sodium-cooled fast reactor cores [4], on advanced safety concepts for SFR [5], and on basic or detailed discussions on the different influencing parameters on the sodium void coefficient [6, 7] and the limited possibilities for enhancing the feedback effects in traditional designs.

Recently, a new proposal has been published. The positive void effect is reduced here in combination with a significant enhancement of the negative fuel temperature effect and a decrease of the positive coolant effect by adding fine distributed moderating material. The study has been focused in the first step on the choice of the ideal moderating material [8, 9] and in a second step on the optimization of the placing

Moderating Material to Compensate the Drawback of High Minor Actinide Containing Transmutation Fuel on the Feedback Effects in SFR Cores

99

of the zirconium hydride to obtain the optimal effect in power distribution and burnup [10, 11] to limit the required changes in the fuel assembly and in the core design.

The calculations are performed with the lattice transport code HELIOS and based detailed full fuel assembly geometry representation. These possibilities are given due to the rapid development of the spectral codes for LWR analysis which solve the integral transport equation in two dimensions on unstructured mesh [12, 13]. These codes offer the chance to investigate the feedback effects on fuel assembly level for different designs in full detail including multigroup visualization of integral and resolved neutron flux and cross sections. The verification of the results for the enhanced feedback effects calculated with HELIOS has been performed by a comparison calculation using MCNP 5 [8, 9].

The insertion of moderating material softens the neutron spectrum of the fast reactor. Nevertheless, for the burning of minor actinides, a hard neutron spectrum is essential due to the fission threshold of these isotopes. Following this knowledge, the effect of the use of a small, fine distributed amount of moderating material on the efficiency of minor actinide burning will be investigated in this publication.

2. Reference System

For the beginning of the study, a reference case based on the European Fast Reactor is defined (see Figure 1). The data is mostly given in the IAEA Fast Reactor Database— 2006 Update [14]. Additional data is taken from Waltar, Reynolds: Fast Breeder Reactors [15] and from European fast reactor (EFR) fuel element design [16]. The following major parameters are used: outer pin diameter 8.5 mm, cladding thickness 0.52 mm, pitch to diameter 1.2, can wall thickness 4.5 mm, wire spacers, and 271 fuel rods per element.

Materials. MOX Fuel with 22.4% Pu fissile and Pu vector from 4% enriched LWR fuel with 50 GWd/tHM burnup. 5-year storage is estimated before reprocessing and 2 years until reuse in the reactor. The Pu-Vektor is (2.6/54.5/23.7/11.3/6.8) and 1.1% Am-241 in depleted uranium (0.3% U-235 content). The smeared fuel density is 9.26 g/cm^3, and the fuel temperature is 900°C. Cladding, wire spacers, and can wall are made from stainless steel 304 along the HELIOS 1.9 definition. The temperatures are 635°C for the cladding and 545°C for wire spacer and can wall. Sodium density is 0.821 g/cm^3 along the formula for liquid saturated sodium at 545°C given in Waltar and Reynolds [15]. The geometric arrangement of the reference system with 10 rings is shown in Figure 1 for 1/6 part of one fuel element. The used power density is 118.8 W/g corresponding to the maximum power density in the EFR. For the calculation HELIOS is used. The internal 112 group fast reactor cross-section library of Studsvik Scandpower is used for the investigation. Only for the calculations of the neutron spectrum the 190 energy group library is used to have a refined energy structure in the thermal range for a better depiction.

Three different minor actinide loadings will be investigated (3% Am, 5% Am, and 2% Np—2% Am) [17]. The Am

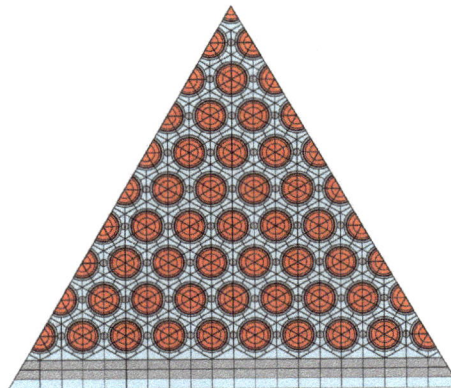

FIGURE 1: 1/6 of the reference fuel element geometry corresponding EFR.

vector is following the same basic definitions as described for the Pu vector mentioned previously. This definition gives 62% Am-241 and 38% Am-243. The data for the used moderators ZrH1.6 and YH is taken from [18].

3. System with Moderating Material

The basic idea to shift a small portion of the neutron spectrum to lower energies is achieved by the introduction of moderating material inside the wire wrapper as is shown in Figure 2. For the visualization purpose, only one unit cell is extracted from the calculated 1/6 of a fuel assembly. The purpose of the moderating material is to force up the absolute value of the negative fuel temperature feedback and to reduce of the positive coolant feedback (consisting of coolant density and temperature effect) coinciding with a reduction of the sodium void effect. The fine distributed placing of the moderating material in each unit cell offers the possibility to keep the original fuel assembly design as well as the power density and the uniform power distribution typical for a homogeneous fast reactor fuel assembly. These three facts ensure that the flow conditions in the fuel assembly can be kept like in the reference design. Thus the flow conditions can still be optimized to keep or improve the safety-related coolability.

The proposed moderating material hydrogen can be inserted in the form of stable metal-hydrogen compounds like zirconium hydride or yttrium hydride; the latter is preferable due to the higher thermal stability. For the manufacturing, it is more suitable to hydride the already produced wire, since it is much easier to handle. Hydride alloys provide a range of hydrogen concentrations in combination with considerable variation in nuclear and mechanical properties. Thus the hydrogen content can be adapted to the special requests. This makes the amount of hydrogen addition into the fuel assembly very flexible; only the maximal values are limited due to the deterioration of the material performance and the solubility limits which decrease with increasing temperatures. "It should not be inferred that the presence of hydrogen in metals is always deleterious. When present

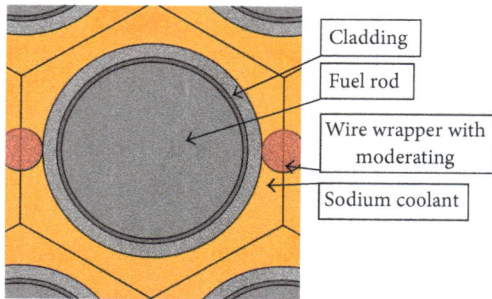

FIGURE 2: Unit cell out of a fuel element based on the European fast reactor (EFR) design with introduced moderation layer.

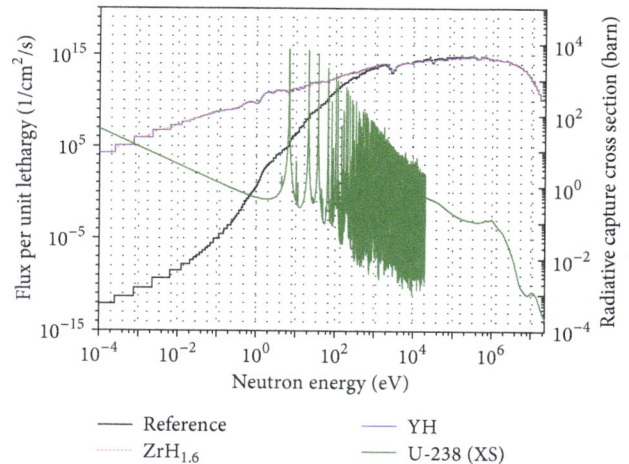

FIGURE 3: Neutron spectrum for the reference fuel assembly and for the fuel assemblies with moderating material based on different hydrogen carrying materials calculated with the 190 group library of HELIOS 1.9.

in amounts less than necessary for embrittlement, hydrogen can cause a noticeable increase in strength so long it can be retained in solution. The important point, again, is that hydrogen presents serious problems only when it is not retained in solution or when its concentration exceeds the solubility limits of the alloy so that hydride precipitates or segregation can occur. Otherwise, hydrogen reacts similarly to other alloying elements in most respects" [18].

Figure 3 gives a general overview on the changes in the neutron flux spectrum due to the insertion of moderating material compared to the reference neutron spectrum. The 190 group HELIOS library is used for these spectral curves to get a sufficiently fine resolution in the thermal groups. The figure shows a significant difference in energy distribution of the neutron flux after the insertion of the moderating material. In the case with the zirconium hydride as wire wrapper, a comparably strong low-energy tail is formed due to the strong moderation effect of the hydrogen atoms in the metallic compound. This low-energy tail has to be seen in conjunction with the radiative capture cross section of U-238 which is added to the figure in green. The insertion of hydrogen causes a significant increase of the share of neutrons in the energies where major capture resonances for the U-238 isotopes appear, especially at energies around 6.67502 eV, 20.8715 eV, 36.6821 eV, 66.0312 eV, 102.559 eV, and 116.8923 eV. Further on, the effectiveness of moderation for the creation of the low-energy tail is compared for a compound based on zirconium and a compound based on yttrium. Both compounds contain exactly the identical amount of hydrogen.

The assembly burnup distribution after 100 GWd/tHM for the reference case and the case with the moderator inside the wire wrapper is given in Figure 4 and compared to the old fashioned solution suggesting pin containing the moderating material. The burnup in the reference fuel assembly (Figure 4(a)) is characterized by a very flat distribution (~1% difference between minimum and maximum) over the fuel assembly as well as over the fuel rods. The reason for this flat burnup distribution can be found in the flat power and neutron flux distribution and in the comparably low total cross sections at the dominating neutron energies. The flat power distribution in the fuel assembly is very favorable, since it disburdens the heat removal. The flat power distribution results in a uniform heat up of the sodium coolant. No hot

spots are created and thus all fuel rods can be operated close to the limit power and until the limit burnup defined in the operating permission.

The burnup distribution for case with fine distributed moderating material (Figure 4(b)) is rather uniform over the fuel assembly, too. A small rim effect appears due to the resonance self-shielding in the U-238 and the increased number of neutrons in the resonance region of the U-238 due to the use of the moderating material (compare Figure 3). A very limited burnup increase occurs at the pins close to the can wall. Nevertheless, the insertion of the moderating material in fine distribution, in layers, and in the wire wrapper does not create any severe deterioration in the power and burnup distribution in the fuel assembly. The evaluation of the pinwise values indicates even a more flat distribution than for the reference case, since the power and burnup at the periphery of the fuel assembly are slightly increased [10]. Only the burnup distribution inside the pin is slightly worsened due to the increased rim effect. This behavior is in strong contrast to the results for the use of moderating material in pins as discussed in several publications [18–20]. A locally concentrated introduction of moderating material like moderator pins causes a significant power and burnup increase in the pins around the moderator pins and a lower burnup in the pins far from the moderator pins (see Figure 4(c)). This uneven power and burnup distributions cause limitations to the obtainable maximal average burnup of the fuel assembly, since comparably low burnt fuel rods appear in areas far from the moderation pins. Additionally, a very strong power increase and a rim effect with all followup problems like hot spots and fuel pellet irritations due to the high burnup appear at the pins next to the moderating pins [10, 11].

The use of fine distributed moderating material offers the possibility to enhance the safety characteristics without changing the major operational and design parameters.

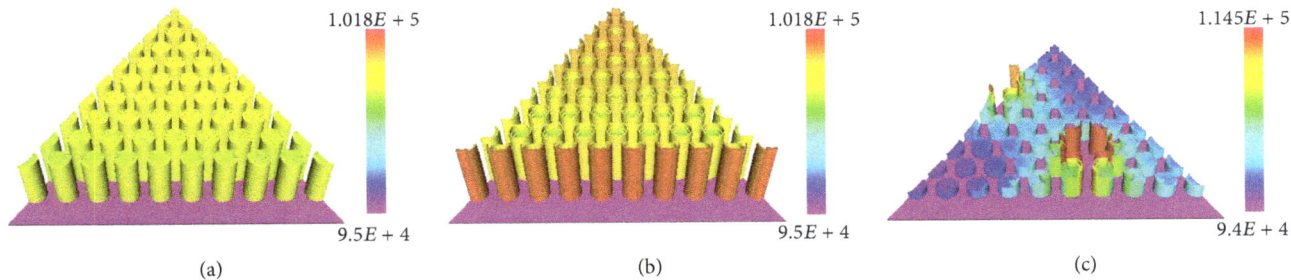

FIGURE 4: Burnup distribution at 100 GWd/tHM for the reference case (a), the case with the moderating material in the wire wrapper (b), and
the case with moderating material arranged in pins (c).

Power density and distribution, fuel configuration and density, fuel assembly geometry, and coolant streaming paths are not changed at all. Only the fissile enrichment has to be increased due to the slightly reduced volume available for the fuel, and the loss of criticality caused by the moderating material has to be compensated.

The idea offers new degrees of freedom for the optimization of the design of the sodium cooled fast reactor core, the cycle strategies, and the transmutation potential. The fine distributed moderating material increases the inherent system stability significantly by reducing the positive feedback effects and enhancing the negative ones. Thus it reduces the probability of reaching sodium voiding in a transient. The interesting point is the possibility of tailoring the feedback coefficients to an ideal value simply by adapting the hydrogen content in the compound. The ideal amount has to be determined from system-specific transient and accident analysis, since strong negative feedback is not desired in all accidental scenarios.

On the suggested moderating materials some comments have to be given. For ZrH, on the one hand, the material is a very efficient moderator, due to the extremely high slowing down power of 1.54 (ZrH$_{1.94}$), a value in the same range as for light water [18]. On the other hand, zirconium hydride of δ or ε phases does not change the associated volume up to well above 1000°C, since these phases are in the single phase region [18]. Thus zirconium hydride does not show the unpleasant swelling behavior of UZrH fuel, which Olander et al. describe as swelling due to void formation around the uranium particles and due to fission gas production [21]. These described processes should not appear in pure zirconium hydride moderator, since both effects are in conjunction with the Uranium component. The equilibrium hydrogen pressure in ZrH increases at high temperatures, especially above 900°C; this is above the sodium voiding temperature [18]. To suppress Hydrogen release, Mueller et al. suggest to use Kanigen nickel which shows promise as a barrier coating for zirconium hydride with low hydrogen content on the basis of hydrogen loss for times up to 100 h at 1300°F in argon [18].

For YH, the slowing down power is comparable, but the thermal stability is significantly higher, up to ~1300°C for yttrium-mono-hydride [18]. Some limited, but very positive, operational experience on the material is available too, since YH has been used in a test assembly for tailoring the neutron

spectrum for tests for possible radioisotope production in the FFTF [22].

For the verification of the very significant results of the influence of moderating material on the feedback effects in general, a basic cross comparison with MCNP for the fuel temperature and moderator effect on k_{inf} has been performed on simplified basis of a unit cell for the reference cases at the beginning of the investigation of the influence of fine distributed moderating materials in SFR cores to assure that the HELIOS results are reliable. Very good agreement for the moderator temperature and density and for the fuel temperature change has been achieved. The deviation between the MCNP perturbation calculation and the HELIOS 1.9 calculation of the coolant effect on k_{inf} was ~2 pcm which indicates that the sodium void effect calculations are reliable too [8, 9].

An investigation of the transferability of the reduction of the sodium void effect from the infinite to the finite system has been carried out based on the basis of an EFR-like core configuration [23]. To avoid any irritations created by group condensation, the same number of groups has been used in DYN3D. The full core calculations have been performed with the nodal code DYN3D [24, 25]. The calculation is based on 47 energy group cross-section sets calculated with HELIOS 1.10 using the given geometry models (compare Figure 1). To avoid any irritations created by group condensation, the same number of groups has been used in DYN3D. The full core calculations have shown that the full gain in sodium void reduction demonstrated in the infinite system can be expected to be reached in the finite system too [8]. Thus, a superposition of the findings in the infinite system and the traditional methods for the sodium void reduction like the pancake core shape or a sodium plenum is possible. Additionally, it has to be mentioned that the influence of sodium void reduction inside the fuel assembly using fine distributed moderating material already comes into action when the first sodium bubble appears. In contrast, the traditional methods to reduce the sodium void by increasing the neutron leakage do not have influence until a significant amount of sodium is voided.

4. Consequences on Minor Actinide Transmutation

Figure 5 shows the influence of an insertion of the minor actinides as well as the moderating material on the infinite

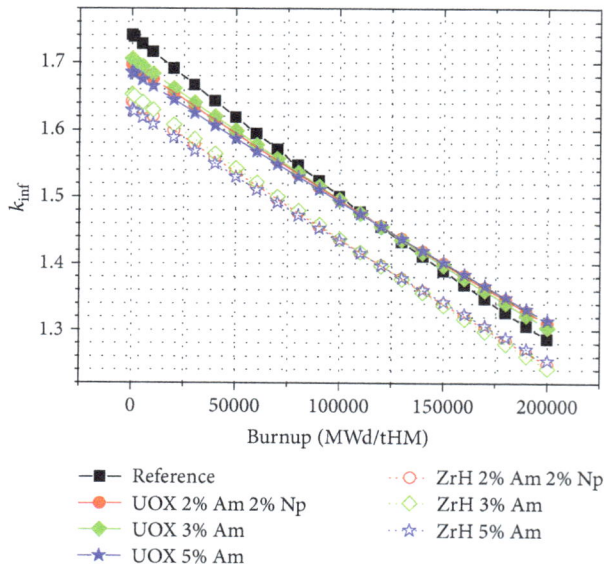

FIGURE 5: Change in the infinite multiplication factor over the burnup for the three different material configurations.

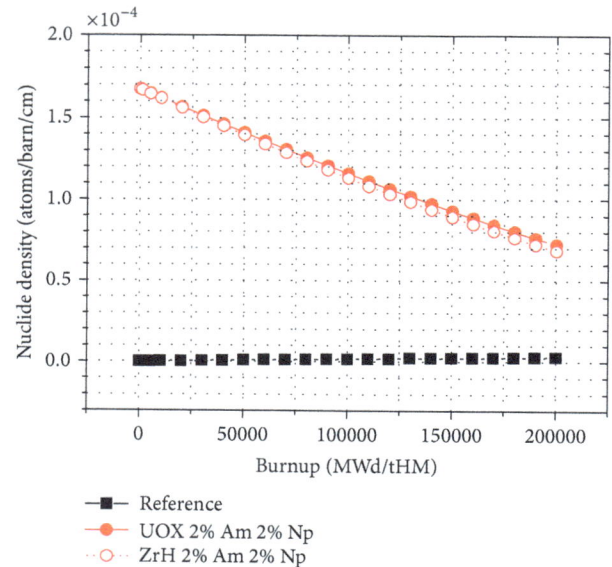

FIGURE 6: Change in the Neptunium-237 content over burnup for the case with and without moderation.

multiplication factor of a fuel assembly over burnup. Three different effects are appearing. On the one hand, the insertion of minor actinides decreases the initial k_{inf} of the fuel assembly, since they cause increased neutron absorption. On the other hand, the insertion of the minor actinides reduces the loss of criticality over burnup, due to the change in the fertile material configuration caused by adding minor actinides as efficient fertile material, replacing the corresponding amount of U-238. The insertion of the moderating material (ZrH1.6) causes in all cases a reduction in the k_{inf} throughout the whole observed burnup period. This reduction of the multiplication factor reduces the possible cycle time. A comparable effect appears for all other, "traditional" methods leading to sodium void reduction too, since they are based on increased neutron leakage. Thus, the breeding is reduced due to the reduced number of available excess neutrons.

The transmutation of Np-237 is shown in Figure 6 over the burnup of the reference fuel assembly. The nuclide density of Np-237 is reduced in both cases, without moderating material as well as with moderating material. In contrast to the reduction of Np-237, in the reference case, a very small amount of Np-237 is formed. The burning of Np-237 in the case with a given begin of life (BOL) Np-237 concentration and the built-up of Np-237 when there is no initial concentration would lead to a comparable asymptotic value for infinitely high burnup. It is visible that the use of the distributed moderating material even enhances the transmutation of Np-237 slightly.

Figure 7 shows the Americium transmutation over burnup for the isotope Am-241 (a) and for Am-243 (b). The Am-241 content in the reference case starts with a finite value, since the defined MOX fuel contains a small fraction of Am-241 as long as Pu-241 appears. The Am-241 is a decay product of Pu-241 and appears already after short storage time of

the separated plutonium used for the production of MOX fuel due to the short half-life of Pu-241 (14 years). In the reference case (full black squares), the initial concentration of Am-241 rises slowly to a maximum around 100 GWd/tHM and decreases after this maximum slightly. In all three cases with an added, higher initial Am-241 content, Am-241 is transmuted. The more efficient the transmutation is, the higher the initial Am-241 content is. An interesting fact is the slightly more efficient transmutation in the cases with distributed moderating material, even with the appearing slightly softer neutron spectrum (see Figure 3). This result is very surprising and it has to be investigated, if there is really more Am-241 burnt, or if the Am-241 is only shifted to the higher element Curium due to absorption processes. All cases, independently of the initial concentration of Am-241, tend to a comparable asymptotic limit for infinite burnup. The transmutation of Am-243 shows a comparable behavior (Figure 7(b)). No Am-243 is in the BOL fuel composition in the reference case; thus there is only the production which occurs during the burnup. In all cases with a BOL Am-243 content higher than 0.75% ($\hat{=}$2% Am content) Am-243 is reduced. The Am-243 transmutation is not significantly influenced by the use of distributed moderating material.

The change in the Curium content during burnup in the minor actinide containing fuel is shown in Figure 8. The Curium isotopes are built up from Americium in all fuel configurations by neutron capture and decay processes. The Cm-242 content (Figure 8(a)) rises in all cases in dependence of the initial Am-241 content to a maximum value. The more the Am-241 is available, the more the Cm-242 is built in the initial phase, but with decreasing Am-241 content, the Cm-242 content starts to be reduced too after reaching the maximum at about 100 GWd/tHM. The insertion of the moderating material leads in all cases to a slightly increased

Moderating Material to Compensate the Drawback of High Minor Actinide Containing Transmutation Fuel on the Feedback Effects in SFR Cores

103

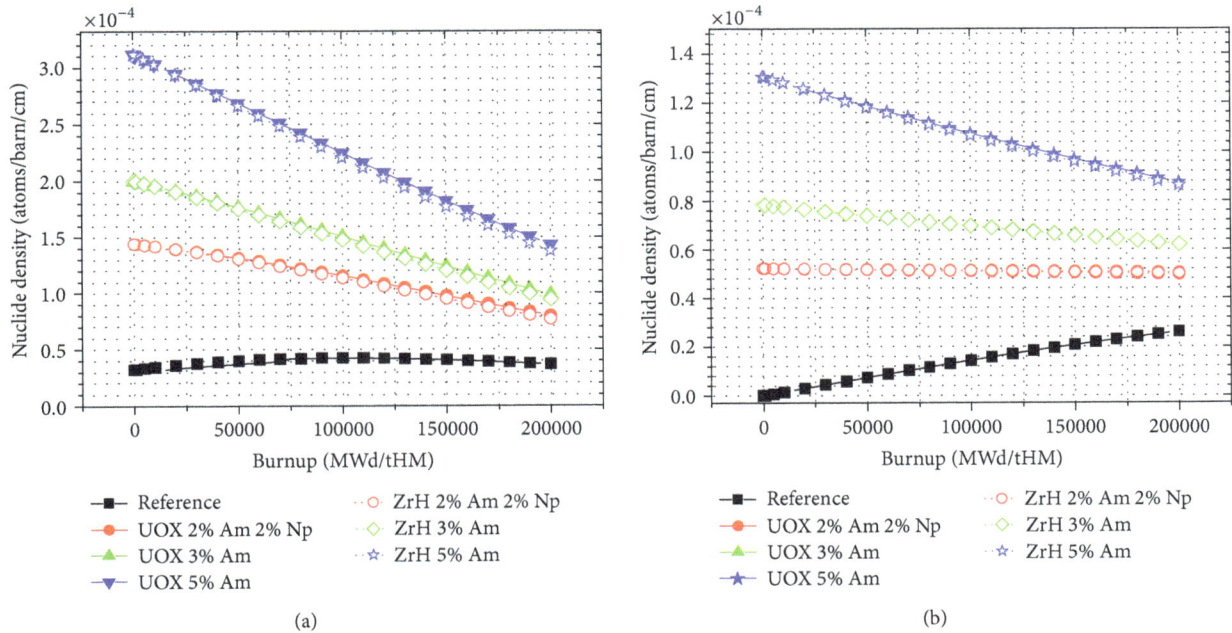

FIGURE 7: Change in the Americium-241 (a) and the Americium-243 (b) content over burnup for the case with and without moderation.

built-up and a slightly higher EOL concentration of Cm-242. The case with the highest Am content (5%) leads after reaching a maximum at ~100 GWd/tHM already to a significant reduction of the Cm-242 content. The Cm-244 content (Figure 8(b)) rises for all cases over the full observed burnup period. The overall Cm-244 content is strongly dependent on the initial Americium content in the fuel. The Cm-244 accumulation is slightly higher for all cases with moderating material.

A more detailed insight into the influence of the fine distributed moderating material on transmutation is given in Tables 1 and 2. The comparison of the transmutation efficiency provides information on which fuel assembly provides a lower content of minor actinides at the end of life (EOL). A positive value indicates an advantageous behavior of the fuel assembly with moderating material; a negative value indicates a better transmutation performance for the assembly without moderating material. The comparison of the transmutation efficiency (see Table 1)

Transmutation efficiency [%]

$$= \left(1 - \frac{\text{isotope number density with moderating material}}{\text{isotope number density without moderating material}}\right)$$
$$\cdot 100 \tag{1}$$

shows that the Am-241 transmutation is roughly 4% more efficient in the fuel assemblies with moderating material for all configurations of transmutation fuel. The Transmutation of Am-243 is slightly less efficient (−1.28%) for a low Americium content and slightly more efficient (1.42%) for high Americium content. In the cases with moderating material,

~2.5% more Cm-242 is produced than that in the reference cases without moderating material. The use of moderating material increases the production of Cm-244 by 7 to 9%. The transmutation of Np-239 is more efficient in the fuel assembly with moderating material. 5.5% more Np-239 is burnt than in the reference case.

The detailed comparison of the differences in the number densities

Transm. efficiency [number density]

= isotope number density with moderating material

− isotope number density without moderating material
$$\tag{2}$$

is given in the lower part of Table 1. A positive number indicates more efficient transmutation in the assembly with moderating material and a negative number more efficient transmutation in the assembly without moderating material. A detailed comparison of the numbers shows that the amount of the Am-241 reduction is higher than the increase in the number densities of the higher isotopes (Am-243, Cm-242, Cm-244) in all cases with moderating material. Thus there is really more Am-241 burnt in the fuel assemblies with fine distributed material than that in the "clean" transmutation fuel assemblies. The effect rises with increasing Americium concentration at BOL. Over all it can be stated that the insertion of the fine distributed moderating material in the wire wrapper does at least not have a negative influence on the transmutation efficiency.

(a)

(b)

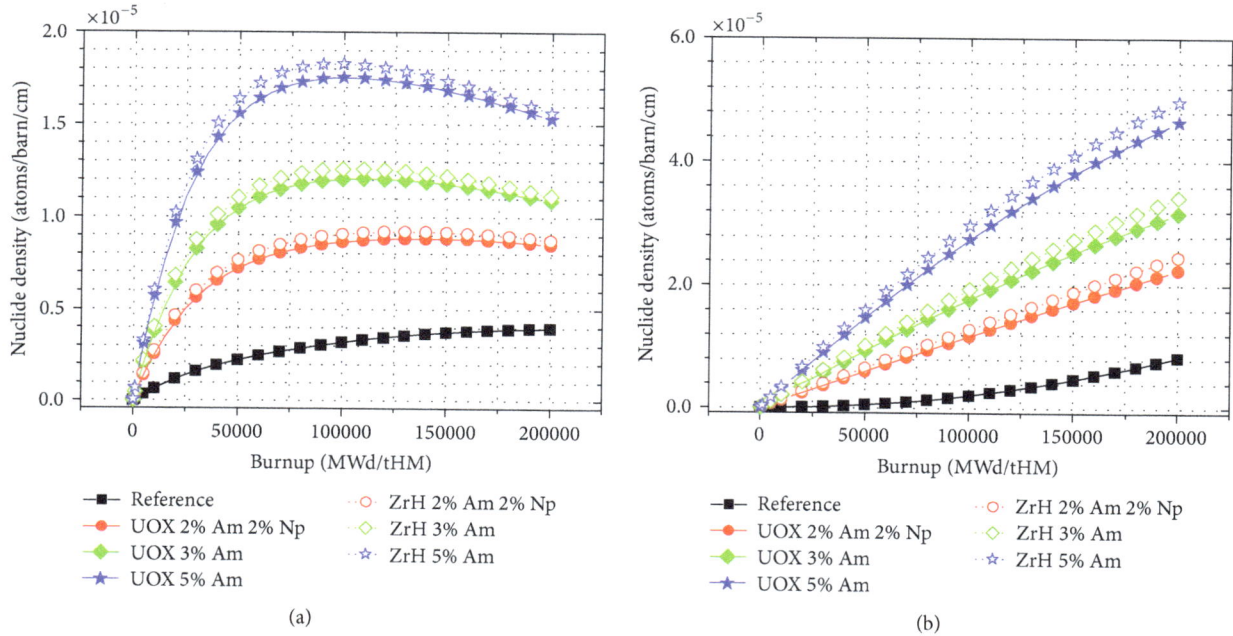

FIGURE 8: Change in the Curium-242 (a) and the Curium-244 (b) content over burnup for the case with and without moderation.

TABLE 1: Comparison of the transmutation efficiency between the reference cases without and the cases with moderating material for the different transmutation fuels.

Transmutation eff. (comparison)	Am-241	Am-243	Cm-242	Cm-244	Np-239
2% Np 2% Am EOL	3.87%	−1.28%	−2.68%	−9.18%	5.65%
3% Am EOL	4.27%	0.02%	−2.43%	−8.24%	
5% Am EOL	4.04%	1.42%	−2.15%	−7.05%	
8% Am EOL	**3.59**%	**2.20**%	**−1.93**%	**−6.30**%	
2% Np 2% Am EOL	−2.9466E − 06	6.0580E − 07	2.1653E − 07	2.0270E − 06	−3.9219E − 06
3% Am EOL	−4.2160E − 06	−1.4000E − 08	2.6400E − 07	2.6060E − 06	
5% Am EOL	−5.7900E − 06	−1.2410E − 06	3.2900E − 07	3.2740E − 06	
8% Am EOL	**−7.6950E − 06**	**−2.8100E − 06**	**4.1970E − 07**	**4.2524E − 06**	

The analysis of the Americium and Neptunium transmutation rate is given in Table 2. Consider

Transmutation rate [%]

$$= \left(1 - \frac{\text{isotope number density at EOL}}{\text{isotope number density at BOL}}\right) \cdot 100. \quad (3)$$

The transmutation rate provides the information on how much of the initially inserted material is transmuted at the end of the lifetime of the fuel assembly. A higher number indicates a better transmutation performance for the corresponding isotope. After a burnup of 200 GWd/tHM, the Am-241 concentration is roughly halved and the transmutation rate is increased with increasing initial Americium content. The transmutation of Am-243 becomes only efficient for high Americium content, at least 3% even better 5%. The transmutation of Np-237 remains nearly constant at a low content of 2% efficient.

The conclusion of this evaluation of the transmutation rate is that transmutation especially of Americium becomes significantly more efficient with increasing initial Americium content. Unfortunately, the permissible Americium content is limited due to the negative influence of the Americium on the feedback effects in the reactor core and due to the negative influence on the fuel behavior caused by the high fission gas release in Americium containing fuel.

The consequence of the insertion of minor actinides on different feedback effects and other safety-related values is given in Table 3. A positive value in the negative Doppler feedback indicates the desired stronger feedback. For the positive coolant effect and void effect, a positive value is desired, since the positive feedback should be reduced to stabilize the reactor. The absorber worth reflects the negative reactivity which is inserted by the addition of a given amount of absorbing material; a value close to zero indicates an unchanged efficiency of the absorber rods; a negative value signs a demand of a higher number of absorber rods; a

Moderating Material to Compensate the Drawback of High Minor Actinide Containing Transmutation Fuel on the Feedback Effects in SFR Cores

105

TABLE 2: Comparison of the transmutation rate for the reference cases and the cases with moderating material for the different transmutation fuels.

Transmutation rate	Reference			With moderator		
	Am-241	Am-243	Np-237	Am-241	Am-243	Np-237
2% Np 2% Am EOL	−47.0%	−9.1%	−58.5%	−49.1%	−8.0%	−60.9%
3% Am EOL	−50.6%	−20.8%		−52.7%	−20.8%	
5% Am EOL	−54.1%	−33.0%		−55.9%	−34.0%	
8% Am EOL	**−55.3%**	**−38.9%**		**−56.9%**	**−40.2%**	

TABLE 3: Feedback effects and absorber worth for the cases with different transmutation fuels.

	UMOX 2% Am 2% Np	+ZrH 2% Am 2% Np	UMOX 3% Am	+ZrH 3% Am	UMOX 5% Am	+ZrH 5% Am	**+ZrH 8% Am**	**1.2*ZrH 8% Am**
Doppler effect T_f + 100 K								
BOL	−30.2%	59.7%	−23.0%	73.8%	−36.4%	49.4%	**18.9%**	**31.9%**
EOL	−26.9%	52.2%	−20.2%	65.2%	−30.9%	46.8%	**18.7%**	**32.3%**
Coolant effect T_c + 50 K								
BOL	4.9%	−8.1%	5.7%	−11.7%	12.2%	−4.6%	**2.1%**	**−0.2%**
EOL	0.9%	−12.0%	2.1%	−11.3%	1.2%	−14.2%	**−13.3%**	**−15.0%**
void effect								
BOL	9.2%	−5.3%	6.9%	−8.1%	10.7%	−3.2%	**2.7%**	**0.9%**
EOL	−0.2%	−11.8%	−0.4%	−12.5%	−0.6%	−12.3%	**−11.9%**	**−13.9%**
Absorber rod worth								
BOL	−7.8%	−3.6%	−6.0%	−1.6%	−9.6%	−5.6%	**−11.1%**	**−10.6%**
EOL	−9.5%	−5.0%	−6.9%	−2.0%	−11.1%	−6.8%	**−13.3%**	**−12.9%**

positive value would allow reducing the number of absorber rods.

The insertion of the minor actinides into the fuel has a strong influence on the safety-related effects of the fuel assembly. The negative Doppler effect is reduced in all cases due to the influence of the minor actinides. The already positive coolant effect (the combination of the coolant temperature and the coolant density effect) becomes more positive with increasing Americium fraction. The positive sodium void effect at BOL increases compared to the reference case in all configurations and the absorber rod worth of the shutdown rods decreases compared to the reference case in all configurations, which requires more shutdown elements for the safe shutdown in hot condition compared to the reference case in all configurations. The parallel insertion of minor actinides and distributed moderating material relaxes the situation significantly. The effect of the moderating material on the Doppler effect is even strong enough to overcompensate the influence of the minor actinides in all cases; thus the Doppler effect is still significantly stronger—more negative—than in the reference case without minor actinides and moderating material. The same behavior can be observed for the coolant effect as well as for the sodium void effect. The influence of the minor actinides is overcompensated in all cases by the insertion of the fine distributed moderating material. Finally, the absorber worth, the reduction of the absorber rod worth is only partly compensated due to the insertion of the moderating material, but the situation is relaxed in all cases due to the insertion of moderating material.

One drawback of the moderating material should not be forgotten; due to the significantly stronger Doppler effect, more shutdown rods are required for reaching a defined cold subcriticality status. The amount of moderating material to be used is sure to be optimized from the point of view of system safety and transient behavior and the thermal stability of the ZrH moderator has to be investigated thoroughly or a more stable material like YH has to be envisaged.

An independent evaluation for the maximum possible amount of minor actinides in combination with the use of fine distributed moderating material closes the study. The strong influence of the minor actinides on the safety relevant feedback effects limits the possible amount of minor actinide insertion. For the given amount of hydrogen insertion, the limit has been found for 8% Americium content where the coolant feedback and the sodium void at BOL become slightly positive (see Table 3 first bold column), but this is still compensated by the enhancement of the Doppler feedback by more than 18%. The consequences of this Americium content can be compensated once more by a slight increase of the hydrogen content. This is studied by a test for 8% Americium content and by a 20% increased $ZrH_{1.6}$ content. The insertion of 20% more hydrogen leads once more to at least balanced or even enhanced feedbacks and thus to a more stable system (see Table 3 second bold column). From the transmutation point of view, the by 3% increased Americium amount (relative 60% more Americium) leads to an increased Am-241 burning by 1% (relative increase ~2%) and an increased Am-243 burning by roughly 6% (relative increase ~18%) (see Table 2 bold row). Thus the expected increase of transmutation efficiency can be found mainly for Am-243 which is more problematic to be transmuted.

Nevertheless, the increase of Americium content cannot be continued ad infinitum, since the acceptable Americium content is limited by the swelling behavior of transmutation fuel too. Experiments have demonstrated that target pellets containing 10–12 wt% ^{241}Am already show swelling of the target pellets by the order of 15%. This is attributed to accumulation of helium, produced by alpha decay of ^{242}Cm that occurs in the transmutation scheme of ^{241}Am [26].

5. Conclusions

The effect of the insertion of fine distributed moderating material on the feedback effects and the void effect in a sodium cooled fast reactor has been investigated in several publications [8–11]. The strong improvement in the fuel temperature and coolant feedback effect and in the sodium void effect has already been demonstrated, but negative consequences to the transmutation efficiency have been assumed due to the softer neutron spectrum.

The effect of fine distributed moderating material on fuel assemblies with a high minor actinide concentration like those foreseen for minor actinide transmutation has been investigated for 3 different fuel compositions: fuel with 2% Np and 2% Am, fuel with 3% Am, and fuel with 5% Am. It is demonstrated that the influence of the minor actinides on the feedback effects can be easily compensated by the insertion of fine distributed moderating material into the wire wrapper. The investigation of the transmutation efficiency has shown that there is only a very limited influence due to the moderating material. On the one hand, the Americium transmutation rate is even slightly higher for the fuel assembly containing moderating material. On the other hand, the Curium production increases slightly. Nevertheless, a detailed comparison of the data shows that the higher Americium efficiency rate is not only because breeding reactions lead to more Curium but also a higher fission rate of Americium is reached. Further, it is shown that the Americium transmutation rate depends strongly on the initial Americium amount in the fuel assembly.

Using fine distributed moderating materials, for example, inside the wire wrapper has the potential to improve the safety of a dedicated core for minor actinide transmutation. Additionally, it is shown that the transmutation rate can be improved since it is possible to increase the BOL amount of minor actinides in the core. This can be reached, since it is possible to eliminate the negative effect of the minor actinides on core stability and transient behaviour up to an Americium content of ~8%. Additionally, it is shown that an increase of the hydrogen content by 20% can even improve all feedback effects in this case with 8% Americium content. Using this configuration allows a relative increase of the Am-243 transmutation rate by nearly 20% which is an important improvement for an isotope which is hard to transmute. Thus the concept of fine distributed moderating material offers the possibility of increased transmutation efficiency in conjunction with an elimination of the negative consequences of transmutation fuel.

The use of fine distributed moderating material can influence the safety of fast reactors strongly—it opens the stage for designable feedback coefficients; thus it creates a new degree of freedom for the optimization of important inherent safety-related parameters in the design of sodium-cooled fast reactor cores.

References

[1] "IAEA TM on Innovative Fast Reactor Designs with Enhanced Negative Reactivity Feedback Features," IAEA Headquarters, Vienna, Austria, 2012, accessed March 11, 2013, http://www.iaea.org/NuclearPower/Meetings/2012/2012-02-27-02-29-TM-NPTD.html.

[2] H. Hummel and D. Okrent, Reactivity Coefficients in Large Fast Power Reactors, ANS, 1970.

[3] R. N. Hill and H. Khalil, "Evaluation of LMR design options for reduction of sodium void worth," in Proceedings of International Conference on the Physics of Reactors, vol. 1, pp. 11–19, Marseille, France, 1980.

[4] G. Rimpault, L. Buiron, P. Sciora, and F. Varaine, "Towards GEN IV SFR design: promising ideas for large advanced SFR core designs," in International Conference on the Physics of Reactors (PHYSOR '08), pp. 2394–2400, Interlaken, Switzerland, September 2008.

[5] B. Carluec and P. L. Pinto, "Lessons learned from Fushima accident as regards the safety approach of the ASTRID project at conceptual design stage," in Proceedings of the IAEA Technical Meeting on Impact of the Fukushima Event on Current and Future Fast Reactor Design, Helmholtz-Zentrum Dresden-Rossendorf, Dresden, Germany, March 2012, accessed March 11, 2013, http://www.iaea.org/NuclearPower/Meetings/2012/2012-03-19-03-23-TM-NPTD.html.

[6] L. Buiron et al., "Innovative core design for generation IV sodium-cooled fast reactors," in Proceedings of the International Congress on Advances in Nuclear Power Plants (ICAPP '10), Nice, France, 2007.

[7] K. Sun et al., "Void reactivity decomposition for the sodium cool fast reactor in equilibrium closed fuel cycle," in Proceedings of the Advances in Reactor Physics to Power the Nuclear Renaissance (PHYSOR '10), Pittsburgh, Pa, USA, 2010.

[8] B. Merk, E. Fridman, and F. P. Weiß, "On the use of zirconium based moderators to enhance the feedback coefficients in a MOX fuelled sodium cooled fast reactor," Nuclear Science and Engineering, vol. 171, 2, pp. 136–149, 2012.

[9] B. Merk, E. Fridman, and F. P. Weiß, "On the use of a moderation layer to improve the safety behavior in sodium cooled fast reactors," Annals of Nuclear Energy, vol. 38, no. 5, pp. 921–929, 2011.

[10] B. Merk and F. P. Weiß, "Analysis of the influence of different arrangements for ZrH moderator material on the performance of a SFR core," Annals of Nuclear Energy, vol. 38, pp. 2374–2385, 2011.

[11] B. Merk and F. P. Weiß, "On the effect of different placing ZrH moderator material on the performance of a SFR core," in Advances in Reactor Physics—Linking Research, Industry, and Education (PHYSOR '12), Knoxville, Tenn, USA, 2012.

[12] E. A. Villarino, R. J. J. Stammler, A. A. Ferri, and J. J. Casal, "HELIOS: angularly dependent collision probabilities," Nuclear Science and Engineering, vol. 112, pp. 16–31, 1992.

[13] R. Sanchez et al., "APOLLO II: a user-oriented, portable, modular code for multigroup transport assembly calculations," *Nuclear Science and Engineering*, vol. 100, article 352, 1988.

[14] IAEA Fast Reactor Database, 2006, http://www.iaea.org/in-isnkm/nkm/aws/frdb/auxiliary/generalInformation.html.

[15] A. E. Waltar and A. B. Reynolds, *Fast Breeder Reactors*, Pergamon Press, New York, NY, USA, 1981.

[16] A. Pay, E. Francillon, B. Steinmetz, D. Barnes, and N. Meda, "European Fast Reactor (EFR) fuel element design," in *Proceedings of the 10th International Conference on Structural Mechanics in Reactor Technology*, Anaheim, Calif, USA, 1989, http://www.iasmirt.org/iasmirt-3/SMiRT10/DC_250515.

[17] J. Carmack and K. O. Pasamehmetoglu:, "Review of transmutation fuel studies," in *INL/EXT-08-13779, or GNEP-FUEL-TD-RT-2008-000050*, U.S. Department of Energy, Transmutation Fuel Campaign, 2008, http://www.inl.gov/technicalpublications/Documents/3901056.pdf.

[18] W. M. Mueller, J. P. Blackledge, and G. G. Libowitz, *Metal Hydrides*, Academic Press, New York, NY, USA, 1968.

[19] K. Tuček, J. Carlsson, and H. Wider, "Comparison of sodium and lead-cooled fast reactors regarding reactor physics aspects, severe safety and economical issues," *Nuclear Engineering and Design*, vol. 236, no. 14–16, pp. 1589–1598, 2006.

[20] S. E. Bays, H. Zhang, and H. Zhao, "The industrial sodium cooled fast reactor," in *ANFM*, INL/CON-09-15519 PREPRINT, 2009, http://www.inl.gov/technicalpublications/Documents/4363828.pdf.

[21] D. Olander, E. Greenspan, H. D. Garkisch, and B. Petrovic, "Uranium-zirconium hydride fuel properties," *Nuclear Engineering and Design*, vol. 239, pp. 1406–1424, 2009.

[22] D. W. Wootan, J. A. Rawlins, L. L. Carter, H. R. Brager, and R. E. Schenter, "Analysis and results of a hydrogen-moderated isotope production assembly in the fast flux test facility," *Nuclear Science and Engineering*, vol. 103, no. 2, pp. 150–156, 1989.

[23] J. C. Lefèvre, C. H. Mitchell, and G. Hubert:, "European fast reactor design," *Nuclear Engineering and Design*, vol. 162, pp. 133–143, 1996.

[24] U. Rohde, U. Grundmann, and S. Kliem, "DYN3D - advanced reactor simulations in 3D," *Nuclear Energy Review*, vol. 2, pp. 28–30, 2007.

[25] C. Beckert and U. Grundmann, "A nodal expansion method for solving the multigroup SP3 equations in the reactor code DYN3D," in *Proceedings of the Joint International Topical Meeting on Mathematics and Computations and Supercomputing in Nuclear Applications (M&C+SNA '07)*, Monterey, Calif, USA, April 2007.

[26] R. J. M. Konings et al., "The EFTTRA-T4 experiment on americium transmutation," *Journal of Nuclear Materials*, vol. 282, no. 2-3, pp. 159–170, 2000.

11

SPES3 Facility RELAP5 Sensitivity Analyses on the Containment System for Design Review

Andrea Achilli,[1] Cinzia Congiu,[1] Roberta Ferri,[1] Fosco Bianchi,[2] Paride Meloni,[2] Davor Grgić,[3] and Milorad Dzodzo[4]

[1] SIET S.p.A., UdP, Via Nino Bixio 27/c, 29121 Piacenza, Italy
[2] ENEA, UTFISSM, Via Martiri di Monte Sole 4, 40129 Bologna, Italy
[3] FER, University of Zagreb, Unska 3, 10000 Zagreb, Croatia
[4] Research and Technology Unit, Westinghouse Electric Company LLC, Cranberry Township, PA 16066, USA

Correspondence should be addressed to Roberta Ferri, ferri@siet.it

Academic Editor: Alessandro Del Nevo

An Italian MSE R&D programme on Nuclear Fission is funding, through ENEA, the design and testing of SPES3 facility at SIET, for IRIS reactor simulation. IRIS is a modular, medium size, advanced, integral PWR, developed by an international consortium of utilities, industries, research centres and universities. SPES3 simulates the primary, secondary and containment systems of IRIS, with 1:100 volume scale, full elevation and prototypical thermal-hydraulic conditions. The RELAP5 code was extensively used in support to the design of the facility to identify criticalities and weak points in the reactor simulation. FER, at Zagreb University, performed the IRIS reactor analyses with the RELAP5 and GOTHIC coupled codes. The comparison between IRIS and SPES3 simulation results led to a simulation-design feedback process with step-by-step modifications of the facility design, up to the final configuration. For this, a series of sensitivity cases was run to investigate specific aspects affecting the trend of the main parameters of the plant, as the containment pressure and EHRS removed power, to limit fuel clad temperature excursions during accidental transients. This paper summarizes the sensitivity analyses on the containment system that allowed to review the SPES3 facility design and confirm its capability to appropriately simulate the IRIS plant.

1. Introduction

The IRIS reactor, with its integral design, is an advanced engineering solution of the latest LWR technology. Medium-sized, safe, modular, and economic, it provides a viable bridge to generation IV and satisfies the GNEP requirements for grid-appropriate NPPs [1–3].

In the frame of an R&D program on nuclear fission, funded by the Italian Ministry of Economic Development, ENEA, as member of the IRIS consortium, is supporting the design, construction, and testing of the SPES3 ITF at SIET laboratories [4–6].

The SPES3 design was carried out following the subsequent steps: (a) definition of a preliminary facility design, based on specified system geometry; (b) setup of the RELAP5 facility model and DBA simulation; (c) comparison of SPES3 and IRIS results against the same transient;

(d) identification of the main differences and understanding of related reasons; (e) FSA application to selected thermo-fluid-dynamic parameters in order to assess and quantify the discrepancies; (f) updating of the SPES3 design to match the IRIS behaviour; (g) final result comparison; (h) final FSA application and assessment of acceptability criteria for considering SPES3 correctly simulating IRIS.

The above-mentioned process allowed to verify the SBLOCA PIRT objectives for the IRIS reactor, as defined by a group of international experts [7]. The Phenomena Identification and Ranking Table put in evidence the thermal-hydraulic phenomena playing an important role in operation of IRIS safety systems. Two figures of merit were considered fundamental for the accident sequence control: containment pressure and reactor vessel mass inventory. Sufficient water in the vessel allows to remove stored energy, and decay heat without fuel clad temperature excursions and adequate heat

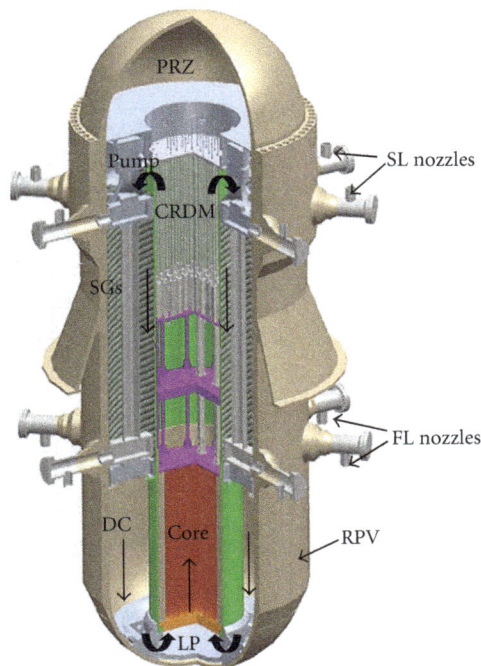

FIGURE 1: IRIS integral layout.

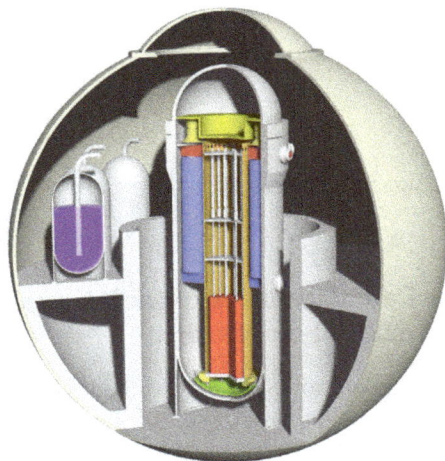

FIGURE 2: IRIS containment systems.

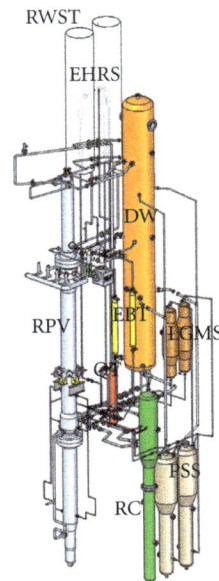

FIGURE 3: SPES3 facility layout.

studies are foreseen for using it in a wider field of application for integral layout SMR simulation [8].

2. IRIS Plant and SPES3 Facility Layout and Nodalization

The IRIS pressure vessel and containment are shown in Figures 1 and 2, whereas the SPES3 facility is presented in Figure 3.

The SPES3 facility reproduces all parts and components of the IRIS plant with 1 : 100 volume scaling factor, 1 : 1 elevation scaling factor, and prototypical fluid and thermal-hydraulic conditions. The reactor vessel includes the internals, consisting of the electrically heated core simulator, the riser with control rod device mechanisms, the pressurizer, the pump suction plenum, the helical coil steam generators, the downcomer, and the lower plenum. Three SGs simulate the eight IRIS SGs. A pump, located outside of the RPV, for room reasons, and connected to it by pipes, simulates eight IRIS pumps. Two emergency boration tanks are simulated and connected to the DVI lines, devoted to direct injection of emergency fluid into the vessel. Three secondary loops simulate four IRIS loops. Each secondary loop is simulated up to the main isolation valves and includes the feed line, the SG, the steam line, and the emergency heat removal system with a vertical tube heat exchanger immersed in a refuelling water storage tank. The IRIS spherical containment compartments are simulated by tanks, connected to each other by pipes and to the RPV by break lines [4, 5]. They include dry-well and reactor cavity, representing the dry zone surrounding the RPV, respectively above and below the mid-deck plane; pressure suppression systems representing the wet zone around the lower part

rejection to the RWST prevents containment overpressurization and contributes to core cooling also thanks to dynamic coupling between the primary and containment systems.

The DBA simulation on the facility allowed to understand the transient plant behaviour and the mutual system interaction. The comparison with the IRIS results led to running many sensitivity cases that required the SPES3 design review for better matching the IRIS transients.

The SPES3 tests will provide a qualified data base for the assessment of codes to be used in the reactor safety analyses.

The SPES3 facility is under construction, based on the IRIS reactor design. The availability of such a complex plant opens the way to other possibilities of exploitation, and

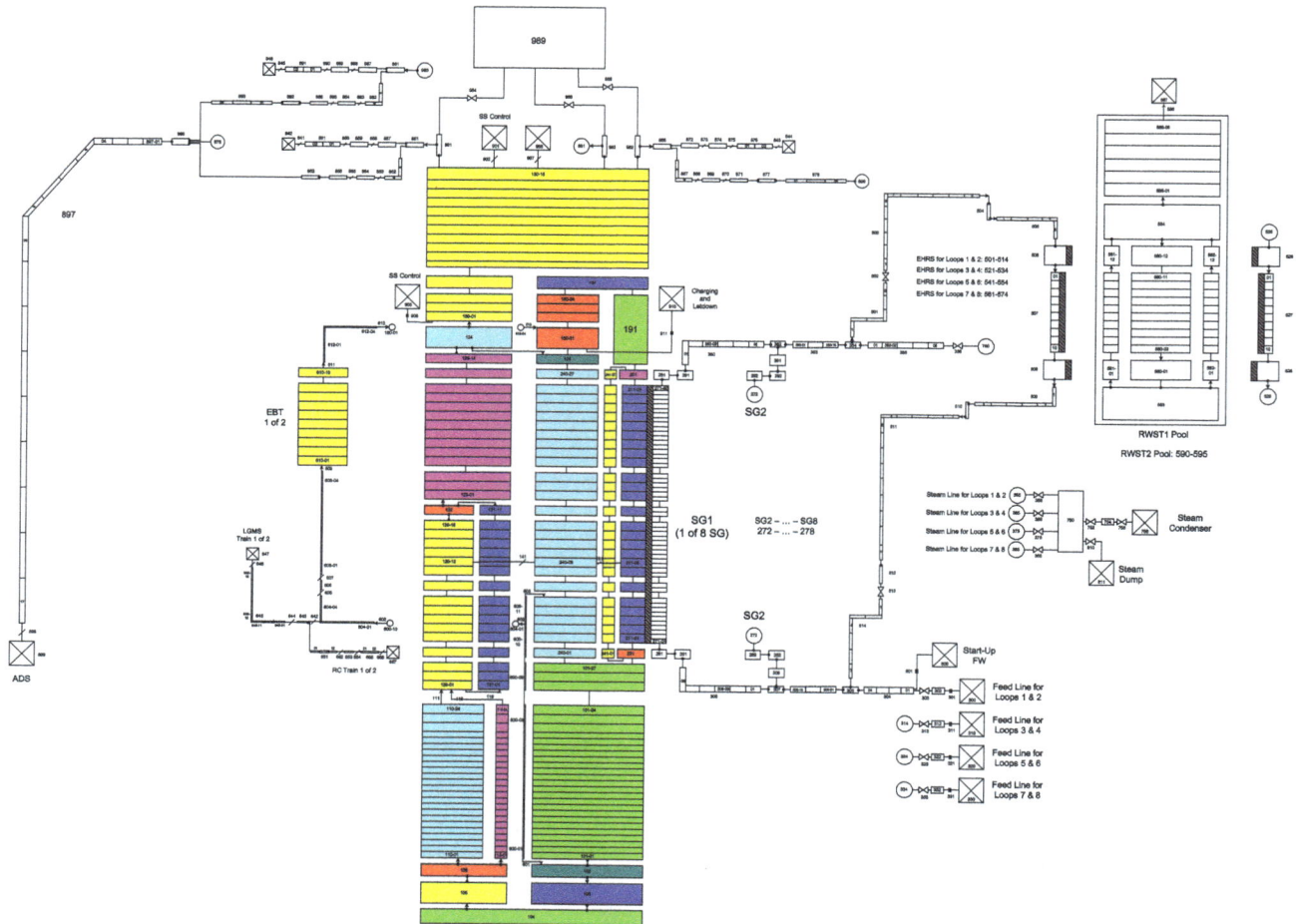

FIGURE 4: IRIS primary and secondary circuit nodalization for RELAP5 code.

of the RPV, suitable to dump pressure in case of containment pressurization; long-term gravity make-up systems representing the cold water reservoir to be poured into the RPV when depressurized. The two stages of the automatic depressurization system are simulated, connected to the pressurizer top, with stage I discharging into the quench tank and stage II directly connecting RPV and DW at high plant elevation.

The facility allows to test both LOCAs and secondary side breaks (DBA and BDBA) as well as to perform separate effect tests on particular components such as SG-EHRS thermally coupled to RWST.

The IRIS nodalization was developed in two parts: the primary and secondary circuits for the RELAP5 code and the containment system for the GOTHIC code (Figures 4 and 5). The RELAP5 nodalization includes 1845 volumes, and 1940 junctions, 1015 heat structures with 8600 mesh points, while the GOTHIC model includes 85 volumes, 28 junctions, and 57 heat structures.

The SPES3 nodalization was completely developed for the RELAP5 code (Figures 6, 7, and 8). It includes 1499 volumes, 1639 junctions, 1839 heat structures, and 19322 mesh points.

3. Design-Calculation Feedback Process for SPES3 Facility Final Design

The RELAP5 model for SPES3 was initially developed on the basis of the preliminary design of the facility, and the steady-state conditions are based on the actual IRIS nominal operation [9]. Five DBAs were simulated with particular attention to the occurring phenomena and sequence of events. In particular, three SB-LOCAs and two secondary side breaks were simulated, according to the specified test matrix [10].

Once the phenomena occurring in the DBAs were investigated, attention was focused on the most challenging transient scenario, the DVI line DEG break, for a direct comparison of the SPES3 and IRIS results.

WEC, in collaboration with the University of Pisa and Politecnico di Milano, developed the Fractional Scaling Analysis for IRIS and SPES3. The method, based on system and time sequence decomposition, allowed to identify the parameters mostly affecting the transient and to quantify the distortions between IRIS and SPES3 simulations introduced by such parameters (e.g., containment tank metal mass, heat transfer at core side wall, etc.).

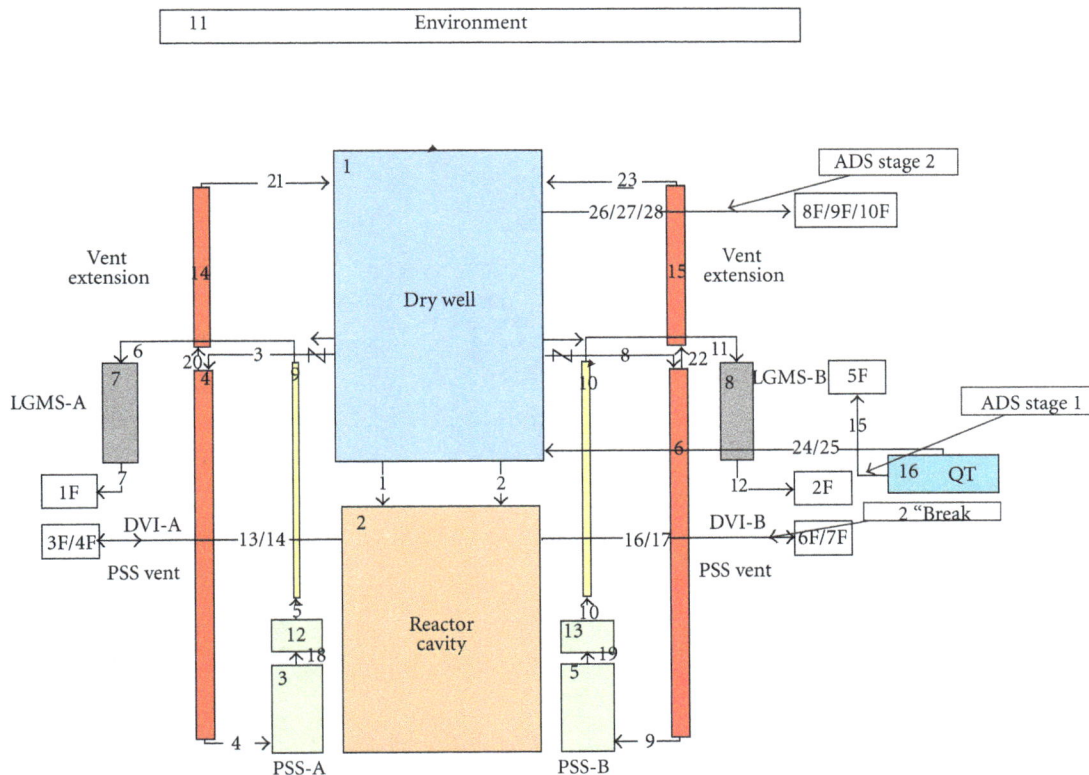

FIGURE 5: IRIS containment nodalization for GOTHIC code.

The first application of the DVI line DEG break evidenced important differences on containment pressure, especially in early phase of the accident, at pressure peak, and also on the long term.

The need of understanding the reasons and reducing discrepancies led to performing a series of sensitivity cases on SPES3 containment, making SPES3 response closer to reactor one, and finalizing the facility design [11, 12].

The main events, identified in the DVI line break transient, are listed below for better understanding all sensitivity analyses and the design-calculation feedback process. Approximate timing of events is reported too. The long-term phase of the transient was simulated to verify the safe, stable plant operation.

(i) The break opening (0 s) causes the RPV blowdown and depressurization, containment pressurization, steam dumping into PSS with air build-up at PSS top, and consequent pressurization;

(ii) the signal of high containment pressure (∼30 s) triggers the reactor scram, the secondary loop isolation, and the actuation of two out of four EHRSs;

(iii) the signal of low PRZ water level (∼120 s) triggers the pump coastdown, and the natural circulation in the core is guaranteed through the check valves, connecting riser and downcomer at one-third of the SG height;

(iv) the signal of low PRZ pressure (∼200 s) actuates the remaining EHRS and triggers the ADS stage I, to help

RPV depressurization, and the EBT intervention, to inject cold water into the primary circuit;

(v) the signal of low differential pressure between RPV and DW (∼2250 s) triggers the LGMS injection into the DVI line and opens the valves connecting RC and DVI line to allow water reverse flow from the containment to the primary side;

(vi) when PSS pressurization is sufficiently high, cold water flows from PSS to DW (3500 s), increasing the RC flooding and allowing water to enter the RPV;

(vii) the signal of low LGMS mass (∼25000 s) opens the ADS stage II with possible reverse steam flow from DW to RPV;

(viii) on the long term (simulation up to 100000 s), the plant is cooled by EHRSs that reject core decay heat to RWST.

The starting point for the sensitivity analyses was the comparison between cases SPES3-97 and IRIS-HT1 results, which showed qualitatively good agreement in occurring phenomena, but also quantitative discrepancies in containment pressure, affecting the sequence of events and transient evolution (Figure 9).

A series of parameters identifying and potentially affecting containment pressure is: (a) SPES3 containment over-volume of about 10% with respect to 1 : 100 scaled IRIS; (b) SPES3 containment metal mass greater than IRIS for mechanical resistance to the same design conditions; (c)

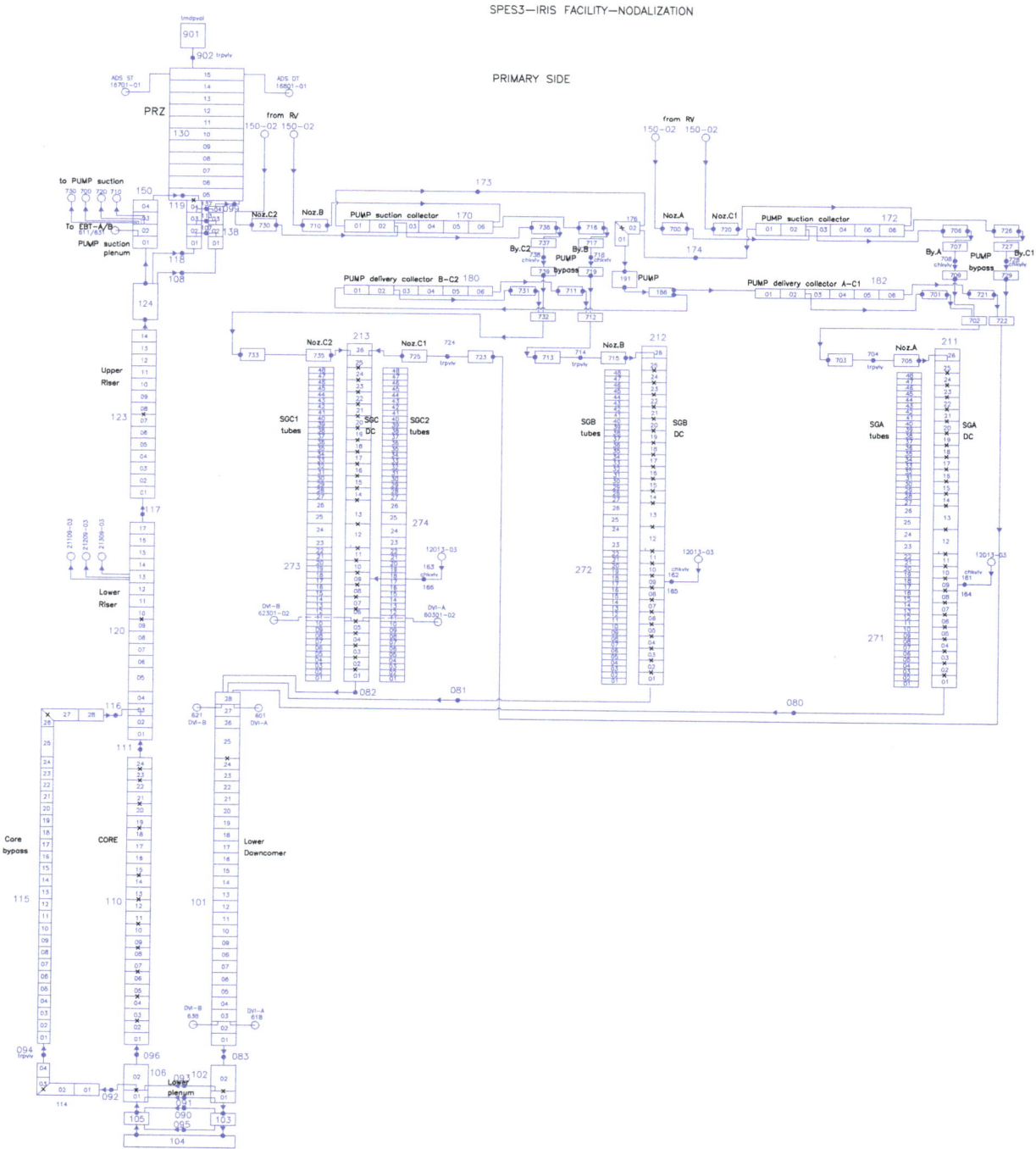

FIGURE 6: SPES3 primary circuit nodalization for RELAP5 code.

SPES3 component surface-to-volume ratio ten times greater than IRIS, due to volume scaling and component height conservation; (d) containment metal structure temperature; (e) containment piping pressure drops; (f) EHRS and RWST modelling and heat transfer coefficients.

A synthesis of the performed sensitivity cases on SPES3 is reported in Table 1, where they are grouped according to the investigated parameters. A synthesis of IRIS cases, utilised for the comparison, is reported in Table 2.

A reduction of DW volume, for correctly scaling IRIS (1 : 100), did not provide great improvements (a few percent) in the containment pressure response. An improvement was observed only in the long term related to lower heat losses to environment due to DW size reduction (Figure 10).

In order to compensate for the extra surface in SPES3, a thermal insulation of DW inner surface was tested with different thickness of Aluminium Silicate Rescor 902. As shown in Figure 11, the introduction of an increasing

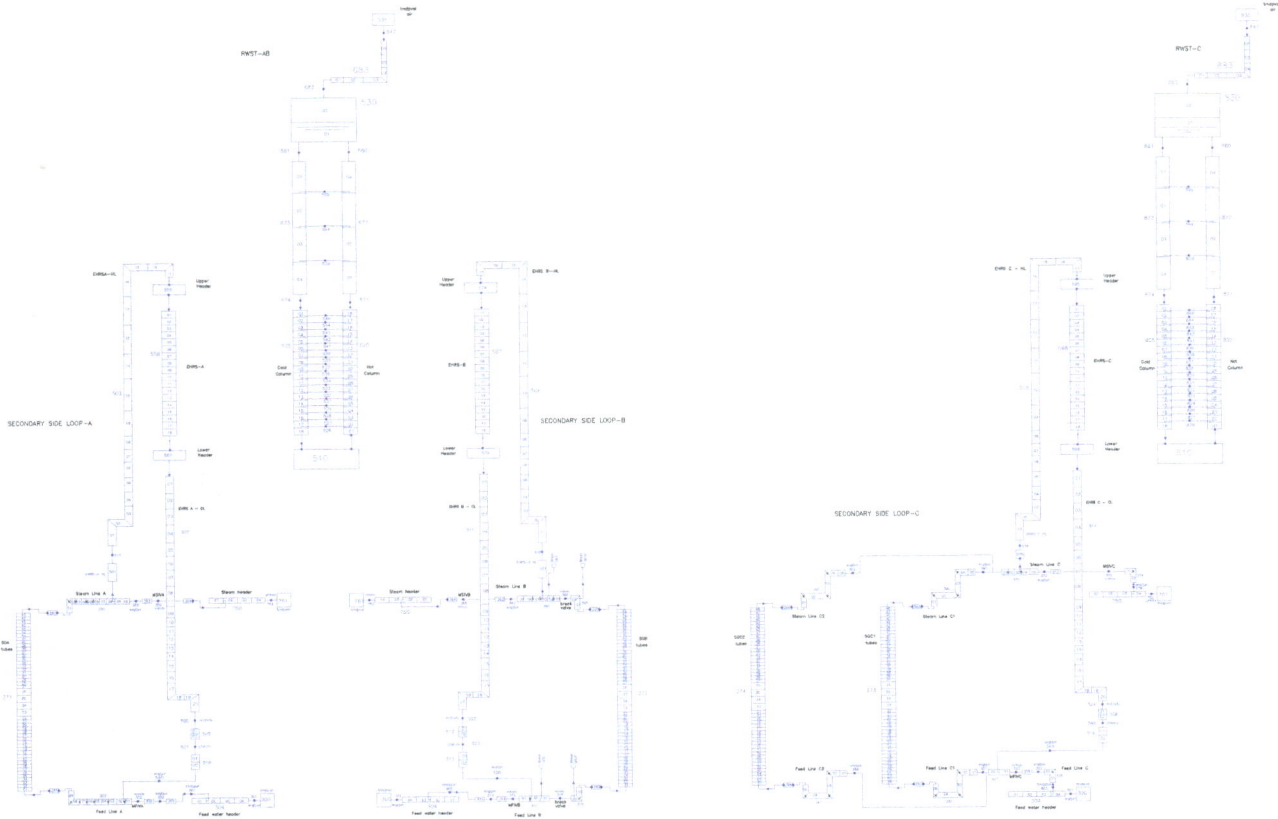

FIGURE 7: SPES3 secondary circuit and EHRS nodalization for RELAP5 code.

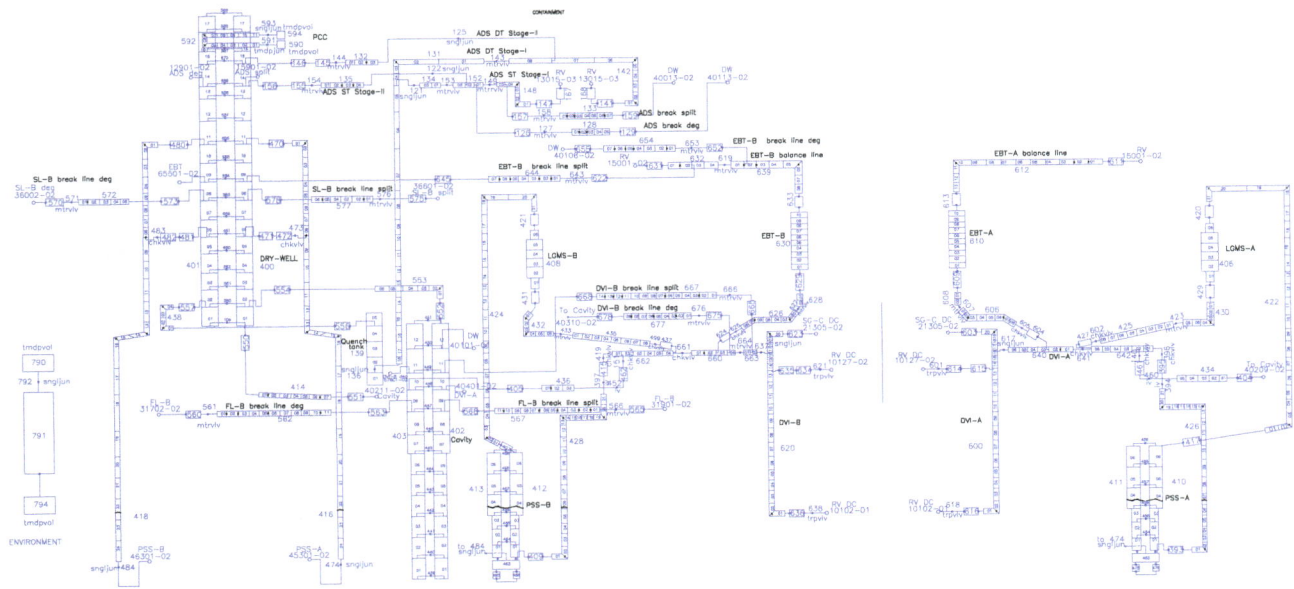

FIGURE 8: SPES3 containment nodalization for RELAP5 code.

thickness of thermal insulation increased pressure in the very short term, but the additional mass reduced the containment pressure peak. The result was that the DW insulation led to worse effects on pressure than with noninsulated DW, showing that masses have larger effects than surfaces.

The influence of DW heat structure mass on containment pressure response was investigated by reducing DW thickness by 40% (25 mm to 15 mm), approximately corresponding to a design pressure of 1.5 MPa, instead of the original 2 MPa. As shown in Figure 12, pressure increase in the early

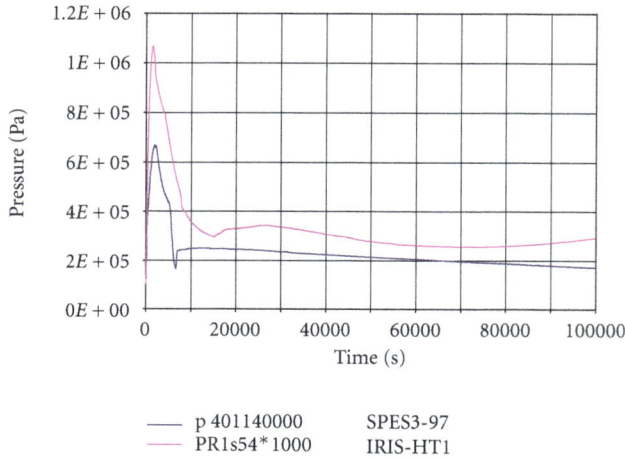

FIGURE 9: SPES3-97 and IRIS-HT1 DW pressure. Note: The IRIS Drywell Volume was subdivided in 4x4x4 sub-volumes in the three directions, marked 1s1, 1s2... up to 1s64. Pressure from a cell at the top of the model was used as reference.

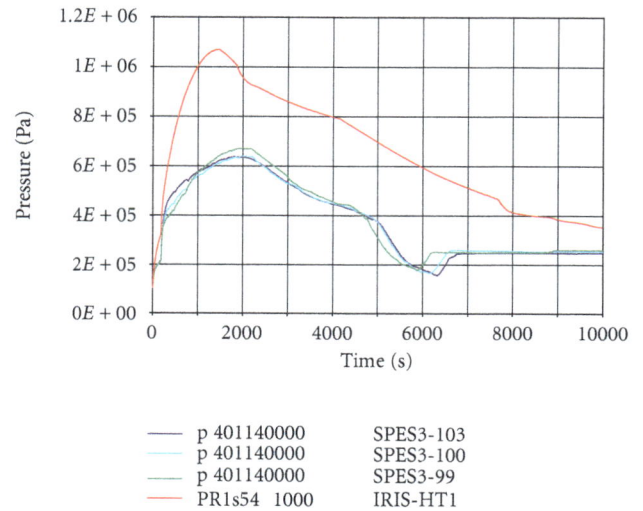

FIGURE 11: SPES3-99, 100, 103 and IRIS-HT1 DW pressure (short term).

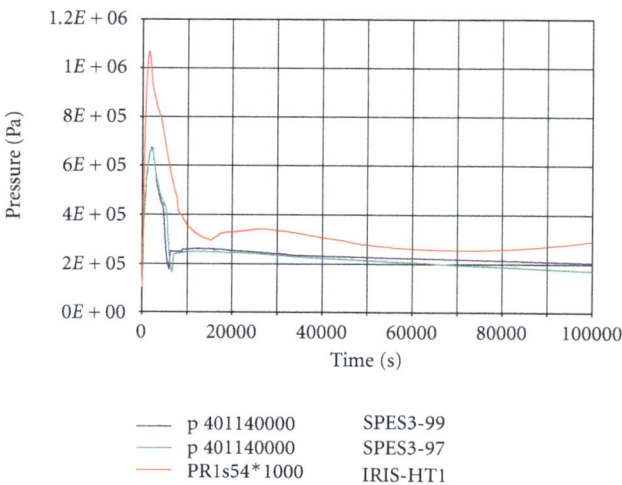

FIGURE 10: SPES3-97, 99 and IRIS-HT1 DW pressure.

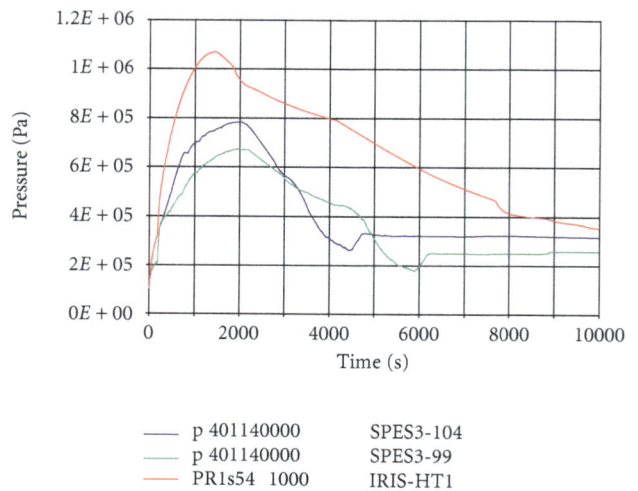

FIGURE 12: SPES3-99, 104 and IRIS-HT1 DW pressure (short term).

phase of the transient was steeper and pressure peak higher. Containment pressure was still below the IRIS one, showing that such DW metal mass reduction was not enough to have the desired pressure response.

A further DW mass reduction was performed in a theoretical case, where concrete plus carbon steel IRIS DW equivalent mass was distributed on SPES3 DW surface, resulting in a thickness of 10 mm AISI 304. As shown in Figure 13, the further DW mass reduction allowed to get pressure values closer to IRIS both in the early phases of the transient and at the pressure peak, but still below IRIS pressure. That showed that other parameters affected the results.

An attempt to investigate how a greater DW volume reduction affects containment pressure was performed by scaling it 1 : 150 with respect to IRIS. Figure 14 shows a pressure gain only in the early phases of the transient

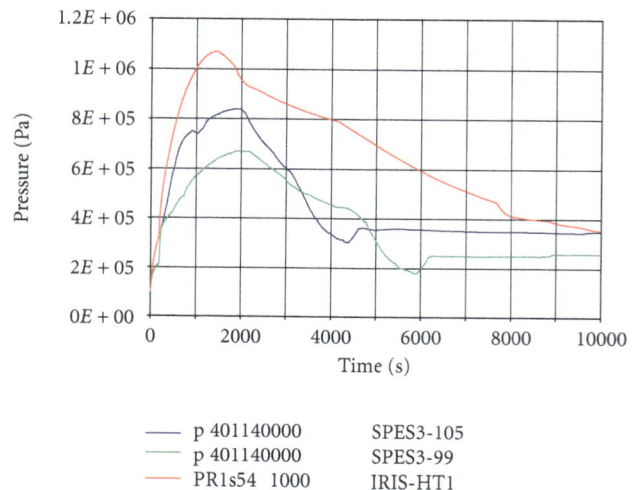

FIGURE 13: SPES3-99, 105 and IRIS-HT1 DW pressure (short term).

TABLE 1: Characteristics of the SPES3 cases.

SPES3 cases	Characteristics	Results
SPES3-97	Containment volume ~110% of IRIS volume; DW wall thickness 25 mm for 2 MPa design pressure; Containment metal wall initial temperature 48.9°C.	Starting point for sensitivity analyses.
effect for mass		
Sensitivity on the DW inner surface insulation		
SPES3-94	1 mm Teflon on the DW inner surface	Little improvements in containment pressure, limited to the early phase of transient.
SPES3-100	DW volume correctly scaled on IRIS volume. 1.5 mm Rescor 902 Aluminium Silicate layer on the DW inner surface	Little improvements in containment pressure. Negative effect for mass addition.
SPES3-103	DW volume correctly scaled on IRIS volume. 3 mm Rescor 902 Aluminium Silicate layer on DW inner surface	Little improvements in containment pressure. Negative effect for mass addition.
Sensitivity on the containment volume		
SPES3-99	DW volume correctly scaled on IRIS volume	Little improvements in containment pressure.
SPES3-106	DW volume scaled 1 : 150 on IRIS volume	Little improvements in containment pressure, limited to the early phase of transient. Modelling due to a component simulation with a different scaling factor.
SPES3-107	Equivalent to SPES3-99 with corrected minor input mistakes	
Sensitivity on the containment metal structure		
SPES3-104	DW wall thickness 15 mm for 1.5 MPa design pressure	Improvement of containment pressure, but not sufficient.
SPES3-105	DW wall thickness 10 mm (equivalent to 1 : 100 IRIS DW mass distributed on SPES3 surface)	Further improvement of containment pressure, but not sufficient.
SPES3-108	DW heat structure directly scaled 1 : 100 in terms of mass and surface (a one-hundredth slice of IRIS structure attributed to SPES3). DW wall material as in IRIS	Improvement of containment pressure response in early phase of the transient, due to DW surface reduction impact. No improvement in later phases, due to other SPES3 containment structures, not simulated in IRIS.
SPES3-111	DW heat structure initial temperature from 48.9°C to 84°C	Heat structure preheating compensates for the extra mass. Containment pressure response not sufficiently improved as other containment heat structures are non-preheated.
SPES3-112	PSS wall eliminated	Containment pressure response improved.
SPES3-115	All containment tanks volume resized to scale IRIS 1 : 100. Thickness reduced to resist 1.5 MPa design pressure. Air space metal structure initial temperature 84°C. PSS main vent pipe no additional restriction. LGMS to DVI line calibrated orifice from 4 mm to 3.2 mm. ADS stage I ST calibrated orifice from 7.019 mm to 5.637 mm. ADS stage I DT calibrated orifice from 9.927 mm to 7.973 mm. EHRS-A and B CL calibrated orifice from 6 mm to 5 mm. EHRS-C CL calibrated orifice from 12 mm to 8.5 mm	Containment pressure similar in SPES3 and IRIS. ADS stage I mass flow correctly reproducing IRIS. LGMS to DVI line and EHRS CL mass flow not correctly reproduced.

TABLE 1: Continued.

SPES3 cases	Characteristics:effect for mass	Results
SPES3-119	Containment volume scaled 1 : 100 on IRIS. DW heat structure directly scaled 1 : 100 in terms of mass and surface. PSS, LGMS, RC thickness 1 mm (to get closer to IRIS nonsimulated structure) LGMS to DVI line calibrated orifice from 2.3 mm to 2.5 mm. PSS vent pipe extension orifice from 5.2 to 7.3 mm. RC to DVI line additional calibrated orifice of 1 mm (original valve D = 10.7 mm) EHRS-C CL calibrated orifice from 6.7 mm to 7 mm	Containment pressure very similar in SPES3 and IRIS. Great importance of correct heat structure simulation.
SPES3-122	As SPES3-120. Containment heat structure initial temperature 48.9°C. As SPES3-130.	For comparison with SPES3-120 to quantify the heat structure preheating influence.
SPES3-132	EHRS-A and B tube 2% additional surface thermally insulated with Teflon (originally 0.6 tubes out of 3 insulated to scale 1 : 100 240 IRIS tubes). EHRS-C 0.2 tubes out of 5 insulated to scale 1 : 100 480 IRIS tubes. Additional 2% tube surface thermally insulated with Teflon	EHRS tube heat transfer surface reduced to correctly simulate IRIS surface.
	Sensitivity on the containment piping pressure drops	
SPES3-109	PSS main vent pipe resizing from 2.5″ to 2″ Sch. 40. PSS vent pipe extension resizing from 3/4″ to 1/2″ Sch. 40	PSS to DW mass flow closer to IRIS one.
SPES3-110	PSS main vent pipe additional restriction at the check valve ($D_{orifice}$ 14.19 mm)	Only early steep but limited containment pressure increase.
SPES3-118	Containment volume scaled 1 : 100 on IRIS. LGMS to DVI line calibrated orifice from 3.2 mm to 2.3 mm. EHRS-C CL calibrated orifice from 8.5 mm to 6.5 mm. PSS vent pipe extension additional restriction ($D_{orifice}$ 5.2 mm). Containment heat structure initial temperature 48.9°C Containment volume scaled 1 : 100 on IRIS. Thickness to resist 1.5 MPa design pressure.	Attempt to match IRIS injection mass flows.
SPES3-120	Air space metal structure initial temperature 84°C. QT initial temperature 48.9°C. LGMS to DVI line calibrated orifice from 2.3 mm to 2.5 mm	IRIS injection mass flow reproduced, but different pressure drops in the pipes.
SPES3-124	PSS main vent pipe resizing from 2″ to 2.5″ Sch. 40 PSS vent pipe extension resizing from 1/2″ to 1″ Sch. 40 LGMS to DVI line calibrated orifice from 2.5 mm to 3.6 mm. PSS vent pipe extension orifice from 7.3 to 19 mm. PSS vent pipe extension connection to DW elevation decrease of 1.5 m to match IRIS. PSS sparger elevation decrease of 0.25 m to match IRIS. PSS bottom modelled with a branch. Containment air space metal structure initial temperature 84°C, water space 48.9°C	Mass flow determined by actual piping pressure drops as in IRIS. Containment pressure response qualitatively and quantitatively close to IRIS. The PSS bottom modelling did not affect the PSS vent pipe emptying mode.

TABLE 1: Continued.

SPES3 cases	Characteristics/effect for mass	Results
SPES3-127	As SPES3-124. PSS bottom modelled with three branches.	The PSS bottom modelling did not strongly affect the PSS vent pipe emptying mode.
SPES3-130	As SPES3-127. RWST top pipe introduction for connection to atmosphere.	Reduced loss of mass at RWST top due to dry air and water contact.
SPES3-135	As SPES3-132 Completely reviewed the EHRS circuits and RWST model: EHRS-A and B tube 4% surface thermally insulated with Teflon other than the originally insulated 0.6 tubes. EHRS-C 4% surface thermally insulated other than the originally insulated 0.2 tubes. EHRS-A and B HL resized from 2″ to 1.25″ Sch. 80. EHRS-A and B CL resized from 1.25″ to 1.5″ Sch. 80. EHRS-C HL resized from 2.5″ to 2″ Sch. 80. EHRS-C CL resized from 1.5″ to 3″ Sch. 80. HL-A and B additional orifice $D = 24$ mm. HL-C additional orifice $D = 17$ mm. CL-A and B orifice resized from 5 mm to 5.9 mm. CL-C orifice resized from 7 mm to 8.3 mm. EHRS tube heat structure fouling factor set to 2.9 left and 2.77 right (original values 2.725 left, 3.54284 right) [13]. RWST-AB and C rising slice area resized from 0.491 m² to 0.119167 m²; recirculation slice resized from 1.217 m² to 1.601169 m² [13]. RWST-AB and C slice side junction area from 0.151 m² to 0.135 m² [13].	The complete revision of the EHRS piping and heat exchanger parameters led to matching transferred energy.
SPES3-146	As SPES3-135. PSS vent pipe extension orifice from 19 mm to 17.5 mm. SG tube inlet orifice from 12.5 mm to 11.7 mm	Good similarity between SPES3 and IRIS BASE CASE for FSA final application.

TABLE 2: Characteristics of the IRIS cases.

IRIS cases	Characteristics	Results
IRIS-HT1	Containment heat structures simulated only for the DW.	Starting point for comparison with SPES3.
IRIS-HT5g	Heat structures added to all containment compartments and secondary sidepiping. PSS main vent pipe connection to DW rise of 4 m. PSS sparger set at 0.75 m from PSS bottom. RWST remodelled with parallel slice approach.	Similar containment pressure response with SPES3.
IRIS-HT6_rwstc	SG tubes inner layer removed which simulated the Fouling. ADS stage II actuation signal corrected to intervene on low LGMS mass. RWST remodelling according to PERSEO area ratio and HTC calibration on experimental data. EHRS heat transfer parameters set as in SPES3 (by multiplier fouling factors) [12, 13]. RWST top pipe introduction for connection to atmosphere. Correction of energy transfer parameters at the GOTHIC and RELAP5 code couplings.	Better matching of EHRS long-term energy transfer to RWST.
IRIS-HT6_rwstc1a	Corrected elevation difference between the RELAP5 and GOTHIC parts of the model: the ADS stage I vent pipe end should be 0.5 m from QT bottom (it was connected to QT top); LGMS tanks rise of 0.75 m.	Results very similar to IRIS-HT6_rwstc BASE CASE for FSA final application.

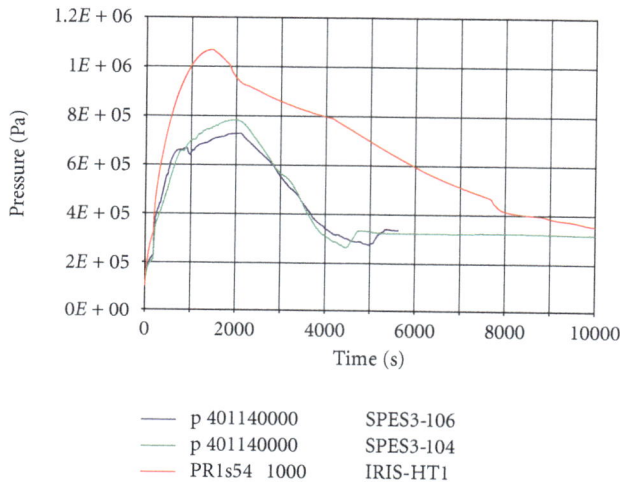

FIGURE 14: SPES3-104, 106 and IRIS-HT1 DW pressure (short term).

—— p 401140000	SPES3-106	
—— p 401140000	SPES3-104	
—— PR1s54 1000	IRIS-HT1	

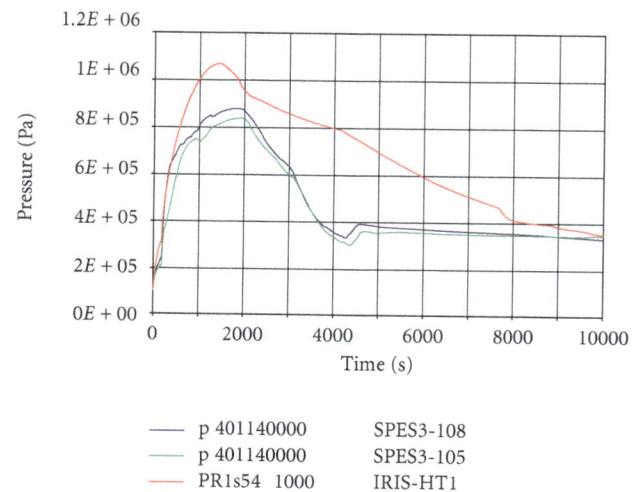

FIGURE 15: SPES3-105, 108 and IRIS-HT1 DW pressure (short term).

—— p 401140000	SPES3-108	
—— p 401140000	SPES3-105	
—— PR1s54 1000	IRIS-HT1	

with a lower peak value. No improvement was obtained by overreducing the DW volume with a scaling factor different from 1 : 100.

A theoretical case, where IRIS DW structures were directly scaled 1 : 100 in mass and surface, was investigated. A one-hundredth vertical slice of IRIS DW structures was attributed to SPES3 DW, maintaining the same thickness and material composition. Figure 15 compares two cases with equivalent heat structure masses, but different surfaces and material properties. Pressure increase was similar in IRIS and SPES3, in the early phase of the transient, when surface has a greater impact, but later energy transfer to heat structures prevailed and SPES3 pressure did not increase as expected, with all SPES3 containment structures being completely simulated against the only IRIS DW structure simulation. Moreover, gas space volume at the PSS and LGMS top was about 14% higher in SPES3 than IRIS, so limiting

the containment pressure peak. Containment pressure trend showed also differences in the depressurization phase, related to steam condensation in RC and DW due to broken loop LGMS water entering the RC (~2000 s in SPES3) through the DVI break line, containment side, and PSS injection into the DW (~3000 s in SPES3). Injection mass flows are shown in Figures 16 and 17, and they depended on different pipe pressure drops and containment pressurization.

An attempt to make closer SPES3 and IRIS PSS to DW injection mass flows was performed: size of PSS main vent pipe and extension was decreased and the results compared with a base case (SPES3-107) equivalent to SPES-99, where minor input mistakes were corrected. PSS injection results were effectively closer to IRIS with consequent slower containment depressurization (Figures 18 and 19).

The attempt was performed to see how a restriction at the PSS main vent pipe check valve affects the steam-air transfer

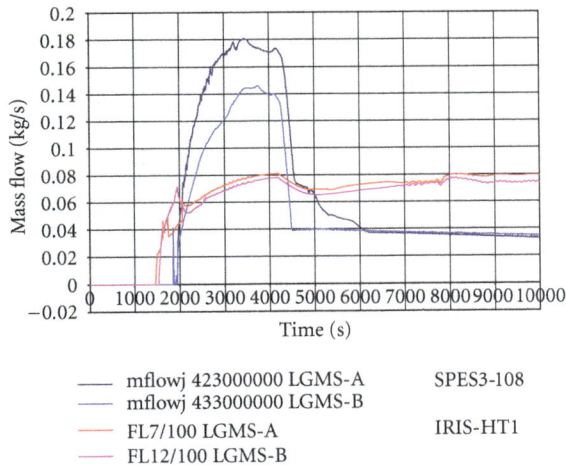

FIGURE 16: SPES3-108 and IRIS-HT1 LGMS to DVI injection mass flow (short term).

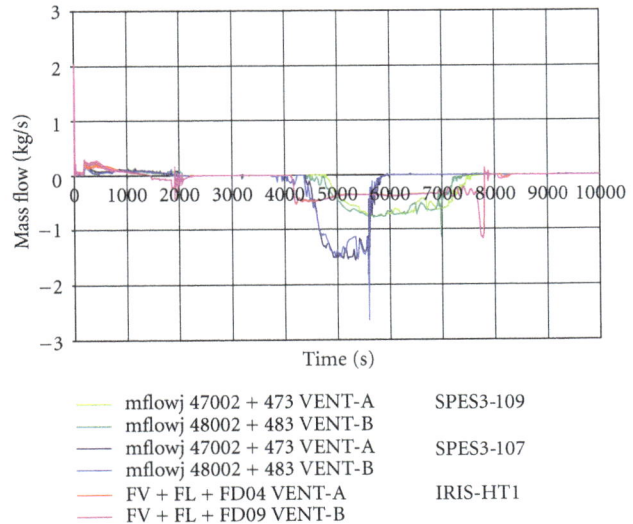

FIGURE 17: SPES3-108 and IRIS-HT1 PSS to DW injection mass flow (short term). Note: for IRIS, the total mass flow (FL + FV + FD) is obtained by the sum of phasic mass flows (liquid, gas and droplets.

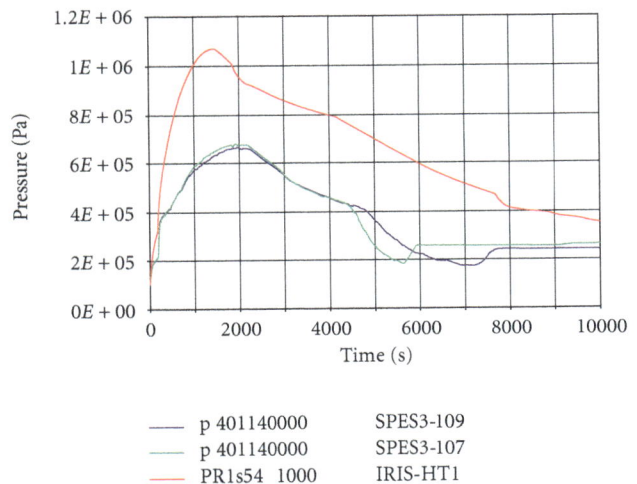

FIGURE 18: SPES3-109, 107 and IRIS-HT1 PSS to DW injection mass flow (short term).

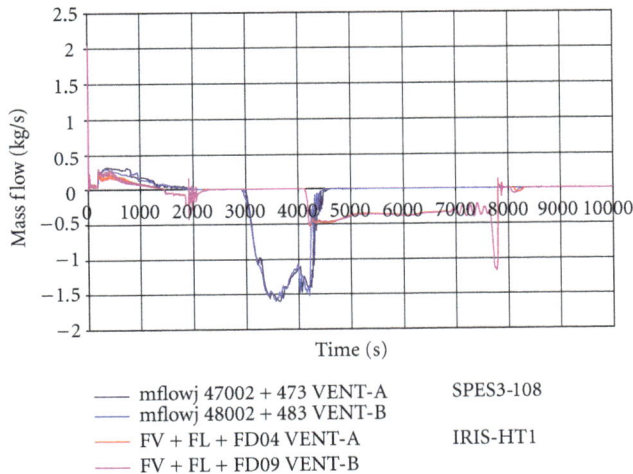

FIGURE 19: SPES3-109, 107 and IRIS-HT1 DW pressure (short term).

from DW to PSS and eventually rise DW pressure. The result was a steep but limited pressure increase (Figure 20).

The impossibility of reducing the DW thickness under 15 mm, to resist 1.5 MPa design pressure, led necessarily to an excess of mass with respect to IRIS 1 : 100 scaled mass. In SPES3-105 case, the IRIS scaled mass was distributed on the SPES3 DW surface obtaining an equivalent thickness of 10 mm. The possibility of compensating for 5 mm extra mass, by preheating the DW heat structures, was investigated. The preheating temperature of 84°C was estimated by an energy balance between the cases with 10 mm and 15 mm thickness from the specified initial temperature of 48.9°C and regime temperature of 172°C after the heat-up transient. Figure 21 compares the cases with 15 mm (SPES3-104), 10 mm (SPES3-105), and 15 mm (SPES3-111) pre-heated DW thickness. The 10 mm and 15 mm pre-heated

DWs are equivalent showing that heat structure preheating compensates for the excess of mass in SPES3. Only DW preheating is not enough to have the same IRIS pressure response.

The comparison between the cases with and without PSS heat structures allowed to quantify the phenomenon of air cooling when steam-air mixture flowed from DW to PSS. The run was interrupted by a nonconvergence error on noncondensable gas properties in the PSS, but available results allowed to evaluate the pressure gain, with respect to the theoretical case with 1 : 100 IRIS DW volume and surface scaled structures, as shown in Figure 22.

In order to reduce distortions on pressure as much as possible, related to scaling mismatching, all the SPES3 containment compartments were scaled 1 : 100 on IRIS volumes and all thicknesses were sized to resist 1.5 MPa

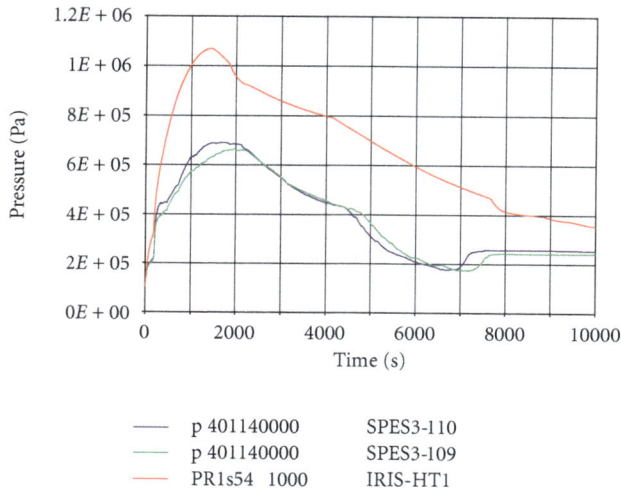

FIGURE 20: SPES3-110, 109 and IRIS-HT1 DW pressure (short term).

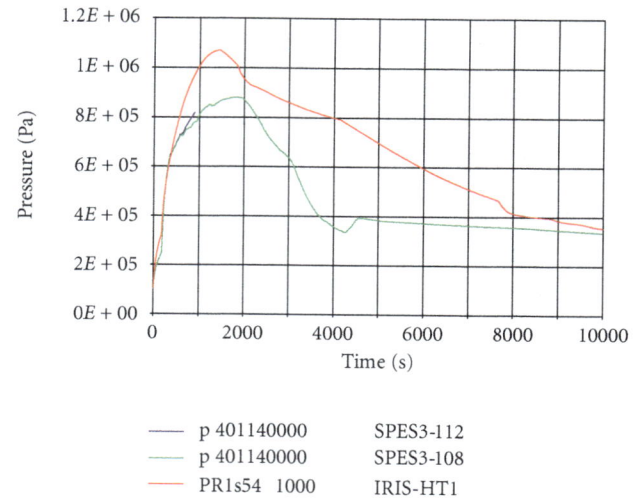

FIGURE 22: SPES3-112, 108 and IRIS-HT1 DW pressure (short term).

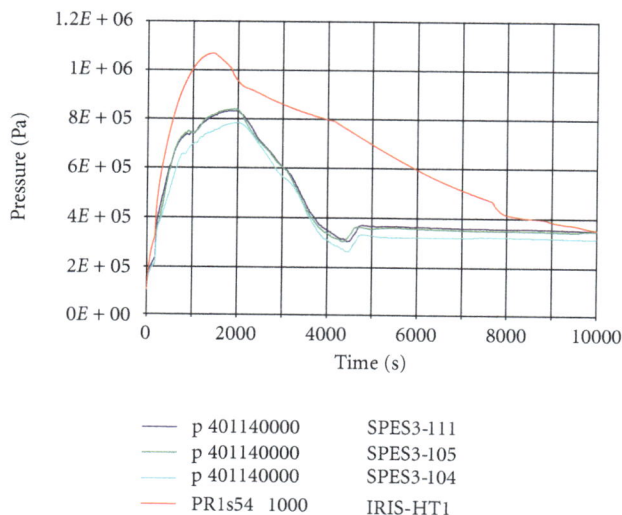

FIGURE 21: SPES3-111, 105, 104 and IRIS-HT1 DW pressure (short term).

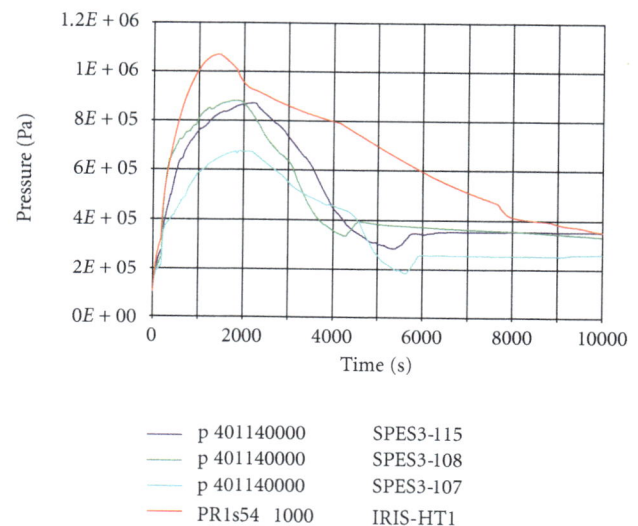

FIGURE 23: SPES3-115, 108, 107 and IRIS-HT1 DW pressure (short term).

pressure, so limiting the thermal inertia of the metal walls. In particular, the containment air zone heat structures were preheated at 84°C, while those in the liquid zone were kept at 48.9°C. Moreover, the calibrated orifices on the LGMS to DVI lines, the ADS stage I, and EHRS-C CL were resized to match IRIS mass flows. Figure 23 shows that, notwithstanding a slower DW pressure increase, that case is similar to the case with IRIS DW scaled 1:100 in mass and surface (SPES3-108), confirming that containment volume resizing and heat structure preheating are good solutions toward IRIS containment pressure response. Orifice resizing was not enough to match the IRIS LGMS to DVI and EHRS-C CL mass flows, while it was correct to scale ADS stage I mass flow (Figure 24), even if a stronger water entrainment was evidenced in IRIS at the second flow peak.

Further calibrations of the injection line orifices were performed that evidenced contact condensation stronger in SPES3 (RELAP5) than in IRIS (GOTHIC).

A case was run where IRIS containment heat structures were reproduced on SPES3 with the DW 1:100 scaled in mass and surface and all other tanks with 1 mm thick walls to avoid code convergence errors in case of complete heat structure removal. Figure 25 shows very similar SPES3 and IRIS containment pressure rising phase and a pressure peak only 0.1 MPa lower in SPES3 than in IRIS, demonstrating the importance of a correct simulation of the heat structures.

The containment piping orifice, sized to match IRIS injection mass flows, allowed a direct comparison between the plants, but it did not meet the piping pressure drop scaling criteria. It allowed to understand two important

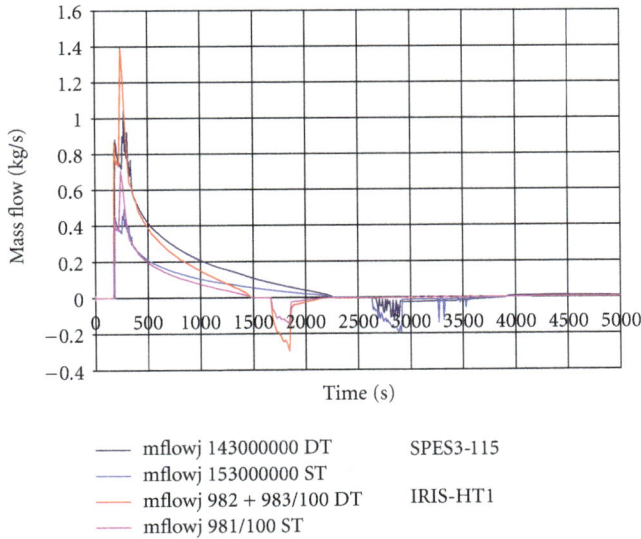

FIGURE 24: SPES3-115 and IRIS-HT1 ADS stage I ST and DT mass flow (short term).

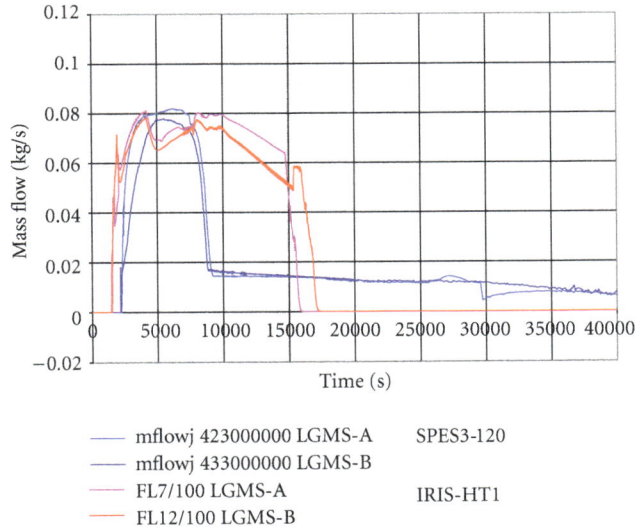

FIGURE 26: SPES3-120 and IRIS-HT1 LGMS injection mass flow.

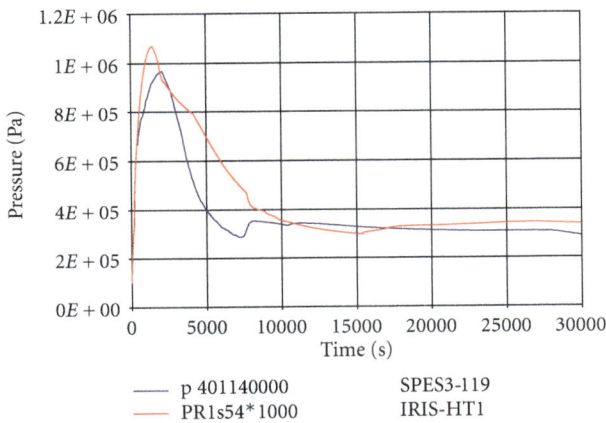

FIGURE 25: SPES3-119 and IRIS-HT1 DW pressure (short term).

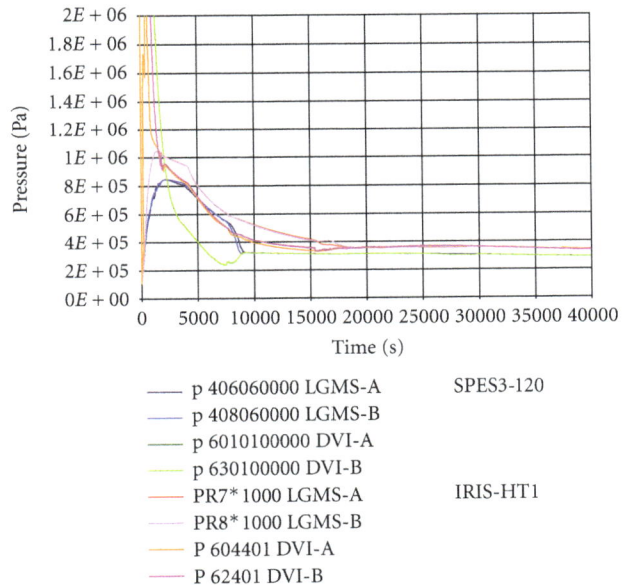

FIGURE 27: SPES3-120 and IRIS-HT1 LGMS and DVI pressure.

differences related to the injection of LGMS into the DVI lines and of PSS into the DW. As shown in Figures 26 and 27, IRIS LGMS injection into the DVI line is always driven by differential pressure between LGMS and DVI line (until about 18700 s), instead in SPES3, such differential pressure extinguishes earlier (around 9200 s) and later LGMS injection is driven only by gravity with a large mass flow decrease. The reason for such early pressure equalization in SPES3 is related to the PSS injection stop, vent pipe emptying, and gas flow from PSS to DW. That phenomenon did not occur in IRIS, where the vent pipes did not empty avoiding air transfer from PSS to DW, keeping the PSS pressurized with respect to DW and DVI (Figure 28).

The sensitivity cases on containment tank geometry, heat structures, and piping pressure drops led to reviewing both the IRIS and SPES3 models in IRIS-HT5g and SPES3-124. All IRIS containment heat structures were simulated and SPES3 piping geometry was adjusted to match IRIS pressure drops.

Figure 29 compares SPES3 and IRIS containment pressures, showing a good qualitative and quantitative agreement. The LGMS to DVI and PSS to DW injection mass flows are shown in Figures 30 and 31. With the same simulated pressure drops in the piping, different values of mass flow are evidenced due to greater differential pressures in SPES3 between LGMS and DVI and PSS and DW. Various reasons could explain these differences and the most likely is the different code simulation of contact condensation with the consequent different pressurization of containment compartments. Remodelling of IRIS RWST led to similar RWST water temperatures, but greater exchanged power in SPES3 caused faster heat-up (Figure 32).

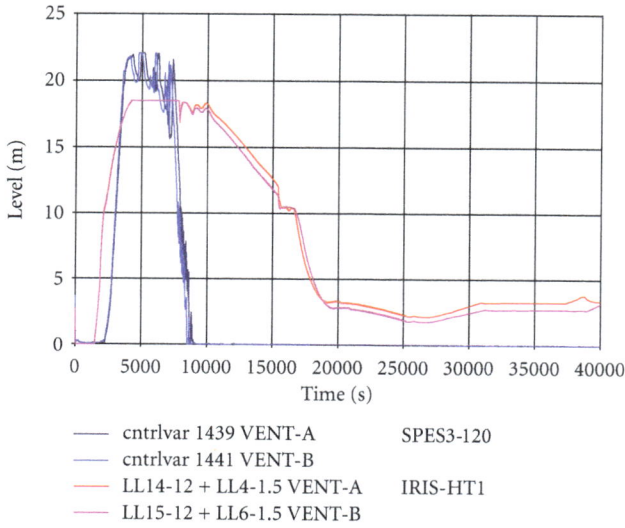

FIGURE 28: SPES3-120 and IRIS-HT1 PSS vent pipe level.

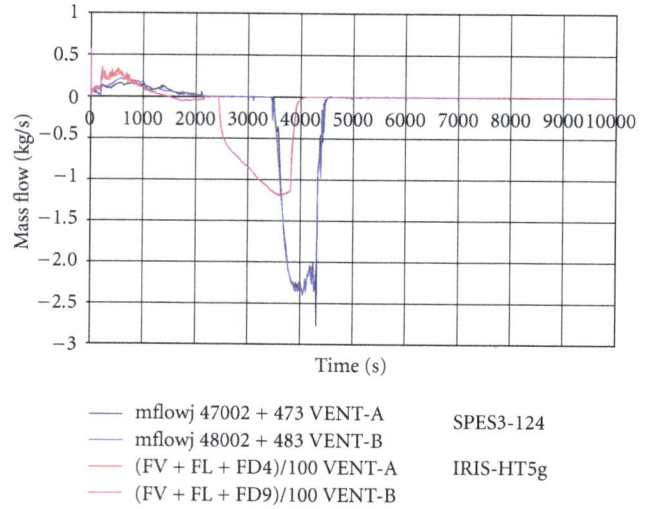

FIGURE 29: SPES3-124 and IRIS-HT5g DW pressure.

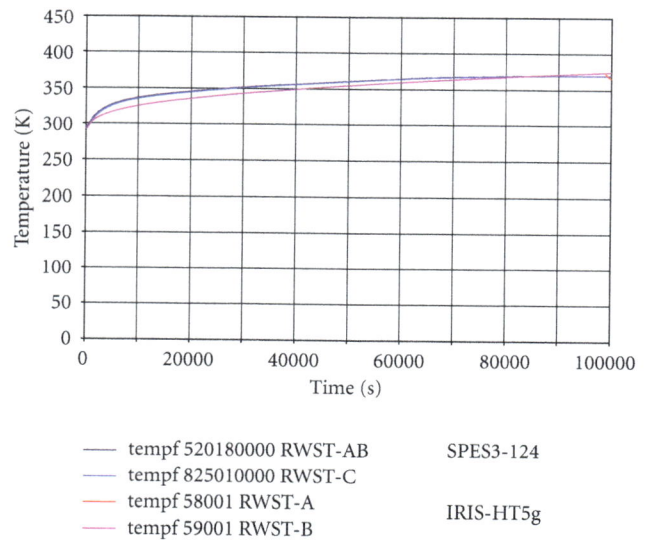

FIGURE 30: SPES3-124 and IRIS-HT5g LGMS to DVI mass flow (short term).

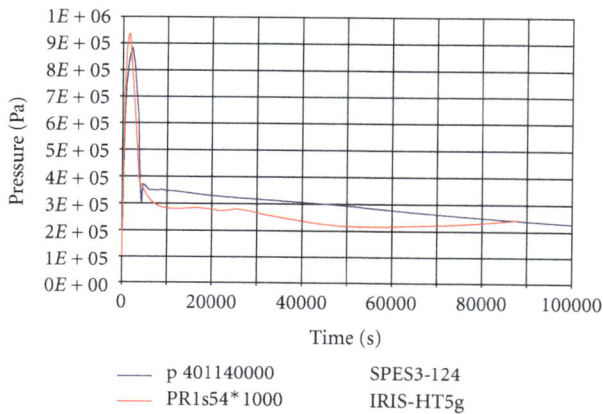

FIGURE 31: SPES3-124 and IRIS-HT5g PSS to DW mass flow (short term).

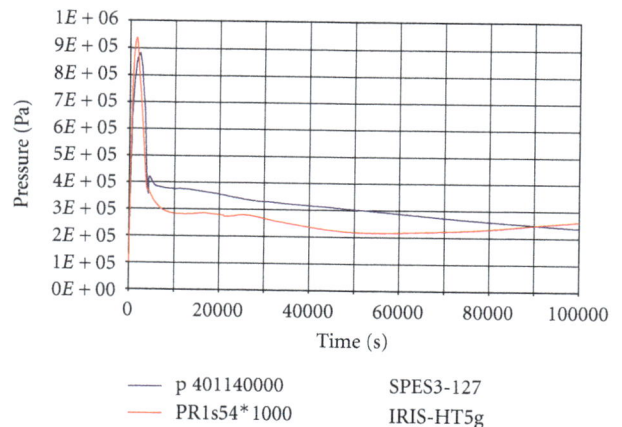

FIGURE 32: SPES3-124 and IRIS-HT5g RWST temperature.

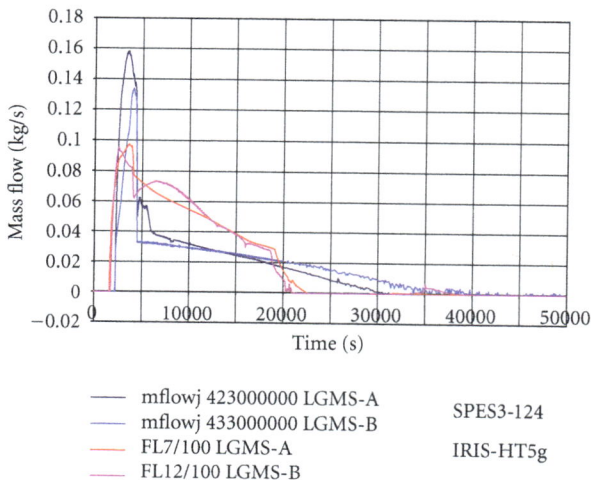

FIGURE 33: SPES3-127 and IRIS-HT5g DW pressure.

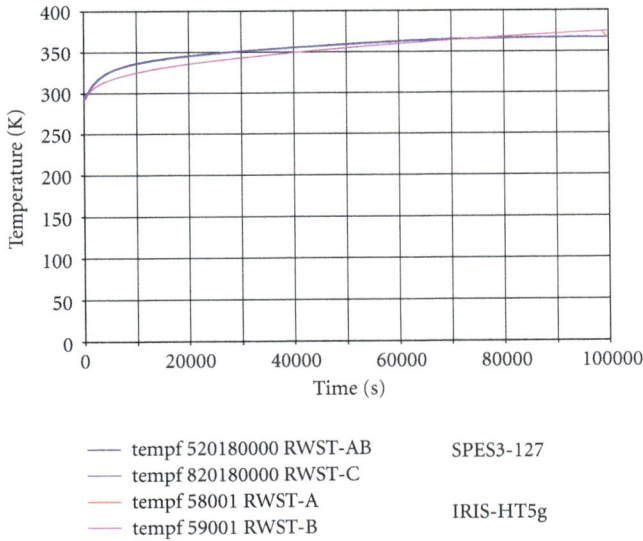

FIGURE 34: SPES3-127 and IRIS-HT5g RWST temperature.

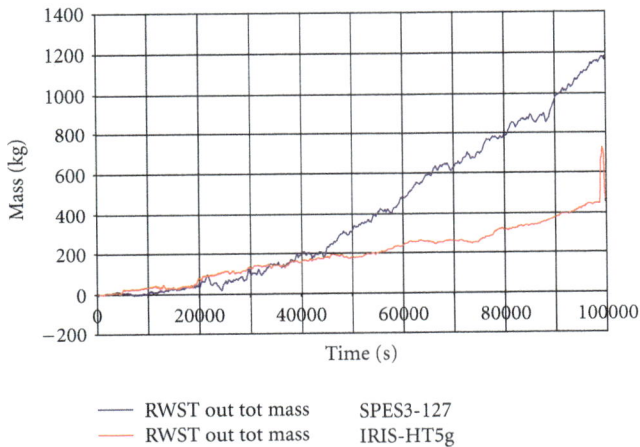

FIGURE 35: SPES3-127 and IRIS-HT5g RWST top integral mass flow.

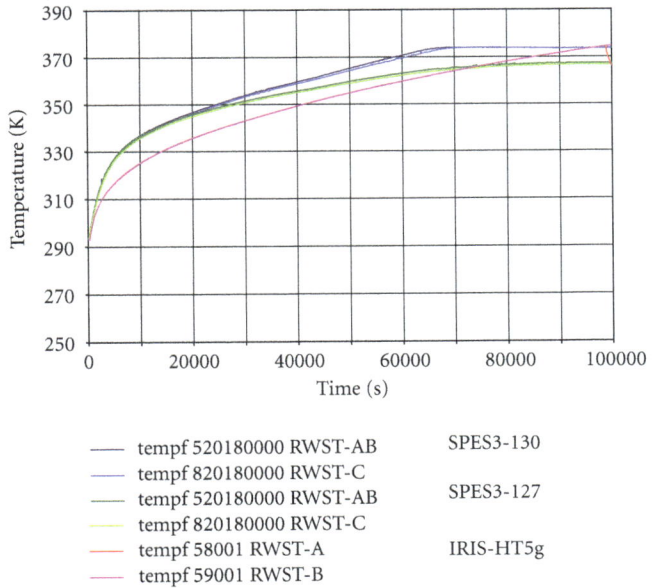

FIGURE 36: SPES3-130, 127 and IRIS-HT5g RWST temperature.

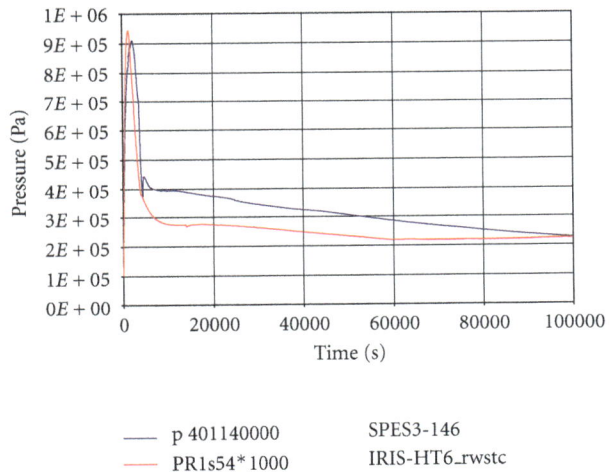

FIGURE 37: SPES3-146 and IRIS-HT6_rwstc DW pressure.

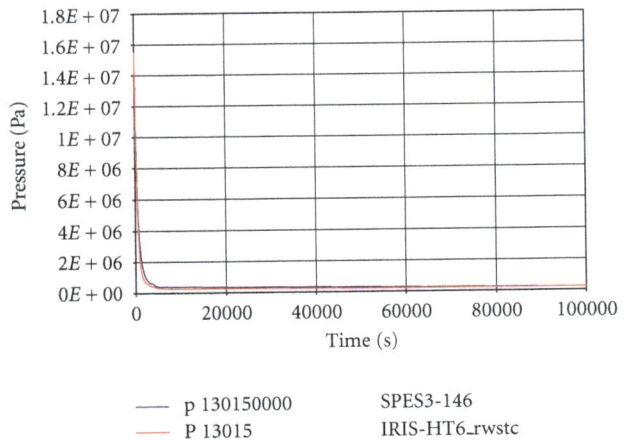

FIGURE 38: SPES3-146 and IRIS-HT6_rwstc PRZ pressure.

The SPES3 PSS bottom remodelling was the further attempt to investigate PSS vent pipe emptying at the end of PSS injection into the DW. No important differences were observed. The SPES3-127 results were used to investigate the EHRS heat transfer to RWST, considered the cause of the different containment pressure trend in the long term, where IRIS increased after 50000 s while SPES3 continued to decrease (Figure 33). RWST temperature in SPES3 did not reach saturation, but established at lower values (Figure 34), due to the direct connection, in the model, of RWST top with the atmosphere control volume, and mass lost through RWST top caused by water solution in dry air with energy removal and temperature limitation (Figure 35).

That phenomenon led to a further modification of the model with the introduction of a discharge pipe at RWST top, limiting the contact surface with air and solving the problem of RWST water temperature that, finally, could

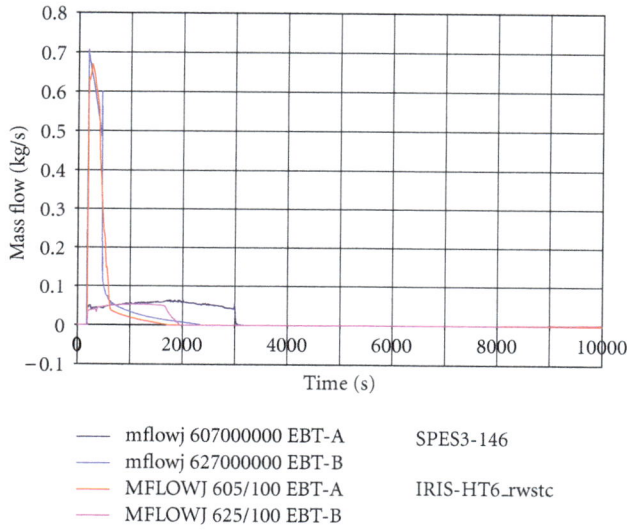

FIGURE 39: SPES3-146 and IRIS-HT6_rwstc EBT injection mass flow (short term).

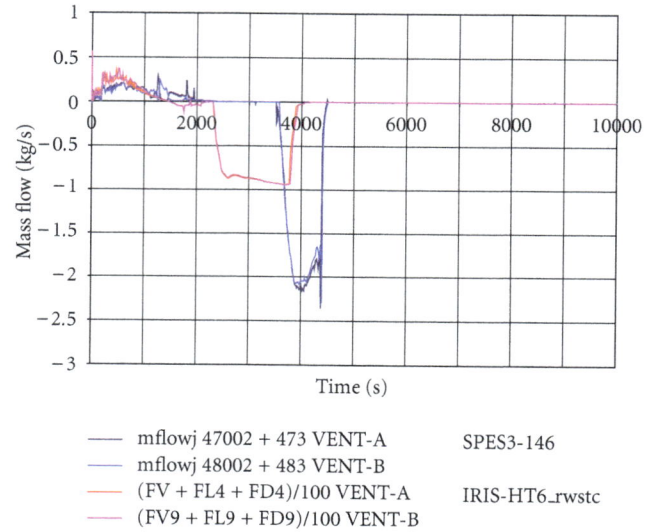

FIGURE 41: SPES3-146 and IRIS-HT6_rwstc PSS to DW injection mass flow (short term).

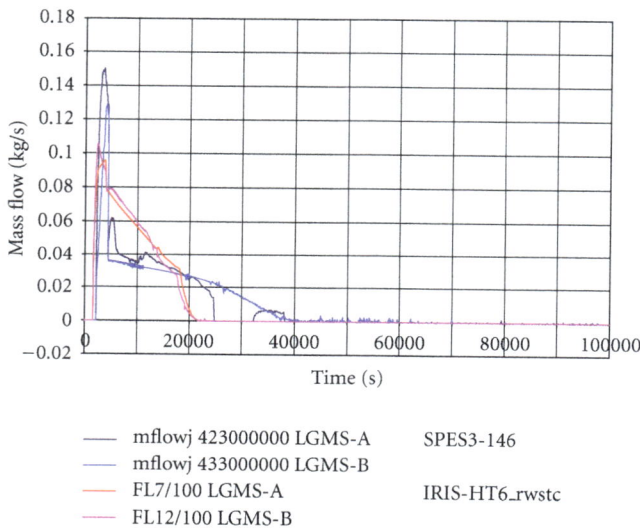

FIGURE 40: SPES3-146 and IRIS-HT6_rwstc LGMS to DVI injection mass flow.

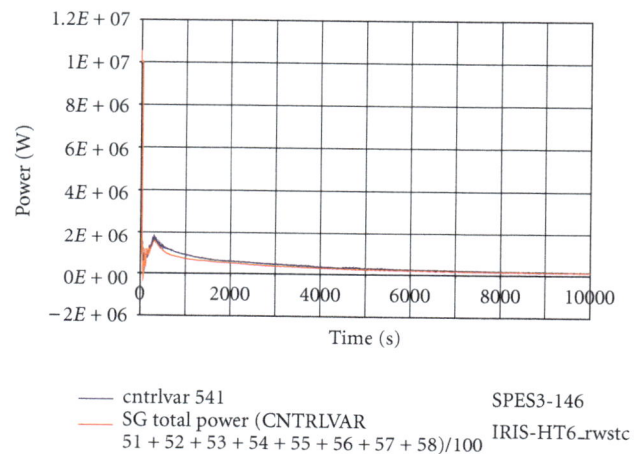

FIGURE 42: SPES3-146 and IRIS-HT6_rwstc SG power (short term).

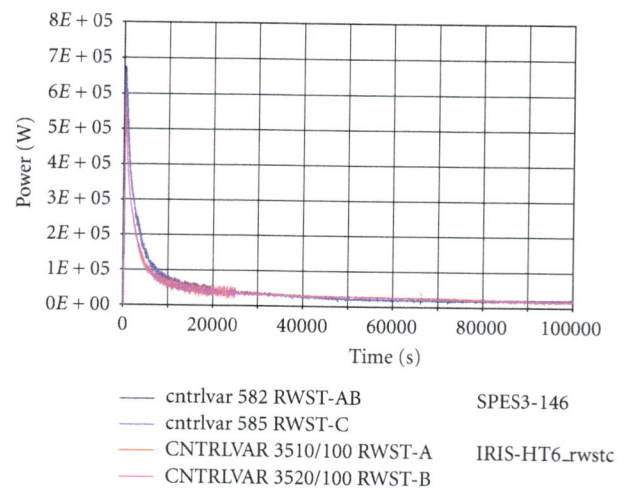

FIGURE 43: SPES3-146 and IRIS-HT6_rwstc RWST power.

reach saturation (Figure 36). Faster water heat-up in SPES3 showed that EHRS energy transfer to the RWST is greater than IRIS. That led to a series of sensitivity cases on a stand-alone model of EHRS-RWST that led to investigating the differences between the models and finding a common modelling approach based on experimental data on an in-pool heat exchanger and literature values of the heat transfer coefficients [13]. The method provided proper multiplying factors for the HTC to be applied to tube heat structures, condensing and boiling side, in the form of fouling factors [14] and a criterion to set the area of the pool slice containing the heat exchanger. IRIS EHRS-RWST was modified accordingly in the IRIS-HT6_rwstc case. A stand-alone model was also utilized to calibrate pressure drops in

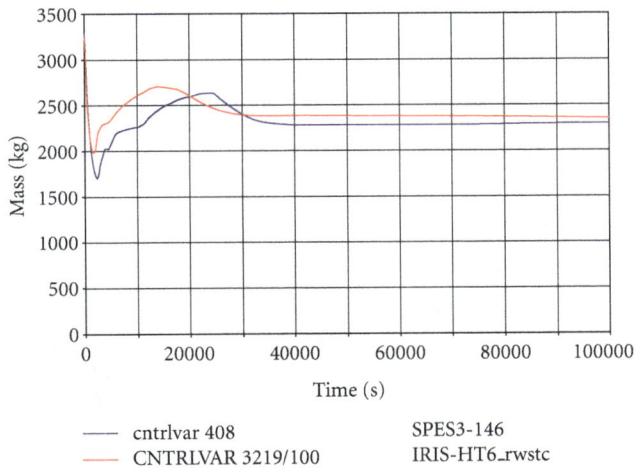

FIGURE 44: SPES3-146 and IRIS-HT6_rwstc RV mass.

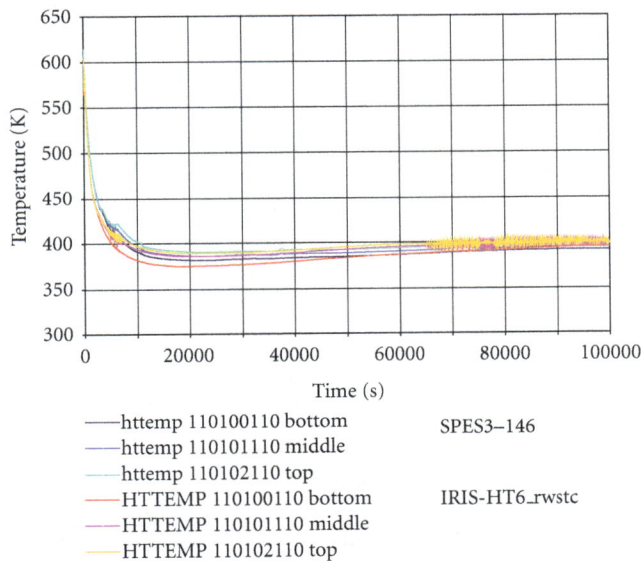

FIGURE 45: SPES3-146 and IRIS-HT6_rwstc core heater rod outer surface temperature.

EHRS hot legs and cold legs to properly reproduce the IRIS loops with adjustment to calibrated orifices. Moreover, the need of thermally insulating 4% of SPES3 HX heat transfer surface was evidenced to compensate for AISI 304 thermal conductivity greater than IRIS Inconel 600.

The SPES3-146 case included all design and model updates previously described and was considered the base case to compare to IRIS-HT6_rwstc. The main quantities of the transient and those that were objective of the SPES3 facility model and design optimization are shown in Figures 37, 38, 39, 40, 41, 42, 43, 44, and 45.

The last IRIS case was successively run to correct some differences, found in RELAP5 and GOTHIC models, about the end elevations of LGMS and ADS stage I lines. Small differences, compared to IRIS-HT6_rwstc results, were found in LGMS flows as well as some changes in ADS stage I

flow, after initial discharge. Such differences do not affect or modify the results of the analysis previously described.

The final FSA is planned to be performed on the SPES3-146 and IRIS-HT6_rwstc1a results.

4. Conclusions

The design of the SPES3 facility was finalized thanks to an iterative calculation-design feedback process that allowed to verify the adequacy of containment pressure and reactor vessel mass inventory simulation, objectives of the SBLOCA PIRT for the IRIS reactor [7].

Since the early simulations, efficiency of IRIS safety systems was demonstrated in coping with SBLOCAs. The comparison with the SPES3 results and the early application of the FSA allowed to identify the main causes of discrepancy between the results and to put in evidence specific phenomena particularly affected by simulation choices. The containment heat structures, the heat transfer from EHRS to RWST, and piping pressure drops were found to be the most affecting parameters in matching the IRIS results. The review of the SPES3 design, in accordance to the above-mentioned parameter optimization, led to demonstrating that the PIRT identified FoMs are satisfied and that the residual discrepancies can be considered conservative: SPES3 RPV mass lower than IRIS mass and SPES3 heater rod temperatures higher than IRIS ones.

Besides the SPES3 design review, the main outcome of this work is the availability of a set of data suitable for the final FSA application, in progress at the moment, and the quantification of SPES3 facility distortions in IRIS simulation.

The SPES3 facility is under construction at SIET laboratories.

Nomenclature

ADS:	Automatic depressurization system
CIRTEN:	Consorzio Interuniversitario per la Ricerca Tecnologica Nucleare (University Consortium for Nuclear Technologic Research)
CL:	Cold Leg
cntrlvar:	(CNTRLVAR) control variable (RELAP5 variable)
CRDM:	Control rod drive mechanism
DBA:	Design basis accident
DC:	Downcomer
DEG:	Double ended guillotine
DT:	Double train
DVI:	Direct vessel injection
DW:	Dry well
EBT:	Emergency boration tank
EHRS:	Emergency heat removal system (EHRS-A, B, C for loops A, B, C)
ENEA:	Agenzia Nazionale per le Nuove Tecnologie, l'Energia e lo Sviluppo Economico Sostenibile (Italian National Agency for New Technologies, Energy and Sustainable Economic Development)

FD: Droplet mass flow rate (GOTHIC variable)
FER: Fakultet Elektrotehnike i Računarstva (Faculty of Electric Engineering and Computing) FL Feed Line
FL: Liquid mass flow rate (GOTHIC variable in the graphs)
FoM: Figure of merit
FSA: Fractional scaling analysis
FV: Gas mass flow rate (GOTHIC variable)
GNEP: Global nuclear energy partnership
GOTHIC: Generation of thermal-hydraulic information for containments
HTC: Heat transfer coefficient
httemp: (HTEMP) heat structure temperature (RELAP5 variable)
HX: Heat exchanger
IRIS: International reactor innovative and secure
ITF: Integral test facility
LGMS: Long-term gravity make-up system
LL: Liquid level (GOTHIC variable)
LOCA: Loss of coolant accident
LP: Lower plenum
LWR: Light water reactor
mflowj: (MFLOWJ) mass flow rate (RELAP5 variable)
NPP: Nuclear power plant
p: (P) Pressure (RELAP5 variable)
PIRT: Phenomena identification and ranking table
PR: Pressure (GOTHIC variable)
PRZ: Pressurizer
PSS: Pressure suppression system
QT: Quench tank
RC: Reactor cavity
RELAP: REactor loss of coolant analysis program
RPV: Reactor pressure vessel
RV: Reactor vessel
RWST: Refueling water storage tank
R&D: Research and development
SB: Small break
SET: Separate effect tests
SIET: Società Informazioni Esperienze Termoidrauliche (company for information and experiences on thermal-hydraulics)
SG: Steam generator
SL: Steam line
SMR: Small and medium-sized reactor
SPES: Simulatore pressurizzato per esperienze di sicurezza (pressurized simulator for safety tests)
ST: Single train
tempf: (TEMPF) liquid temperature (RELAP5 variable)
WEC: Westinghouse Electric Company LLC.

References

[1] M. D. Carelli, L. E. Conway, L. Oriani et al., "The design and safety features of the IRIS reactor," *Nuclear Engineering and Design*, vol. 230, no. 1–3, pp. 151–167, 2004.

[2] M. D. Carelli, B. Petrović, L. E. Conway et al., "IRIS design overview and status update," in *Proceedings of the 13th International Conference on Nuclear Engineering (ICONE13-50442 '05)*, Beijing, China, May 2005.

[3] B. Petrović, M. D. Carelli, and N. Cavlina, "IRIS—international reactor innovative and secure: progress in development, licensing and deployment activities," in *Proceedings of the 6th International Conference on Nuclear Option in Countries with Small and Medium Electricity Grids*, Dubrovnik, Croatia, May 2006.

[4] M. D. Carelli, B. Petrović, M. Dzodzo et al., "SPES-3 experimental facility design for IRIS reactor integral testing," in *Proceedings of the European Nuclear Conference (ENC '07)*, Brussels, Belgium, September 2007.

[5] M. Carelli, L. Conway, M. Dzodzo et al., "The SPES3 experimental facility design for the IRIS Reactor simulation," *Science and Technology of Nuclear Installations*, vol. 2009, Article ID 579430, 12 pages, 2009.

[6] R. Ferri, A. Achilli, C. Congiu et al., "SPES3 facility and IRIS reactor numerical simulations for the SPES3 final design," in *Proceedings of the European Nuclear Conference (ENC '10)*, Barcelona, Spain, May June 2010.

[7] T. K. Larson, F. J. Moody, G. E. Wilson et al., "Iris small break loca phenomena identification and ranking table (PIRT)," *Nuclear Engineering and Design*, vol. 237, no. 6, pp. 618–626, 2007.

[8] IAEA-TECDOC 1536, "Status of small reactor designs without on-site refuelling (IAEA '07)," 2007.

[9] R. Ferri and C. Congiu, SPES3-IRIS facility nodalization for RELAP5 Mod.3.3 code and steady state qualification. SIET 01 423 RT 08 Rev.0, ENEA FPN-P9LU-017, 2009.

[10] R. Ferri and C. Congiu, SPES3-IRIS facility RELAP5 base case transient analyses for design support. SIET 01 489 RT 09 Rev.0., ENEA FPN-P9LU-035, 2009.

[11] R. Ferri and C. Congiu, SPES3-IRIS facility RELAP5 sensitivity analyses of the Lower Break transient for design support. SIET 01 499 RT 09 Rev.0., FPN- P9LU-040, 2009.

[12] R. Ferri, SPES3-IRIS facility RELAP5 sensitivity analyses on the containment system for design review. SIET 01 526 RT 09 Rev.0., ENEA NNFISS-LP2-017, 2010.

[13] R. Ferri and P. Meloni, Approach for a correct simulation of the SPES3-IRIS Emergency Heat Removal System with the RELAP5/MOD3 Code. SIET 01 745 RT 11 Rev.0. Piacenza, Italy, 2011.

[14] RELAP5 code manual. NUREG/CR-5535/Rev.1 Idaho National Engineering Laboratory (USA), 2001.

12

Countercurrent Air-Water Flow in a Scale-Down Model of a Pressurizer Surge Line

Takashi Futatsugi,[1] Chihiro Yanagi,[2] Michio Murase,[2] Shigeo Hosokawa,[1] and Akio Tomiyama[1]

[1] *Department of Mechanical Engineering, Faculty of Engineering, Kobe University, 1-1 Rokkodai, Nada, Hyogo Kobe, 657-8501, Japan*
[2] *Institute of Nuclear Safety System, Inc. (INSS), 64 Sata, Mihama-cho, Mikata-gun, Fukui 919-1205, Japan*

Correspondence should be addressed to Akio Tomiyama, tomiyama@mech.kobe-u.ac.jp

Academic Editor: Thomas Hoehne

Steam generated in a reactor core and water condensed in a pressurizer form a countercurrent flow in a surge line between a hot leg and the pressurizer during reflux cooling. Characteristics of countercurrent flow limitation (CCFL) in a 1/10-scale model of the surge line were measured using air and water at atmospheric pressure and room temperature. The experimental results show that CCFL takes place at three different locations, that is, at the upper junction, in the surge line, and at the lower junction, and its characteristics are governed by the most dominating flow limitation among the three. Effects of inclination angle and elbows of the surge line on CCFL characteristics were also investigated experimentally. The effects of inclination angle on CCFL depend on the flow direction, that is, the effect is large for the nearly horizontal flow and small for the vertical flow at the upper junction. The presence of elbows increases the flow limitation in the surge line, whereas the flow limitations at the upper and lower junctions do not depend on the presence of elbows.

1. Introduction

The mid-loop operation is to be conducted during plant refueling and maintenance of a PWR (Pressurized Water Reactor). In this operation, the reactor coolant level is kept around the primary loop center, and decay heat is removed by RHR (Residual Heat Removal) systems. If the loss of cooling systems such as RHR and/or other cooling systems takes place, cooling water in the reactor core may be heated up to boil and the top of the fuel assembly can be exposed to the air. In such an event, reflux cooling by the steam generators (SG) is regarded as one of the possible and effective core cooling methods. The reflux cooling is a way of core cooling by making use of water condensed in SGs. The steam generated in the reactor core and water condensed in the SG form a countercurrent flow in the hot leg. The authors therefore measured CCFL (Countercurrent Flow Limitation) characteristics in a scale-down model of a hot leg using air and water [1] and reported that CCFL can be accurately evaluated based on a one dimensional momentum balance for air-water two-phase flow [2]. In addition to this CCFL, the steam generated in the reactor core and water condensed in the pressurizer due to heat transfer to the vessel wall may also form a countercurrent flow in a surge line which connects the hot leg and the pressurizer. The ROSA-IV/LSTF (Rig-of-Safety-Assessment No. 4/Large Scale Test Facility) experiment [3], which simulated the loss of RHR systems during mid-loop operation, reported that water actually accumulated in the pressurizer due to CCFL in the surge line. When the core coolant moves to the primary coolant system and remains there, the reactor core water level decreases. Thus, characteristics of CCFL in the surge line must be well understood for safety evaluation of the mid-loop operation.

Takeuchi et al. [4] calculated CCFL characteristics for a slightly inclined surge line of an AP600 using the momentum equations for steam and water. They reported that (1) CCFL in a vertical pipe is more dominant than that in a slightly inclined pipe, (2) the horizontal elbow increases the falling water volume, and (3) CCFL in the vertical pipe is the most dominant among various CCFLs taking place at different locations in the surge line. Although their prediction overestimated the falling water volume compared with the small break LOCA (Loss of Coolant Accident) data conducted at the AP600-scale test facility (APEX) [5, 6], there are no experimental data for validating their predictions. The surge line consists of a vertical pipe, a vertical elbow and an inclined pipe with several elbows. The flow in the surge line is very complicated due to its complex geometry, and therefore, it is difficult to apply the data and knowledge of CCFL obtained in a simple geometry such as straight pipes and ducts to CCFL in the surge line.

In this study, we carried out experiments using air and water in a 1/10-scale model and measured CCFL characteristics in the surge line. Effects of inclination angle and elbows in the surge line on CCFL characteristics were also investigated.

2. Experimental Setup

Figure 1 shows the experimental setup. It consists of the lower tank corresponding to a reactor vessel, the surge line, the upper tank simulating a pressurizer, and the air and water supply systems. The surge line is made of acrylic resin for the observation of flow pattern in the pipe. The internal diameter is 30 mm. The geometry of the surge line is shown in Figure 2. Air is supplied through the sidewall of the lower tank. Water is supplied through the bottom face of the upper tank. They form a countercurrent flow in the surge line. The elbow is made of two acrylic blocks with semicircular grooves to keep the channel cross-section circle. At a constant flow rate Q_{Lin} of water supplied to the upper tank, the flow rate Q_L of water falling into the lower tank was measured at each gas flow rate Q_G to obtain a relationship between Q_L and Q_G. The Q_L was measured not only by increasing Q_G but also by decreasing Q_G to check a possibility of hysteresis in CCFL. The experimental ranges were $J_{Lin}(= 4Q_{Lin}/\pi D^2) = 0.02$–$0.12$ m/s and $J_G(= 4Q_G/\pi D^2) = 0$–5.5 m/s. CCFL data were plotted by using the dimensionless gas and liquid volumetric fluxes, J_G^* and J_L^*, given by [6]

$$J_k^* = J_k \left\{ \frac{\rho_k}{gD(\rho_L - \rho_G)} \right\}^{1/2}, \quad (k = G, L), \quad (1)$$

where J is the superficial velocity, ρ the density, g the acceleration of gravity, and D the pipe diameter. The subscripts G and L denote the gas and liquid phases, respectively. The inclination angle θ of the surge line was changed from 0.0 to 5.0 deg. (0.0, 0.6, 1.0, 2.0 and 5.0 deg.) to investigate the effects of θ on CCFL characteristics. We also measured CCFL characteristics by replacing the surge line with the straight pipe shown in Figure 3 to examine effects of elbows.

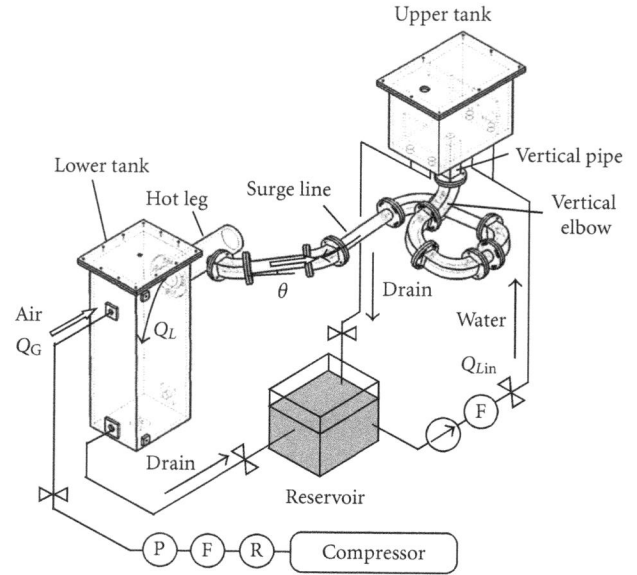

(F: flow meter, P: pressure gage, R: regulator)

FIGURE 1: Experimental setup (Surge line).

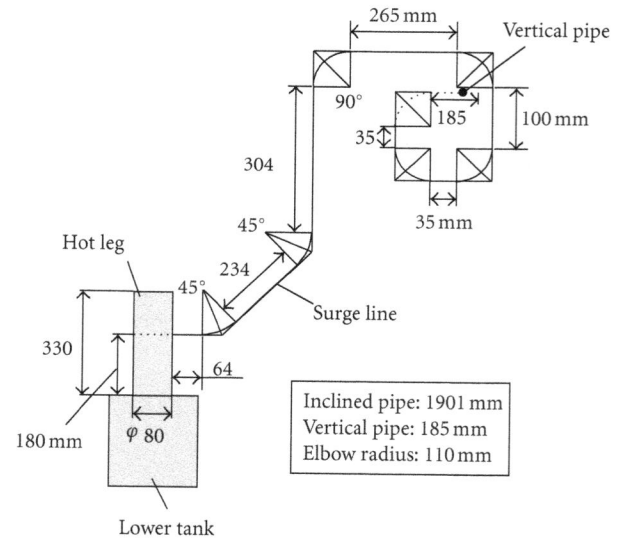

FIGURE 2: Geometry of surge line.

3. Results and Discussion

3.1. Classification of CCFL. Depending on the inclination angle θ and J_G, CCFL took place at three different locations, that is, at the upper junction, in the surge line, and at the lower junction as shown in Figure 4. Hereafter, CCFL at the upper junction between the surge line and the upper tank, in the surge line and that at the lower junction between the surge line and the hot leg will be referred to as CCFL-U, CCFL-S, and CCFL-L, respectively.

In CCFL-U, the flow limitation occurs only at the upper junction of the surge line as shown in Figure 5(a), and

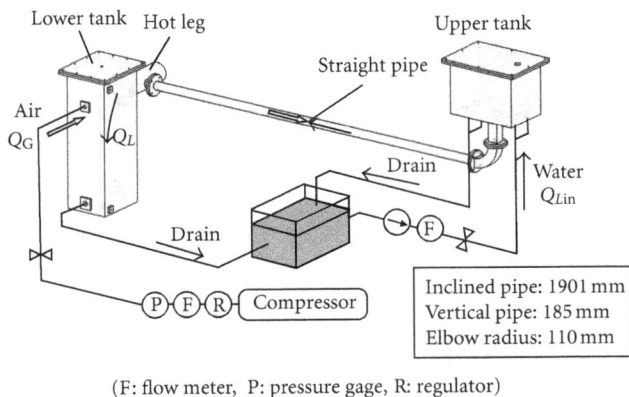

(F: flow meter, P: pressure gage, R: regulator)

FIGURE 3: Experimental apparatus with straight pipe.

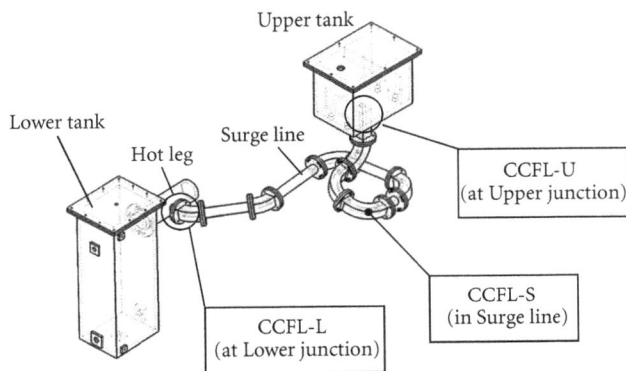

FIGURE 4: Flow limitation locations and CCFL classification.

therefore, water in the surge line and at the lower junction smoothly flows toward the lower tank. To the contrary, the flow limitation occurs not only at the upper junction but also in the surge line in CCFL-S as shown in Figure 5(b). Waves are generated in the surge line and move toward the upper junction. In CCFL-L, the flow limitation occurs at the lower junction as well as at the upper junction. Water is accumulated near the lower junction to form large liquid slugs and periodically flows back toward the upper junction as shown in Figure 5(c).

3.2. CCFL Characteristics at Reference Condition.

Figure 6 shows CCFL characteristics at J_{Lin} = 0.07 m/s and θ = 0.6 deg. As J_G^* increases from zero, flow limitation takes place when J_G^* reaches a certain critical value (point A in Figure 6). At point A, waves form in the surge line and move toward the upper junction. At the same time, J_L^* suddenly decreases to the flooding point (point B in Figure 6), that is, CCFL-S occurs. Further increase in J_G^* reduces J_L^*, and all the water returns to the upper tank at the flow reversal point (point C in Figure 6). When J_G^* is decreased from the flow reversal point C, J_L^* gradually increases as shown in Figure 6. The minimum J_G^* observed in the decreasing process is smaller than J_G^* at the flooding point B. Thus, hysteresis exists in the CCFL characteristics. The hysteresis is caused by the difference in the presence of initial waves on the

gas-liquid interface. There is, however, no difference in the dependence of J_L^* on J_G^* between the processes of increasing and decreasing J_G^*.

3.3. Effects of θ.

Figure 7 shows CCFL characteristics at various θ. At θ = 0.0 and 0.6 deg., only CCFL-S takes place at any values of J_G. At θ = 1.0 deg., CCFL disappears in the surge line, and therefore, it is classified as CCFL-U when J_G is low, whereas CCFL-S occurs at high J_G. This disappearance of CCFL in the surge line is due to the enhancement of water drainage by increasing θ. At θ = 2.0 and 5.0 deg., CCFL-L appears instead of CCFL-S at high J_G. This indicates that the flow limitation at the lower junction becomes dominant because CCFL in the surge line is mitigated by the increase in θ. At low J_G, CCFL-U occurs not only for θ = 2.0 and 5.0 deg. but also for θ = 1.0 deg. These results show that type of CCFL depends on θ and J_G^*, and the dependence of the relation between J_G^* and J_L^* on θ is different among CCFL-S, CCFL-U, and CCFL-L.

Figure 8 shows characteristics of CCFL-S at various θ. A small change in θ causes a large change in the falling water flow rate, that is, the dependence of CCFL-S on the inclination angle is very large. The increase in θ results in the mitigation of flow limitation in the surge line due to the enhancement of water drainage. Hence, CCFL-S occurs only at low θ.

Figure 9 shows characteristics of CCFL-L at various θ. CCFL-L is also affected by θ, and the flow limitation becomes weaker as θ increases. The dependence of CCFL-L on the inclination angle, however, is weaker than that of CCFL-S. Since the holdup at the lower junction depends not only on the water velocity along the surge line but also on the velocity of water falling into the lower tank, the weak dependence of the falling water velocity on θ might be a cause of the small dependency of CCFL-L on θ.

Figure 10 shows characteristics of CCFL-U at various θ. CCFL-U is the limitation at the upper junction between the upper tank and the vertical pipe, and the gravity force acting on the water along the surge line (the vertical pipe) is $\rho g \cos \theta$. Hence, CCFL-U has a very weak dependence on θ at small θ as shown in Figure 10.

These experimental results confirm that the effect of θ on CCFL depends on the flow direction, that is, the effect is large for the nearly horizontal flow in the surge line, small for the vertical flow at the upper junction, and intermediate for the flow at the lower junction at which the flow changes its direction from horizontal to vertical directions. The CCFL in the surge line is determined by the most strong flow limitation among CCFL-S, CCFL-U and CCFL-L.

3.4. Onset of Flooding.

The gas volumetric flux at the onset of flooding is important information when designing surge lines. Figure 11 shows the onset of flooding measured by increasing the gas volumetric flux. Flooding-S, Flooding-U, and Flooding-L in Figure 11 represent that the flooding takes place in the surge line, at the upper junction and at the lower

Upper junction

Inclined pipe

(a) $\theta = 5.0\,\text{deg}$, $J_{Lin} = 0.07\,\text{m/s}$, $J_G = 3.2\,\text{m/s}$

Upper junction

Inclined pipe

(b) $\theta = 0.6\,\text{deg}$, $J_{Lin} = 0.07\,\text{m/s}$, $J_G = 3.8\,\text{m/s}$

Upper junction

Inclined pipe

(c) $\theta = 5.0\,\text{deg}$, $J_{Lin} = 0.07\,\text{m/s}$, $J_G = 5.5\,\text{m/s}$

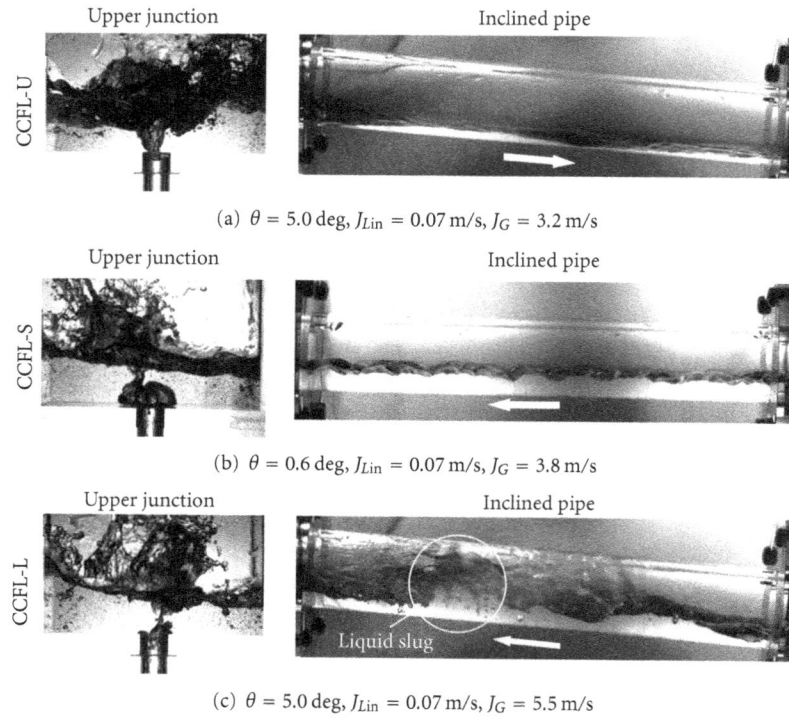

FIGURE 5: Typical flow pattern at CCFL conditions.

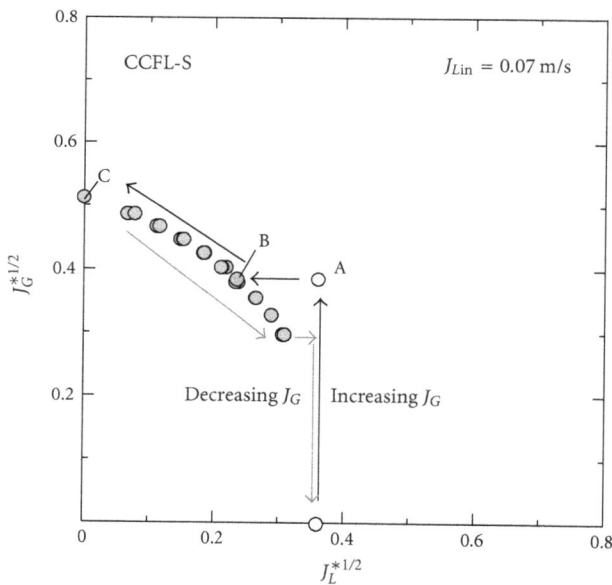

FIGURE 6: CCFL characteristics ($\theta = 0.6$ deg.).

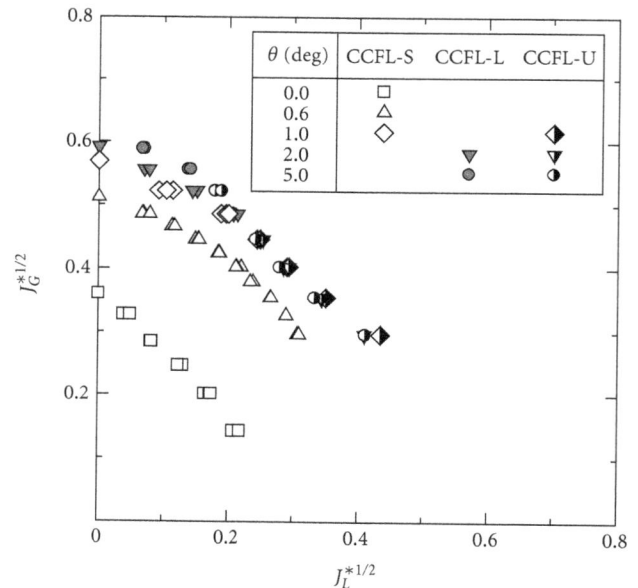

FIGURE 7: CCFL characteristics (effects of θ).

junction, respectively. The line in Figure 11 is drawn by using the Wallis's model [7–9]:

$$\sqrt{J_G^*} + \sqrt{J_{Lin}^*} = 1. \qquad (2)$$

The J_G^* at the onset of flooding decreases as J_{Lin}^* increases, and J_G^* increases with θ. The flooding always occurs in the surge line at low θ. On the other hand, at high θ, it occurs at the lower junction when J_{Lin}^* is low and at the upper junction at high J_{Lin}^*. Since the flooding at the upper junction is similar to that in a vertical pipe, the points of Flooding-U are not far from (2).

3.5. Effects of J_{Lin}. Figure 12 shows the CCFL characteristics under three different J_{Lin} conditions ($J_{Lin} = 0.02$, 0.07, and

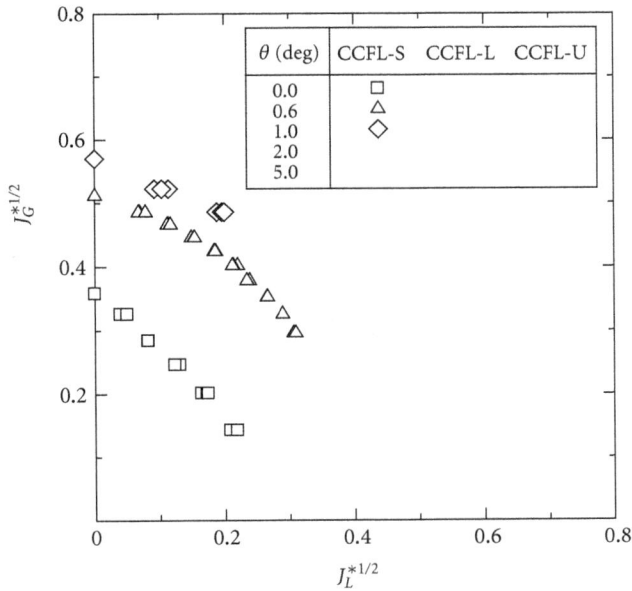

FIGURE 8: CCFL-S characteristics (effects of θ).

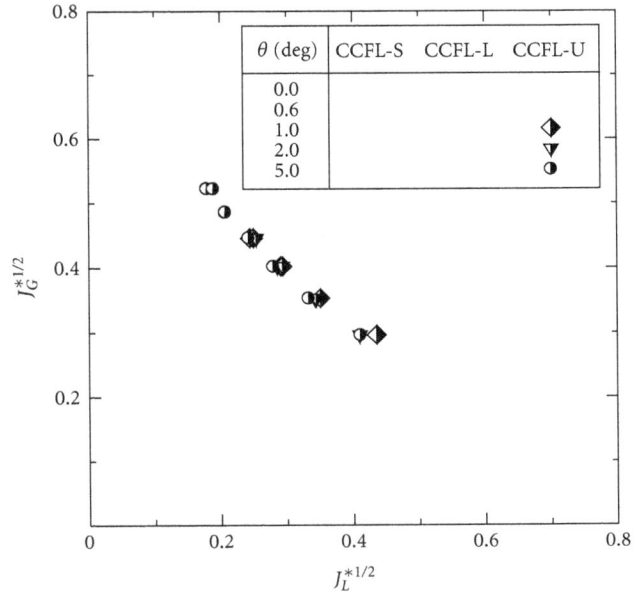

FIGURE 10: CCFL-U characteristics (effects of θ).

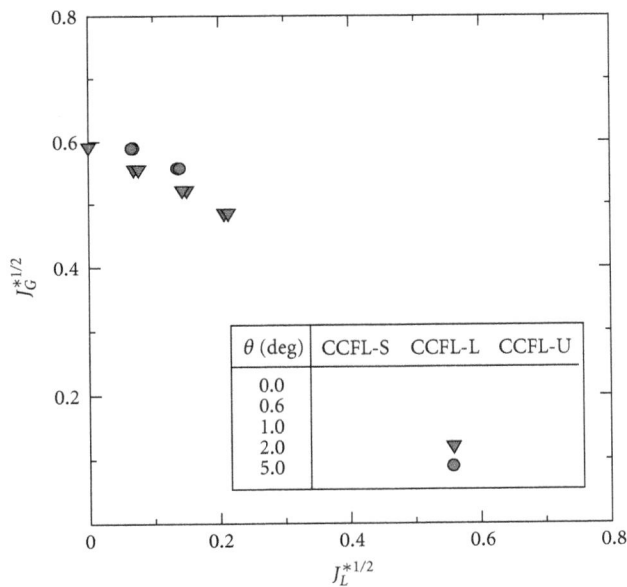

FIGURE 9: CCFL-L characteristics (effects of θ).

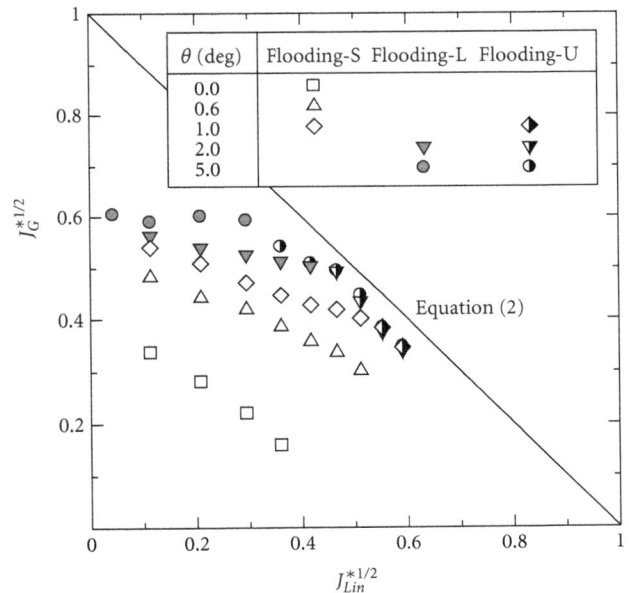

FIGURE 11: Onset of flooding.

0.12 m/s) for θ = 0.6 and 5.0 deg. The volumetric flux J_L^* of the falling water does not depend on the volumetric flux J_{Lin} of the supplied water in the upper tank, irrespective of θ and a type of CCFL (CCFL-S, CCFL-U, and CCFL-L). This is because the water level in the upper tank depends not on J_{Lin} but on the height of the partition in the upper tank when the flow limitation takes place.

3.6. Effects of Elbows. Figure 13 shows comparisons of CCFL characteristics between the surge line and the straight pipe. CCFL in the straight pipe is also classified into CCFL-S,

CCFL-U, and CCFL-S. CCFL-S occurs at low θ whereas CCFL-U, and CCFL-L appear at high θ. The flow limitation in the surge line is stronger than that in the straight pipe as shown in Figure 13(a). The elbows, therefore, enhance the flow limitation in the surge line, which contradicts the predictions obtained by Takeuchi et al. [4]. They explained that centrifugal force in the elbow section stabilizes the gas-liquid interface and inhibits the flow limitation. However, the centrifugal force would make the liquid film thinner and increase the wall friction. In addition, the presence of elbows would increase pressure drop in the line, in other words,

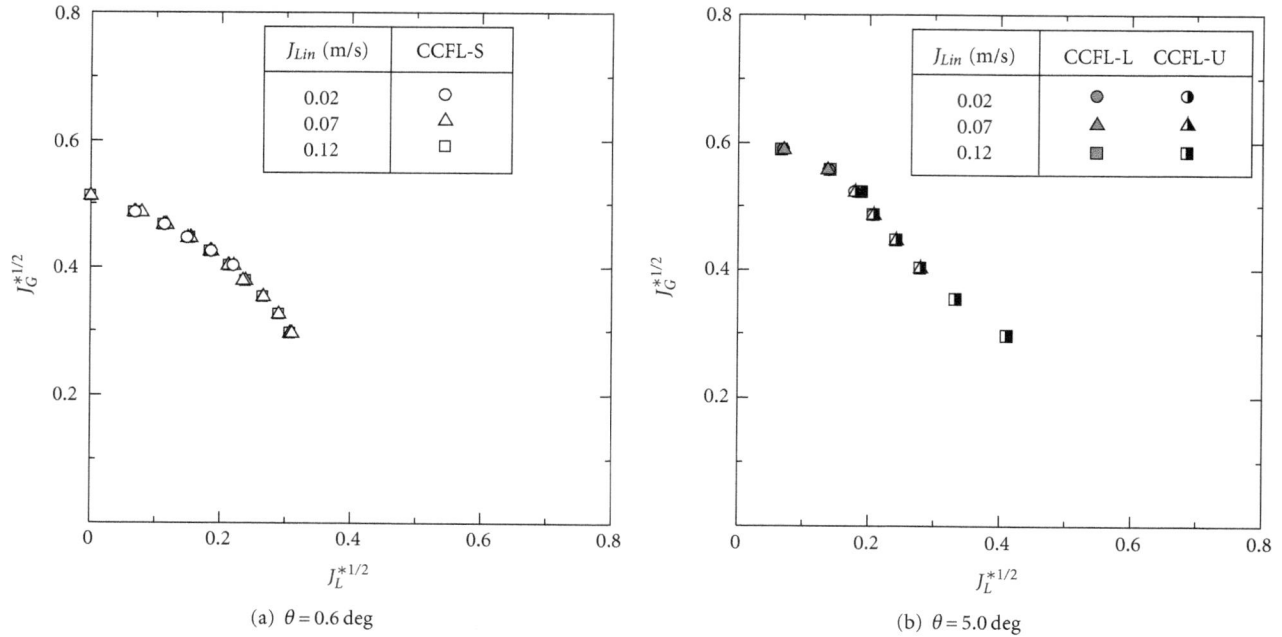

FIGURE 12: CCFL characteristics (effects of J_{Lin}).

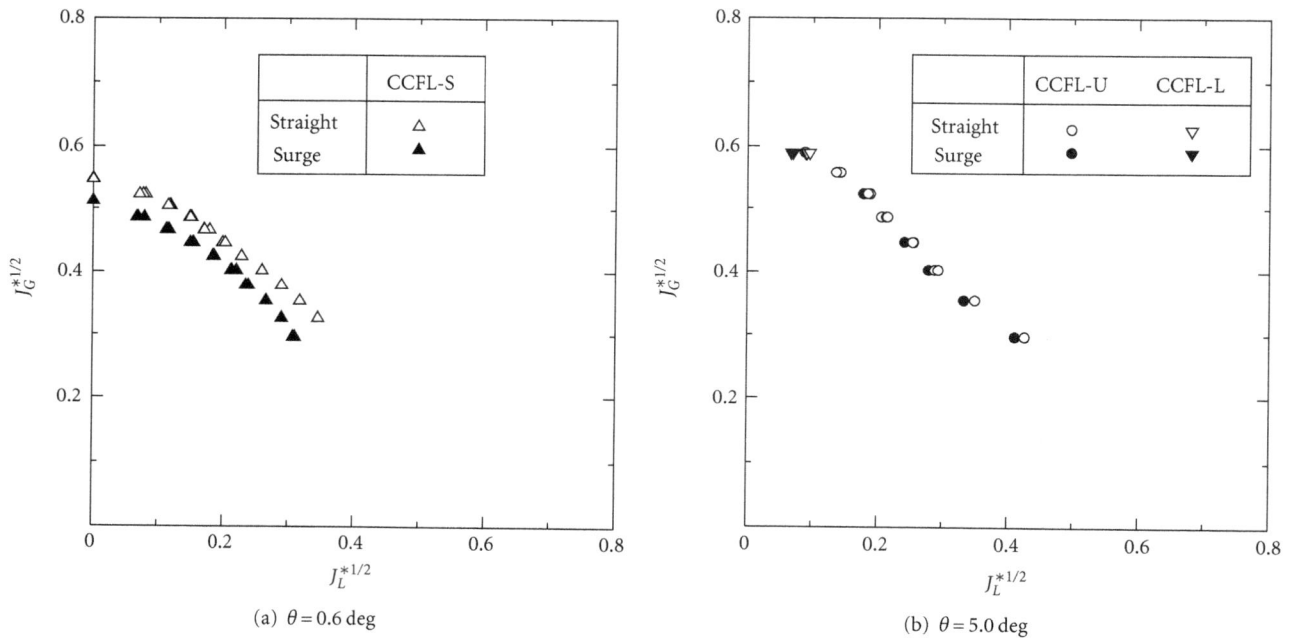

FIGURE 13: Effect of elbows on CCFL characteristics.

increases the force acting on water in the upstream direction. These effects would result in enhancement of flow limitation. The present result, therefore, supports the latter speculation rather than Takeuchi's one. On the other hand, CCFL-U and CCFL-L do not depend on the presence of elbows as shown in Figure 13(b). This is because the flow limitation occurs at the junctions, and therefore, it has no relation with the elbows in the surge line.

3.7. Discussion on Effects of Size and Fluid Properties on CCFL Characteristics. Minami et al. [10] measured CCFL characteristics in a scale-down model of PWR hot leg and confirmed thorough comparisons with literature [11–14] that the effects of the size and fluid properties are small. Since CCFL-L and CCFL-S in the surge line are similar to CCFL in the hot leg, this result implies that their dependence on the size and fluid properties is also small. On the other hand, CCFL-U is similar to CCFL in a vertical pipe. Many researches, which are summarized in textbooks [7, 15, 16], have been carried out for CCFL in a vertical pipe. These researches indicate that the Kutateladze number is more appropriate than the dimensionless volumetric flux J_k^* for large diameter tubes, and that the fluid properties can be taken into account by using the Bond number, viscosity ratio and/or Grashof number. This kind of knowledge can be utilized when applying the present results to a system with different pipe sizes or different fluid properties.

4. Conclusions

Countercurrent air-water flow in a scale-down model of a PWR pressurizer surge line was measured to understand characteristics of countercurrent flow limitation, CCFL. As a result, the following conclusions were obtained.

(1) CCFL takes place at three different locations, that is, at the upper junction, in the surge line, and at the lower junction. CCFL characteristics are governed by the most dominating flow limitation among the three.

(2) CCFL characteristics depend on the inclination angle of the surge line and the air flow rate. The effects of inclination angle on CCFL depend on the flow direction, that is, the effect is large for the nearly horizontal flow in the surge line, small for the vertical flow at the upper junction, and intermediate for the flow at the lower junction at which the flow changes its direction from horizontal to vertical directions.

(3) The presence of elbows enhances the flow limitation in the surge line, whereas the flow limitations at the upper and lower junctions do not depend on the presence of elbows.

Nomenclature

D: Pipe diameter [m]
g: Acceleration of gravity [m/s^2]
J: Volumetric flux [m/s]
J^*: Dimensionless volumetric flux
Q: Volume flow rate [m^3/s]
ρ: Density [kg/m^3]
θ: Angle of inclination [deg.].

Subscripts

G: Gas phase
L: Liquid phase
Lin: Liquid phase supplied to the upper tank.

References

[1] N. Minami, D. Kataoka, A. Tomiyama, S. Hosokawa, and M. Murase, "Countercurrent gas-liquid flow in a rectangular channel simulating a PWR hot leg (1) flow pattern and CCFL characteristics," *Japanese Journal of Multiphase Flow*, vol. 22, no. 4, pp. 403–412, 2008 (Japanese).

[2] N. Minami, M. Murase, D. Nishiwaki, and A. Tomiyama, "Countercurrent gas-liquid flow in a rectangular channel simulating a PWR hot leg (2) analytical evaluation of countercurrent flow limitation," *Japanese Journal of Multiphase Flow*, vol. 22, no. 4, pp. 413–422, 2008 (Japanese).

[3] H. Nakamura, J. Katayama, and Y. Kukita, "Loss of residual heat removal (RHR) event during PWR mid-loop operation: ROSA-IV/LSTF experiment without opening on primary loop pressure boundary," *American Society of Mechanical Engineers, Fluids Engineering Division (Publication) FED*, vol. 140, pp. 9–16, 1992.

[4] K. Takeuchi, M. Y. Young, and A. F. Gagnon, "Flooding in the pressurizer surge line of AP600 plant and analyses of APEX data," *Nuclear Engineering and Design*, vol. 192, no. 1, pp. 45–58, 1999.

[5] L. E. Hochreiter, S. V. Fanto, L. E. Conway, and L. K. Lau, "Integral testing of the AP600 passive emergency core cooling systems," *Journal of Power and Energy*, vol. 207, no. 4, pp. 259–268, 1993.

[6] J. N. Reyes, "Scaling Analysis for the OSU AP600 Integral Systems and Long Term Cooling Facility," OSU-NE-9204, 1992.

[7] G. B. Wallis, *One Dimensional Two-Phase Flow*, McGraw Hill, New York, NY, USA, 1969.

[8] S. Levy, *Two-Phase Flow in Complex Systems*, Wiley Interscience, 1999.

[9] G. F. Hewitt and G. B. Wallis, *ASME Multi-Phase Flow Symposium*, ASME, Philadelphia, Pa, USA, 1963.

[10] N. Minami, D. Nishiwaki, T. Nariai, A. Tomiyama, and M. Murase, "Countercurrent gas-liquid flow in a PWR hot leg under reflux cooling (I) air-water tests for 1/15-scale model of a PWR hot leg," *Journal of Nuclear Science and Technology*, vol. 47, no. 2, pp. 142–148, 2010.

[11] H. J. Richter, G. B. Wallis, K. H. Carter et al., "Deentrainment and Countercurrent Air-Water Flow in a Model PWR Hot-Leg," NRC-0193-9, U.S. Nuclear Regulatory Commission, 1978.

[12] A. Ohnuki, "Experimental study of counter-current two-phase flow in horizontal tube connected to an inclined riser," *Journal of Nuclear Science and Technology*, vol. 23, no. 3, pp. 219–232, 1986.

[13] A. Ohnuki, H. Adachi, and Y. Murao, "Scale effects on countercurrent gas-liquid flow in a horizontal tube connected to an inclined riser," *Nuclear Engineering and Design*, vol. 107, no. 3, pp. 283–294, 1988.

[14] F. Mayinger, P. Weiss, and K. Wolfert, "Two-phase flow phenomena in full-scale reactor geometry," *Nuclear Engineering and Design*, vol. 145, no. 1-2, pp. 47–61, 1993.

[15] P. B. Whalley, *Boiling Condensation and Gas-Liquid Flow*, Oxford University Press, New York, NY, USA, 1987.

[16] J. M. Delhaye, M. Giot, and M. L. Riethmuller, *Thermohydraulics of Two-Phase Systems for Industrial Design and Nuclear Engineering*, Hemisphere, New York, NY, USA, 1981.

13

Squeezing Force of the Magnetorheological Fluid Isolating Damper for Centrifugal Fan in Nuclear Power Plant

Jin Huang,[1,2] **Ping Wang,**[1] **and Guochao Wang**[1]

[1] *Chongqing Institute of Automobile, Chongqing University of Technology, Chongqing 400054, China*
[2] *The Key Laboratory of Manufacture and Test Techniques for Automobile Parts, Chongqing University of Technology, Chongqing 400054, China*

Correspondence should be addressed to Jin Huang, jhuangcq@163.com

Academic Editor: Yan Yang

Magnetorheological (MR) disk-type isolating dampers are the semi-active control devices that use MR fluids to produce controllable squeezing force. In this paper, the analytical endeavor into the fluid dynamic modeling of an MR isolating damper is reported. The velocity and pressure distribution of an MR fluid operating in an axisymmetric squeeze model are analytically solved using a biviscosity constitutive model. Analytical solutions for the flow behavior of MR fluid flowing through the parallel channel are obtained. The equation for the squeezing force is derived to provide the theoretical foundation for the design of the isolating damper. The result shows that with the increase of the applied magnetic field strength, the squeezing force is increased.

1. Introduction

With the continuous construction of nuclear power plants in China, more and more attentions are paid to safety and reliability of the auxiliary equipments in nuclear power plant. There is a pressing need to researching earthquake resistance of auxiliary equipment in the nuclear power plant. The centrifugal fan in the nuclear power plant must have good aseismatic performance that is the important condition for the safe operation of nuclear power plants in the earthquake. However, there have been few reports on vibration insulator for centrifugal fan [1, 2]. Earthquake response attenuation of centrifugal fans can be improved by connecting them with isolator. In this paper, we propose an isolating damper based on magnetorheological (MR) for centrifugal fan.

MR fluids are materials of micron-sized and magnetized particles in a carrier fluid. In the absence of an applied magnetic field, MR fluids flow freely. The fluids exhibit Newtonian-like behavior. Upon application of a magnetic field, these fluids exhibit viscoplastic behavior with yield strength [3]. Altering the strength of an applied magnetic field will precisely control the shear yield stress of the fluid. Based on the mechanical characteristics, the fluids can be used in the magnetically controlled devices such as brakes [4] and dampers [5, 6].

The MR isolating damper is one such device that provides controllable squeezing force. Altering the strength of an applied magnetic field will change the squeezing force of the MR isolating damper [7]. The research of MR dampers and their applications have been done in many different ways. In the field of the flow model description for MR damper, Boelter and Janocha [8] analyzed the working mode (shear mode, flow mode, squeeze mode) in MR damper. Mcmanus et al. [9] studied the squeeze flow mode in accompany with shear in MR damper, analyzed the damper force-velocity characteristics under different values of magnetic field strength. For the research of application in building structures, Motra et al. [10] researched the response attenuation of seismically excited adjacent buildings connected by a MR damper. P. Y. Lin and T. K. Lin [11] introduced a bridge isolation system that combines the rolling pendulum system (RPS) and the MR damper. Dragasius et al. [12] researched the resistance force generated by linear hydrocylinder-type magnetorheological fluid (MRF) damper acting in different regimes. It was defined that the force increases practically linearly when increasing strength of the magnetic field and

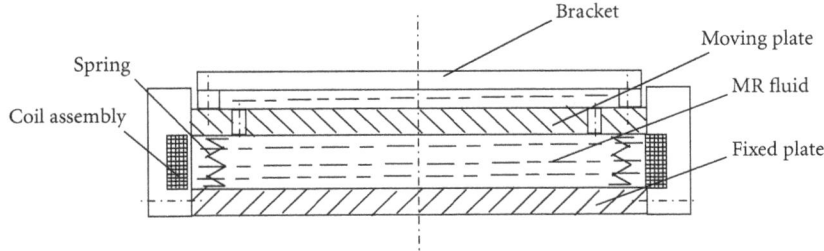

FIGURE 1: Operational principle of a circular plate MR isolating damper.

when increasing piston speed. Erkus and Johnson [13] investigated the dissipativity and performance characteristics of the semiactive control of the base-isolated benchmark structure with MR dampers. Rajamani and Larparisudthi [14] proposed a vibration control model for MR isolating damper.

In this paper, biviscosity model is used to describe the constitutive characteristics of MR fluids subject to an applied magnetic field. The operational principle of the MR isolating damper is introduced. Analytical solutions for the axisymmetric squeeze flow behavior of a biviscosity fluid are obtained. The velocity equation and the location of the unyield flow region are obtained. The expression for the squeezing force is derived to provide the theoretical foundation for the design of the isolating damper. The result shows that with the increase of the applied magnetic field strength, the dynamic yield stress of the MR fluid goes up rapidly, and the squeezing force is increased.

2. Operational Principle

The schematic configuration of the proposed circular plate MR isolating damper is shown in Figure 1. The MR isolating damper consists of an MR fluid, moving plate, working gap, electromagnet coil, flux guide, and housing. The MR fluid fills the working gap between the moving plate and housing. The electromagnet coil in the housing provides the magnetic field in the working gap. During relative vertical motion between the moving plate and housing, MR fluid is squeezed in the working gap. Thus, the pressure drop due to flow resistance of MR fluid in the working gap is induced. The MR isolating damper produces a controllable squeezing force due to the yield stress of the MR fluid, if a certain level of magnetic field is applied through the working gap.

3. Modeling and Analysis

3.1. Analysis of Flow Velocity in Working Gap. Figure 2 shows a schematic of the axisymmetric squeeze flow model. The gap between the disks is filled with the MR fluid. The bottom disk is fixed, while the top disk approaches the bottom disk with a constant velocity U. Gravity is neglected, so that the $z = h/2$ plane is a plane of symmetry for the flow. The force acting on bottom disk is

$$F = \int_0^R 2\pi r p \, dr, \qquad (1)$$

where, p is the pressure, R is radius of the disks.

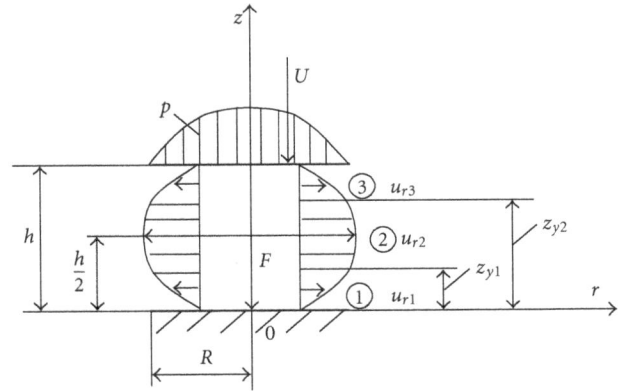

FIGURE 2: Squeeze model of MR fluid between two parallel disks.

To circumvent the "squeeze-flow paradox" of the Bingham model [15] and to facilitate an analytical analysis, the biviscosity constitutive model is employed in this study to describe the behavior of the MR fluid, see in Figure 3. It is known that the biviscosity model can provide very useful and convenient ways for the calculation of materials' behavior with yield, particularly from an analytical perspective.

The constitutive equation of the biviscosity model can be represented by the following two expressions [16]:

$$\tau = \tau_0 + \eta \frac{du_r}{dz}, \qquad |\tau| \geq \tau_y(H), \qquad (2a)$$

$$\tau = \eta(H) \frac{du_r}{dz}, \qquad |\tau| \leq \tau_y(H), \qquad (2b)$$

where τ is the shear stress, $\tau_y(H)$ is the dynamic yield stress developed in response to an applied magnetic field H, τ_0 is the intercept stress, η is the viscosity of the fluid when stress is higher than the dynamic yield stress, $\eta(H)$ is the viscosity coefficient subjected to a stress lower than the dynamic yield stress, it's a function of the magnetic field strength. Viscosity ratio $\varepsilon = \eta/\eta(H)$ is an important parameter for the biviscosity constitutive model, $\tau_0 = (1 - \varepsilon)\tau_y(H)$. Note that when $\varepsilon \to 0$, (2a) and (2b) are the Bingham constitutive model, whereas when $\varepsilon = 1$ they are the Newtonian constitutive model.

It is evident from the constitutive equation that the biviscosity constitutive model can be divided into two regions based on the dynamic yield stress. The MR fluid to be yielded

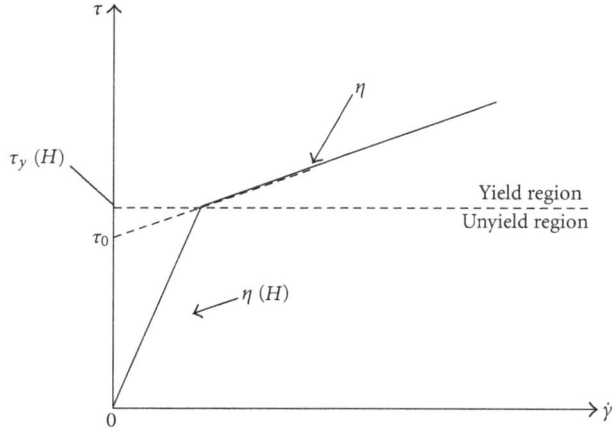

FIGURE 3: Biviscosity model of MR fluid.

when the magnitude of the fluid's internal stress is greater than the yield stress $\tau_y(H)$, the fluid exhibits typical Newtonian behavior. The MR fluid is unyield when the magnitude of the fluid's internal stress is smaller than $\tau_y(H)$, the fluid flows slowly with highly viscous.

For the disc-type squeeze flow model in Figure 2, assuming steady state condition ($\partial/\partial t = 0$), symmetry ($\partial/\partial\theta = 0$), no tangential and axial velocity ($u_\theta = u_z = 0$) but only radial, no body force, and no pressure gradient in thickness direction, the Navier-Stokes equation in the r direction for the cylindrical coordinates (r, θ, z) is simplified to

$$\eta \frac{d^2 u_r}{dz^2} = -m, \tag{3}$$

where u_r is the radial velocity, $m = -dp/dr$, dp/dr is the pressure gradient in radial direction.

By integrating (3), the velocity profile can be easily obtained as

$$u_r = -\frac{m}{2\eta} z^2 + A_1 z + A_2, \tag{4}$$

where A_1 and A_2 symbolize the integral constants.

The MR fluid exhibits Newtonian behavior in the absence of an applied magnetic field, applying boundary conditions of $u_r = 0$ at $z = 0$ and $z = h$, the flow velocity u_{r0} can be obtained as follows:

$$u_{r0} = \frac{m}{2\eta} z(h - z). \tag{5}$$

According to the assumptions, the momentum equation in the r-direction is

$$\frac{d\tau_{zr}}{dz} = -m. \tag{6}$$

By integrating (6) along the thickness direction and applying boundary condition of $\tau_{zr} = 0$ at $z = h/2$, the following equation for the shear stress can be obtained:

$$\tau_{zr} = m\left(\frac{h}{2} - z\right). \tag{7}$$

Assume that z_{y_1} and z_{y_2} denote the position of the unyield region boundaries measured from the bottom wall, respectively, see Figure 2. The yield surface is the locus of points where the shear stress is equal to the yield stress ($|\tau| = \tau_y(H)$), the location of the unyield region can be determined by satisfying the conditions such that

$$z_{y_1} = \frac{h}{2} - \frac{\tau_y(H)}{m}, \tag{8a}$$

$$z_{y_2} = \frac{h}{2} + \frac{\tau_y(H)}{m}. \tag{8b}$$

As seen in Figure 2, the flow is composed of three regions: two yield regions ($0 \le z \le z_{y_1}$, $z_{y_2} \le z \le h$) contacting the bottom and top walls and an unyield region ($z_{y_1} \le z \le z_{y_2}$).

In yield region $0 \le z \le z_{y_1}$, the two boundary conditions are

$$u_r = 0 \quad \text{at } z = 0, \tag{9a}$$

$$\frac{du_r}{dz} = 0 \quad \text{at } z = z_{y_1}. \tag{9b}$$

Using the boundary conditions (9a) and (9b), determining the integral constant in (4), the velocity profile in yield region $0 \le z \le z_{y_1}$ can be obtained as

$$u_{r1} = \frac{m}{2\eta}(hz - z^2) - \frac{\tau_y(H)}{\eta} z, \quad 0 \le z \le z_{y_1}. \tag{10}$$

In unyield region $z_{y_1} \le z \le z_{y_2}$, the two boundary conditions are

$$u_r = u_{r1} \quad \text{at } z = z_{y_1}, \tag{11a}$$

$$\frac{du_r}{dz} = 0 \quad \text{at } z = \frac{h}{2}. \tag{11b}$$

The velocity profile in unyield region $z_{y_1} \le z \le z_{y_2}$ can be obtained by using the boundary conditions (11a) and (11b) as

$$u_{r2} = \frac{m\varepsilon}{2\eta}(hz - z^2) + m\frac{1-\varepsilon}{2\eta}\left(hz_{y_1} - z_{y_1}^2\right) - \frac{\tau_y(H)}{\eta} z_{y_1},$$

$$z_{y_1} \le z \le z_{y_2}. \tag{12}$$

Similar to the previous analysis, by using the boundary conditions of $u_r = 0$ at $z = h$ and $du_r/dz = 0$ at $z = z_{y_2}$, the velocity profile in yield region $z_{y_2} \le z \le h$ can be obtained as

$$u_{r3} = \frac{m}{2\eta}(hz - z^2) - \frac{\tau_y(H)}{\eta}(h - z), \quad z_{y_2} \le z \le h. \tag{13}$$

3.2. *Pressure Distribution.* The pressure gradient dp/dr can be determined by using the law of conservation of mass

$$\frac{D}{Dt}\int_V \rho \, dV = 0, \tag{14}$$

where, ρ is the density of MR fluid, V is the volume of MR fluid. Equation (14) indicates that the derivative of liquid mass is zero

$$\int_V \frac{\partial \rho}{\partial t} dV + \int_s \rho u_j n_j ds = 0, \tag{15}$$

where, s is the integral area, u_j and n_j denote the velocity tensor and direction cosines.

When $z_{y_1} = h/2 - \tau_y(H)/m \leq 0$, the MR fluid flows slowly with highly viscous, the pressure gradient can be expressed as

$$\frac{dp}{dr} = \frac{6\eta(H)r}{h^3} U. \tag{16}$$

When $z_{y_1} = h/2 - \tau_y(H)/m > 0$, the MR fluid flows with a floating core, submitting (10), (12), (13) into (15) is

$$\pi r^2 U + 2\pi r \left(\int_0^{z_{y_1}} u_{r1} dz + \int_{z_{y_1}}^{z_{y_2}} u_{r2} dz + \int_{z_{y_2}}^h u_{r3} dz \right) = 0, \tag{17}$$

where

$$\int_0^{z_{y_1}} u_{r1} dz = \frac{mh}{4\eta} z_{y_1}^2 - \frac{m}{6\eta} z_{y_1}^3 - \frac{\tau_y(H)}{2\eta} z_{y_1}^2,$$

$$\int_{z_{y_1}}^{z_{y_2}} u_{r2} dz = 2 \int_{z_{y_1}}^{h/2} u_{r2} dz$$

$$= \frac{mh\varepsilon}{2\eta} \left(\frac{h^2}{4} - z_{y_1}^2 \right) - \frac{m\varepsilon}{3\eta} \left(\frac{h^3}{8} - z_{y_1}^3 \right)$$

$$+ \frac{1-\varepsilon}{\eta} mh z_{y_1} \left(\frac{h}{2} - z_{y_1} \right) - 2m z_{y_1} \left(\frac{h}{2} - z_{y_1} \right),$$

$$- \frac{2\tau_y(H) z_{y_1}}{\eta} \left(\frac{h}{2} - z_{y_1} \right),$$

$$\int_{z_{y_2}}^h u_{r3} dz = \int_0^{z_{y_1}} u_{r1} dz. \tag{18}$$

So

$$h^3 m^3 + \left[6Ur\eta - (1-\varepsilon)3h^2 \tau_y(H) \right] m^2 + 4(1-\varepsilon)\tau_y^3(H) = 0. \tag{19}$$

3.3. Squeezing Force. The force acting on the bottom disc can be obtained by integrating the pressure along the radial direction:

$$F = \int dF = \int_0^R 2\pi r p \, dr. \tag{20}$$

The subsection integral is used to get the force as

$$F = \left[\pi p r^2 \right]_0^R - \int_0^R \pi r^2 \frac{dp}{dr} dr. \tag{21}$$

The boundary conditions of pressure are

$$\frac{dp}{dr} = 0 \quad \text{at } r = 0, \qquad p = 0 \quad \text{at } r = R. \tag{22}$$

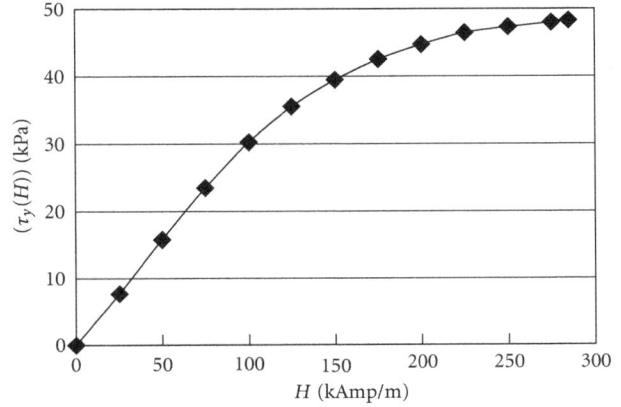

FIGURE 4: Yield strength versus magnetic field strength.

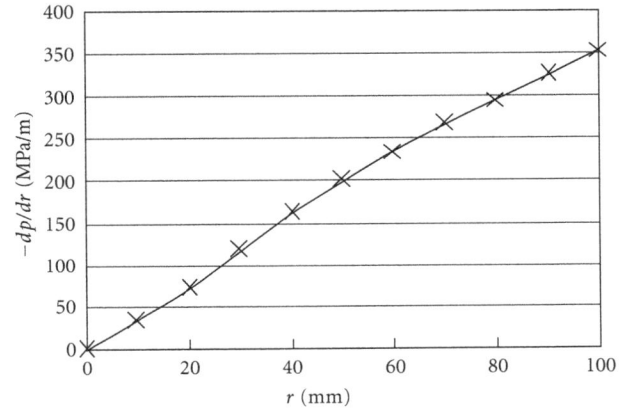

FIGURE 5: The pressure gradient versus radius.

Substituting (21) into (20), the force acting on the bottom disc can be obtained as

$$F = -\pi \int_0^R r^2 m \, dr. \tag{23}$$

4. Results and Discussion

A typical MR fluid is used in this paper. Figure 4 shows the yield strength of MR fluid under different magnetic field strength, measured by experiment. For the purpose of illustration, the following parameters are given: $h = 1$ mm, $R = 100$ mm, $U = 5$ m/s, $\eta = 0.092$ Pa·s, and $\varepsilon = 0.001$.

The relationship between the pressure gradient and the radius with apparent yield stress value of 30 kPa is shown in Figure 5 under the application of magnetic field strength of 100 kAmp/m by solving (16) and (19). The absolute value of the pressure gradient increases with the radius in linear, approximately.

When the strength of magnetic field is 100 kAmp/m, the yield surface determined by (8a) and (8b) with apparent yield stress value of 30 kPa along the radial direction is shown in Figure 6. When $r \leq 17.7$ mm, the MR fluid is unyield totally,

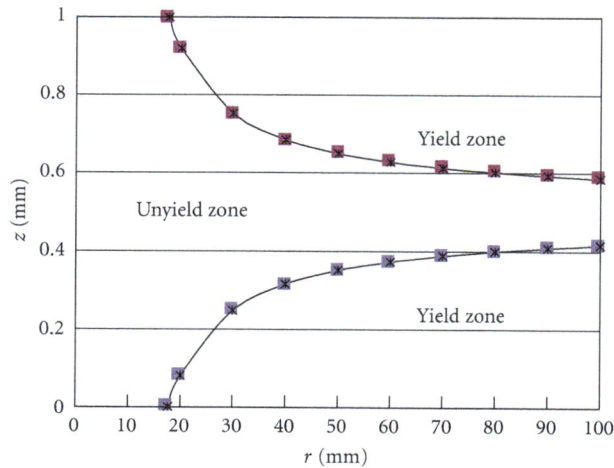

FIGURE 6: The yield surface along the radial direction.

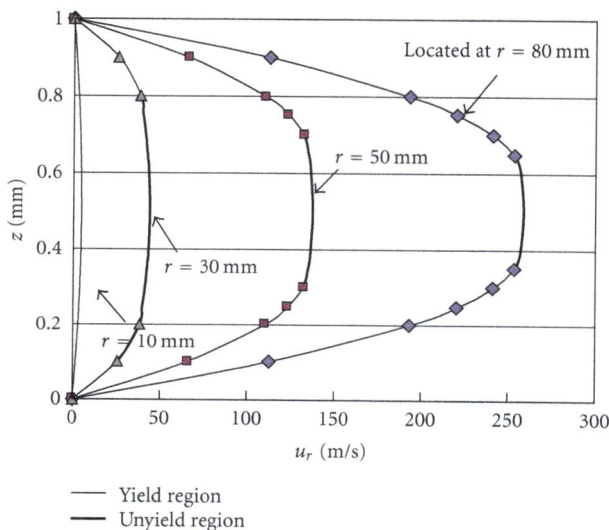

FIGURE 7: Velocity profiles for different location of radius.

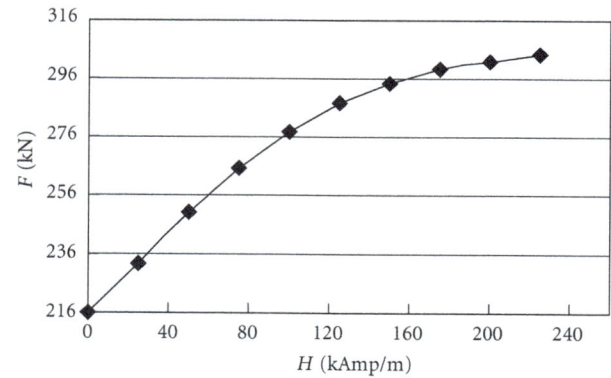

FIGURE 8: Squeezing force versus magnetic field strength.

5. Conclusions

The flow behaviors of MR fluid in circular plate MR isolating damper are investigated theoretically in this paper. The equations for the velocity and the squeezing force are derived to provide the theoretical foundation for the design of the MR isolating damper. The unyield region of MR fluid tends to move toward the symmetry plane of working gap as the radius increases. With the increase of the applied magnetic field strength, the squeezing force increased.

Acknowledgments

This work is supported by Project 51175532 by the National Natural Science Foundation of China and key Project 2011BA4028 by Natural Science Foundation Project of CQ CSTC.

flowing slowly with highly viscous. When $r > 17.7$ mm, the MR fluid is yielded partially. The location of yield surface tends to move toward the symmetry plane of working gap as the radius increases.

The velocity profiles can be obtained by (10), (12), and (13) with apparent yield stress value of 30 kPa, shown in Figure 7. The velocity profiles for the squeeze flow of MR fluid appear to be parabolic curves with respect to z, and their curvature tends to decrease with radius, the unyield region decreases accordingly.

The squeezing force versus magnetic field strength is shown in Figure 8. In the absence of an applied magnetic field, the force is 216 kN. The forces are 250 kN, 278 kN, and 302 kN at the strength of magnetic field of 50 kAmp/m, 100 kAmp/m, and 200 kAmp/m, respectively. The results indicate that with the increase of the velocity and the applied magnetic field, the squeezing force is increased.

References

[1] B. L. Wu, "Simply design for absorbing base of centrifugal fan," *Compressor Blower & Fan Technology*, no. 6, pp. 46–47, 2000 (Chinese).

[2] H. Higashimori, K. Kuma, M. Goto, K. Kimura, and M. Koga, "Centrifugal fan has two sets of inlet damper blades, with one set controlled to form opening different from that formed by other set," *US patent*, JP2005188390-A, 2005.

[3] J. Huang, J. Q. Zhang, and J. N. Liu, "Effect of magnetic field on properties of MR fluids," *International Journal of Modern Physics B*, vol. 19, no. 1–3, pp. 597–601, 2005.

[4] J. Huang, J. Q. Zhang, Y. Yang, and Y. Q. Wei, "Analysis and design of a cylindrical magneto-rheological fluid brake," *Journal of Materials Processing Technology*, vol. 129, no. 1–3, pp. 559–562, 2002.

[5] K. K. Ahn, D. Q. Truong, and M. A. Islam, "Modeling of a magneto-rheological (MR) fluid damper using a self tuning fuzzy mechanism," *Journal of Mechanical Science and Technology*, vol. 23, no. 5, pp. 1485–1499, 2009.

[6] J. Huang, J. He, and G. Lu, "Analysis and design of magnetorheological damper," *Advanced Materials Research*, vol. 148-149, pp. 882–886, 2011.

[7] G. Yang, B. F. Spencer, J. D. Carlson, and M. K. Sain, "Large-scale MR fluid dampers: modeling and dynamic performance

considerations," *Engineering Structures*, vol. 24, no. 3, pp. 309–323, 2002.

[8] R. Boelter and H. Janocha, "Design rules for MR fluid actuators in different working modes," in *Smart Structures and Materials 1997: Passive Damping and Isolation*, vol. 3045 of *Proceedings of SPIE*, pp. 148–159, March 1997.

[9] S. J. McManus, K. A. St. Clair K.A., P. É. Boileau, J. Boutin, and S. Rakheja, "Evaluation of vibration and shock attenuation performance of a suspension seat with a semi-active magnetorheological fluid damper," *Journal of Sound and Vibration*, vol. 253, no. 1, pp. 313–327, 2002.

[10] G. B. Motra, W. Mallik, and N. K. Chandiramani, "Semi-active vibration control of connected buildings using magnetorheological dampers," *Journal of Intelligent Material Systems and Structures*, vol. 22, no. 16, pp. 1811–1827, 2011.

[11] P. Y. Lin and T. K. Lin, "Control of seismically isolated bridges by magnetorheological dampers and a rolling pendulum system," *Structural Control and Health Monitoring*, vol. 19, no. 2, pp. 278–294, 2012.

[12] E. Dragasius, V. Grigas, D. Mazeika, and A. Sulginas, "Evaluation of the resistance force of magnetorheological fluid damper," *Journal of Vibroengineering*, vol. 14, no. 1, pp. 1–6, 2012.

[13] B. Erkus and E. A. Johnson, "Dissipativity analysis of the base isolated benchmark structure with magnetorheological fluid dampers," *Smart Materials & Structures*, vol. 20, no. 10, Article ID 105001, 2011.

[14] R. Rajamani and S. Larparisudthi, "On invariant points and their influence on active vibration isolation," *Mechatronics*, vol. 14, no. 2, pp. 175–198, 2004.

[15] S. D. R. Wilson, "Squeezing flow of a Bingham material," *Journal of Non-Newtonian Fluid Mechanics*, vol. 47, pp. 211–219, 1993.

[16] K. Zhu, R. Ge, and B. Xi, "Squeezing flow of electrorheological fluid between two circular plates," *Journal of Tsinghua University*, vol. 39, no. 8, pp. 80–83, 1999 (Chinese).

14

MR Continuously Variable Transmission Driven by SMA for Centrifugal Fan in Nuclear Power Plant

Jianzuo Ma,[1] Hongyu Shu,[1] and Jin Huang[2, 3]

[1] College of Mechanical Engineering, Chongqing University, Chongqing 400044, China
[2] Chongqing Automobile College, Chongqing University of Technology, Chongqing 400054, China
[3] The Key Laboratory of Manufacture and Test Techniques for Automobile Parts,
  Chongqing University of Technology, Chongqing 400054, China

Correspondence should be addressed to Jianzuo Ma, mjzcqu@163.com

Academic Editor: Yan Yang

The running efficiency of centrifugal fan affects the economical efficiency of the ventilation system. In this paper, we proposed a continuously variable transmission system based on magnetorheological fluid and shape memory alloy for improving the operating efficiency of the centrifugal fan. The equation of transmission torque developed by magnetorheological fluid is derived to compute the torque transmission ability in the continuously variable transmission system. A shape memory alloy spring actuator is designed to control the electric current in coil assembly. The results indicate that the change of temperatures has a tremendous influence on the electric current in coil assembly, the transmission torque of the continuously variable transmission system changes rapidly according to the temperatures acting on shape memory alloy spring actuator, and the output angular velocity of the centrifugal fan can be adjusted continuously.

1. Introduction

As an important component of ventilation system in nuclear power plant, the centrifugal fan plays an important role to ensure that the ventilation system is running reliably. The running efficiency of centrifugal fan affects the economical efficiency of the ventilation system. In a ventilation system, the flow rate or pressure should be timely changed according to the working conditions, so centrifugal fan should correspondingly adjust its velocity for improving the operating efficiency of the centrifugal fan. In order to solve the problem of traditional fan-driver which cannot export consecutive velocity, a magnetorheological (MR) continuously variable transmission system which is driven by shape memory alloy (SMA) is provided in this paper.

MR fluids and SMAs are known as smart materials for their properties can change rapidly on different external conditions. The yield stress of MR fluids changes rapidly and reversibly when an external magnetic field is applied [1–3].

MR fluids are useful for the efficient control of the torques and forces transmission; they can be used in clutches [4–8], brakes [9, 10], shock absorbers [11, 12], valves [13], and so on. SMAs may undergo mechanical shape changes at relatively low temperatures and retain them until heated, then coming back to the initial shape [14, 15]. This makes SMAs unique compared to other smart materials that can be used for actuator applications [16–18].

An MR continuously variable transmission system transmits torque by the shear stress of the MR fluids from the driving shaft to driven shell. The MR continuously variable transmission system has the property that its transmitting torque changes quickly in response to an external magnetic field. Huang et al. [19] proposed the possibility of application of the MR fluids to variable speed transmission. Jiang et al. [20] gave a new type of self-pressurized structure of magnetorheological fluids continuously variable transmission (MRFCVT) with V-shape working gap. Ma et al. [21] derived the necessary working gap and the volume of MR fluid for the MR fan clutch based on MR fluid properties, the desired control torque ratio, the angular velocity, and load torque of the clutch.

In this paper, Herschel-Bulkley model is used to describe the constitutive characteristics of MR fluids subject to an applied magnetic field. The operational model of the continuously variable transmission system is established to derive the formula for the torque transmitted by MR fluids. A sliding mode SMA actuator is proposed to modify the magnetic field acting on working gap under thermal effect. The properties of the MR continuously variable transmission system are studied in detail to provide an effective approach for improving the operating efficiency of the centrifugal fan.

2. Operational Principle

The MR continuously variable transmission system relies on MR fluid as a transmission medium to transmit torque. The operational principle of the MR continuously variable transmission system which is driven by SMA is shown in Figure 1. Transmission shaft and driving disc are initiative members, and shell is a driven one. The initiative members rotate at definite velocity, ω_1. The MR fluid fills the working gap between the driving disc and shell. The shell is joined to centrifugal fan. In the absence of magnetic field, MR fluid keeps flowing, so its transmission torque is only a very small viscous torque. However, a magnetic flux path is formed when electric current is put through the excitation coil. As a result, the magnetic particles in MR fluid are gathered to form chain-like structures, in the direction of the magnetic flux path. These chain-like structures restrict the motion of the MR fluid, thereby increasing the shear stress of the fluid. When the shear stress is large enough, initiative and driven members can finish a synchronous rotation.

Yield strength of MR fluid is a function of magnetic field strength [1], thus by changing magnetic field strength, the shear stress of MR fluid can be adjusted. So the transmitted torque of the MR continuously variable transmission system and the output rotate speed of driven shell can be modulated by varying the current in coil. The current in coil can be modulated by a sliding mode SMA actuator which alters the sliding distance according to the temperature acting on SMA spring, shown in Figure 2. The SMA helical spring works against a conventional steel spring (referred here as the "biasing" spring). At low temperatures, the steel spring is able to completely deflect the SMA spring to its compressed length. When increasing the temperature of the SMA spring, it expands, compressing the steel spring and moving the push rod.

3. Properties of SMA and MR Fluid

The most commonly used SMA elements for actuators are helical springs, for this form produces a large displacement. The force that a spring of any material produces at a given deflection depends linearly on the shear modulus of the material. SMAs exhibit a large temperature dependence on

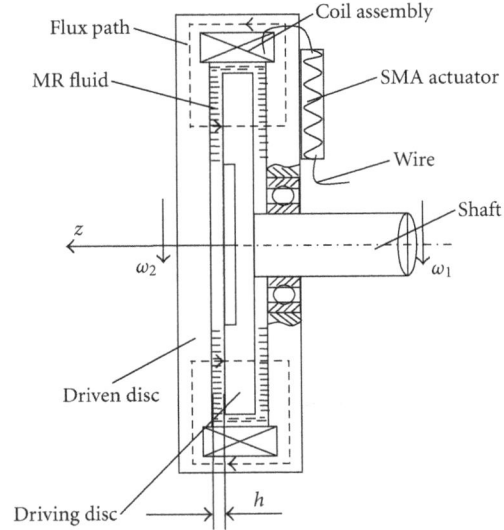

FIGURE 1: Operational principle of the MR continuously variable transmission system.

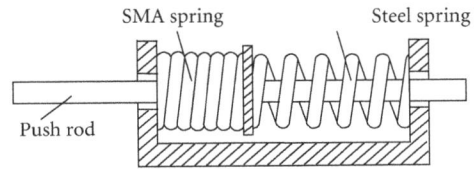

FIGURE 2: Operational principle of the SMA actuator.

the material shear modulus. The relationship between shear modulus and temperature for SMAs is given by

$$G = \begin{cases} G_M & \text{when } T < M_f,\, T < A_s \\ G(T) & \text{when } M_f \le T \le A_f \\ G_A & \text{when } T > A_f,\, T > M_s, \end{cases} \quad (1)$$

where G is the shear modulus of SMA, T is temperature, M_s, M_f, A_s, and A_f are the start and finish transformation temperatures of martensite and austenite, respectively, G_M and G_A are the shear modulus of martensite and austenite, respectively. When $M_f \le T \le A_f$, in absence of stress, shear modulus of SMA can be expressed approximately as:

$$G(T) = G_M + \frac{G_A - G_M}{2}[1 + \sin\phi(T - T_m)]. \quad (2)$$

In the process of heating, $T_m = (A_s + A_f)/2$, $\phi = \pi/(A_f - A_s)$; in the process of cooling, $T_m = (M_s + M_f)/2$, $\phi = \pi/(M_s - M_f)$.

MR fluids exhibit a controllable yield stress-like behavior in shear, whereby the application of a magnetic field transverse to the flow creates a resistance to flow which increases with an increasing magnetic field. To accommodate the shearing thinning observed in MR fluids, the Herschel-Bulkley model [22] can be used to describe the flow behavior of MR fluid:

$$\tau_2 = \tau_y(H) + K\left|\dot{\gamma}_2\right|^m \text{sgn}(\dot{\gamma}_2) \quad \tau_2 \geq \tau_y(H)$$
$$\dot{\gamma}_2 = 0 \qquad \tau_2 < \tau_y(H), \tag{3}$$

where τ_2 is the total shear stress of MR fluid, $\tau_y(H)$ is the yield strength caused by the applied magnetic field, $\dot{\gamma}_2$ is the shear rate of MR fluid, and m, K are constants. In the Herschel-Bulkley model, the constants m, K and the function $\tau_y(H)$ are empirically determined from experiments.

4. Analysis of SMA Spring Actuator

The scheme of the proposed actuator with aSMA spring and conventional steel against-spring is illustrated in Figure 2, where at low temperature the SMA spring will be compressed and when heated will extend with a pushing actuation.

The expression for shear stress in aSMA spring is described as

$$\tau_1 = \kappa \frac{8FD}{\pi d^3} = \kappa \frac{8FC}{\pi d^2}. \tag{4}$$

Here, the axial load is F, D is the average diameter of the spring, d represents the wire diameter, C is the spring index, $C = D/d$, and κ is known as the Wahl correction factor applied:

$$\kappa = \frac{4C-1}{4C-4} + \frac{0.615}{C}. \tag{5}$$

The relationship between compressed length δ and shear strain γ_1 for SMA spring is given by

$$\delta = \frac{n\pi D^2}{d}\gamma_1, \tag{6}$$

where n is the number of turns in the spring.

The wire diameter for the actuator can be obtained from (4) for acceptable values of C ranging from 3 to 12:

$$d = \sqrt{\kappa \frac{8FC}{\pi \tau_1}}. \tag{7}$$

The number of turns in the spring can be obtained from (6):

$$n = \frac{\Delta\delta d}{\pi\Delta\gamma_1 D^2}, \tag{8}$$

where $\Delta\delta$ represents the stroke of the actuator, and $\Delta\gamma_1$ is the strain difference at high and low temperatures:

$$\Delta\gamma_1 = \gamma_L - \gamma_H. \tag{9}$$

For SMA spring actuator in Figure 2, the axial load F has the relationship with the compressed length of SMA spring δ as below:

$$\frac{F(T)}{\delta(T)G(T)} = \frac{F_L}{\delta_L G_L},$$
$$F(T) = F_L + \frac{F_H - F_L}{\Delta\delta}S(T), \tag{10}$$

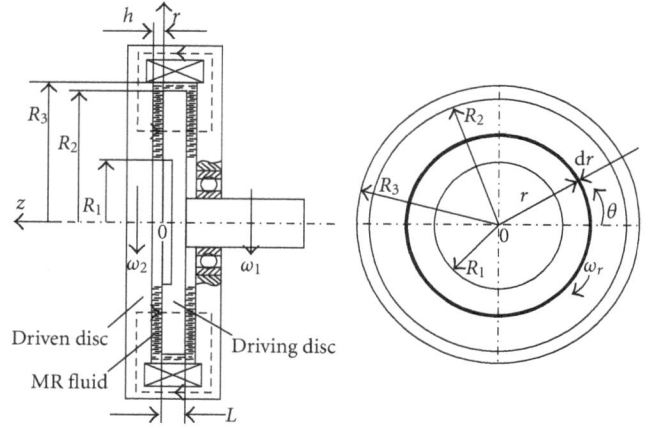

FIGURE 3: Circular flow mode of MR fluid between the two paralleled discs.

where $F(T)$, $\delta(T)$, and $G(T)$ are the axial load, compressed length, and shear modulus of SMA spring at temperature T, respectively, F_L, δ_L, and G_L are the axial load, compressed length, and shear modulus of SMA spring at low temperature, respectively, F_H is the axial load at high temperature, and $S(T)$ is the output displacement of SMA spring actuator:

$$S(T) = \delta_L - \delta(T). \tag{11}$$

The output displacement of SMA spring actuator can be obtained from (1), (2), (6), (10), and (11):

$$S(T) = \frac{(G(T) - G_L)\Delta\delta F_L \gamma_L}{(d/n\pi D^2)G(T)\Delta\delta F_L + (F_H - F_L)G_L\gamma_L}. \tag{12}$$

5. Analysis of Transmission Torque

Figure 3 shows the flow behavior of MR fluid in the working gap between driving disc and shell. In order to determine the fluid flow between driving disc and shell, the following assumptions are given: the fluid is incompressible. There is no flow in radial direction and axial direction, but only tangential flow. The flow velocity of MR fluid is a function of radius. The pressure in the thickness direction of MR fluid is constant. The strength of magnetic field in the gap of the activation region is well distributed.

The angular velocity of MR fluid in the working gap can be obtained as follows.

At the range of $0 \leq z \leq h$, $R_1 \leq r \leq R_2$:

$$\omega_r = \omega_1 + \frac{(\omega_2 - \omega_1)z}{h}. \tag{13a}$$

At the range of $R_2 < r \leq R_3$:

$$\omega_r = \frac{R_3^2 R_2^2}{R_3^2 - R_2^2}\left[\left(\frac{R_3^2 - r^2}{R_3^2 r^2}\right)\omega_1 + \left(\frac{r^2 - R_2^2}{r^2 R_2^2}\right)\omega_2\right]. \tag{13b}$$

The fluid shear strain rate in (3) may be approximated by:

$$\dot{\gamma}_2 = \begin{cases} -r\dfrac{d\omega_r}{dz} & R_1 \leq r \leq R_2 \\[2mm] r\dfrac{d\omega_r}{dr} & R_2 \leq r \leq R_3. \end{cases} \tag{14}$$

Taking a microunit at distance r location in a circle, microshear torque of the unit imposed on disc is as follows:

$$dJ = dF \cdot r = (\tau_2 \cdot dS) \cdot r. \tag{15}$$

The total transmission torque is:

$$J = \int_r dJ. \tag{16}$$

Apply the boundary conditions of the continuously variable transmission system: $\omega_r = \omega_1$, at $z = 0$ and $R_1 \leq r \leq R_2$; $\omega_r = \omega_2$, at $z = h$ and $R_1 \leq r \leq R_2$; $\omega_r = \omega_1$, at $r = R_2$; $\omega_r = \omega_2$, at $r = R_3$. The transmission torque could be achieved from (13a), (13b), (14), (15), and (16):

$$J = \frac{2\pi}{3}\left(R_2^3 - R_1^3\right)\tau_y(H) + \frac{2\pi K}{m+3}\left(R_2^{m+3} - R_1^{m+3}\right)\left(\frac{\omega_1 - \omega_2}{h}\right)^m$$

$$+ K\left[4\pi L\frac{(\omega_1 - \omega_2)R_2^2 R_3^2}{R_3^2 - R_2^2}\right]^m\left[\frac{\pi L}{2}(R_2 + R_3)^2\right]^{1-m}. \tag{17}$$

6. Computational Results and Discussions

Figure 4 shows the relation, obtained from the experiment, between the dynamic yield stress and the magnetic field strength for a typical MR fluid. From the figure we can find that the dynamic yield stress is proportional to the square of the magnetic field strength. MR fluid exhibits dynamic yield stresses of 0~30 kPa for the applied magnetic field strength of 0~175 kAmp/m. The ultimate strength of MR fluid is limited by magnetic saturation. The result shows that, with the increase of the applied magnetic field strength, the dynamic yield stress goes up rapidly.

According to (12) and (2), the effect of temperature in output displacement of SMA spring actuator can be analyzed, show as in Figure 5. In this study, Ti-49.8 at. % Ni SMA wire is used, its start and finish temperatures of the martensitic and austenitic phase transformation are $M_s = 78°C$, $M_f = 50°C$, $A_s = 74°C$, and $A_f = 95°C$, respectively. The shear moduli of martensite and austenite are $G_M = 7.5$ GPa and $G_A = 25$ GPa, respectively. The axial loads of SMA spring at low and high temperatures are $F_L = 15$ N and $F_H = 40$ N, respectively. The stroke of the actuator is $\Delta\delta = 20$ mm. Assume that the low temperature shear strain is $\gamma_L = 1.5\%$ and the value of spring index is $C = 7$. The wire diameter of SMA spring for the actuator which can be obtained from (7) is $d = 1.7$ mm, the number of turns which can be obtained from (8) is $n = 25$. As shown in Figure 5, the output displacement of SMA spring actuator increases with the increasing of temperature that can be controlled by temperature.

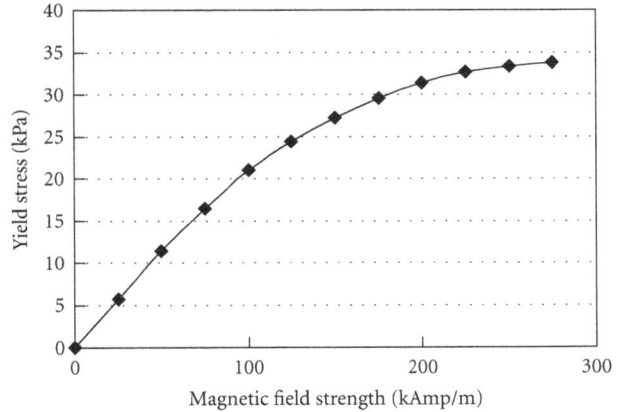

FIGURE 4: Yield stress versus magnetic field strength.

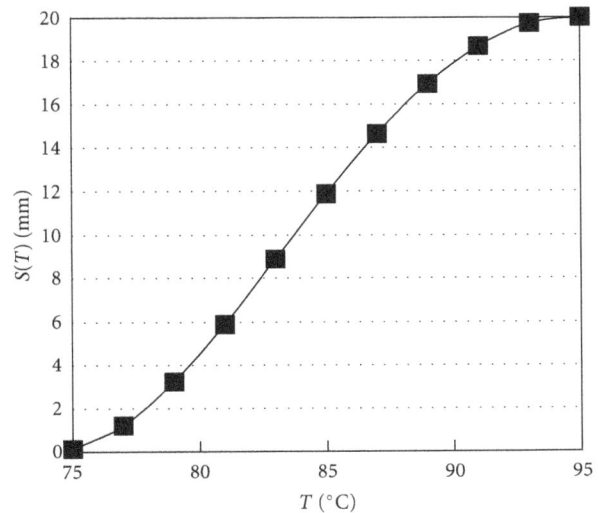

FIGURE 5: The output displacement versus temperature.

According to (17), the effect of magnetic field strength in transmission torque of the MR continuously variable transmission system is analyzed, as shown in Figure 6. The viscosity of the typical MR fluid is 0.042 Pa·s. The transmission torque is mainly produced by yield stress of MRF and viscosity torque is very small, so it can be assumed that $m = 1$, $K = 0.042$ Pa.s in the Herschel-Bulkley model. Geometric parameters of the continuously variable transmission system are inner radius $R_1 = 60$ mm, outer radius $R_2 = 120$ mm, $R_3 = 122$ mm, working gap $h = 1$ mm, $L = 4$ mm. The maximum input angular velocity is $\omega_{1\max} = 233$ rad/s. The transmission torques are 72.5 N·m, 130.4 N·m, 173.1 N·m, and 212.8 N·m at the magnetic field strength of 50 kAmp/m, 100 kAmp/m, 150 kAmp/m, and 200 kAmp/m, respectively. The results indicate that with the increase of magnetic field strength the transmission torque is increased.

The output angular velocity versus various temperatures is shown in Figure 7. It is assumed that the torque of lord

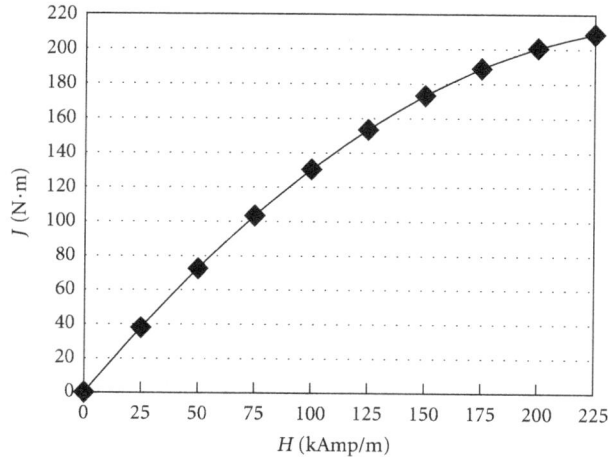

FIGURE 6: Transmission torque under different magnetic field strength.

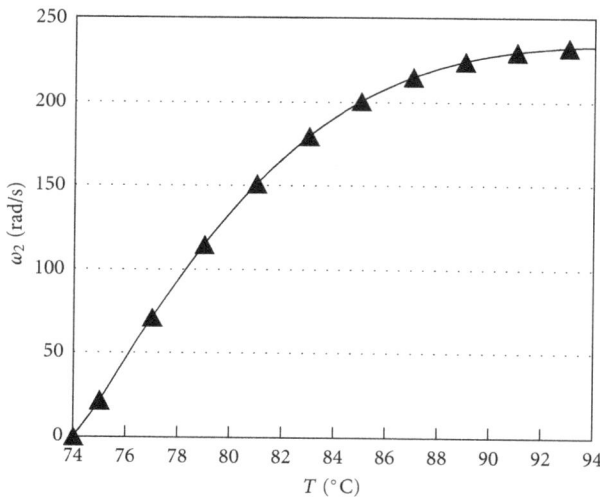

FIGURE 7: The output angular velocity versus temperature.

can be calculated from the empirical formula as $(J_L = 5.5) \times (10^{-5}) \times (30\omega_2/\pi)^2$ approximately, the current in excitation coil increases by 0.1A as push rod of SMA actuator goes ahead 1 mm, and magnetic field strength produced by excitation coil increases 10 kAmp/m. The result shows that with the increase of the temperature, the output angular velocity is increased.

7. Conclusions

The design method of an MR continuously variable transmission system driven by SMA is investigated theoretically in this paper. The equation of transmission torque developed by the MR fluid is derived. The transmission torque of the continuously variable transmission system under different magnetic field strength is analyzed. An SMA spring actuator is designed to control the transmission torque of the continuously variable transmission system. The output angular velocity of the transmission system versus temperatures is analyzed. With the increase of the applied magnetic field strength, the transmission torque of the continuously variable transmission system is increased. The output displacement of SMA spring actuator is controlled by temperature. The output angular velocity of the transmission system increases with the increasing of temperature acting on SMA actuator, rapidly and adaptively.

Acknowledgments

This work was supported by project 51175532 supported by the National Natural Science Foundation of China and key Project 2011BA4028 supported by Natural Science Foundation Project of CQ CSTC.

References

[1] X. Q. Peng, F. Shi, and Y. F. Dai, "Magnetorheological fluids modelling: without the no-slip boundary condition," *International Journal of Materials and Product Technology*, vol. 31, no. 1, pp. 27–35, 2008.

[2] J. Zhang, J. Q. Zhang, and J. F. Jia, "Characteristic analysis of magnetorheological fluid based on different carriers," *Journal of Central South University of Technology*, vol. 15, no. 1, pp. 252–255, 2008.

[3] M. S. Kim, Y. D. Liu, B. J. Park, C. Y. You, and H. J. Choi, "Carbonyl iron particles dispersed in a polymer solution and their rheological characteristics under applied magnetic field," *Journal of Industrial and Engineering Chemistry*, vol. 18, no. 2, pp. 664–667, 2012.

[4] A. S. Shafer and M. R. Kermani, "On the feasibility and suitability of MR fluid clutches in human-friendly manipulators," *IEEE/ASME Transactions on Mechatronics*, vol. 16, no. 6, pp. 1073–1082, 2011.

[5] T. Kikuchi, K. Otsuki, J. Furusho et al., "Erratum: development of a compact magnetorheological fluid clutch for human-friendly actuator (Advanced Robotics (2010) 24 (1489-1502))," *Advanced Robotics*, vol. 25, no. 9-10, p. 1363, 2011.

[6] P. Kielan, P. Kowol, and Z. Pilch, "Conception of the electronic controlled magnetorheological clutch," *Przeglad Elektrotechniczny*, vol. 87, no. 3, pp. 93–95, 2011.

[7] A. L. Smith, J. C. Ulicny, and L. C. Kennedy, "Magnetorheological fluid fan drive for trucks," *Journal of Intelligent Material Systems and Structures*, vol. 18, no. 12, pp. 1131–1136, 2007.

[8] Z. Herold, D. Libl, and J. Deur, "Design and testing of an experimental magnetorheological fluid clutch," *Strojarstvo*, vol. 52, no. 6, pp. 601–614, 2010.

[9] J. Huang, J. Q. Zhang, Y. Yang, and Y. Q. Wei, "Analysis and design of a cylindrical magneto-rheological fluid brake," *Journal of Materials Processing Technology*, vol. 129, no. 1–3, pp. 559–562, 2002.

[10] A. Farjoud, N. Vahdati, and Y. F. Fah, "Mathematical model of drum-type MR brakes using herschel-bulkley shear model," *Journal of Intelligent Material Systems and Structures*, vol. 19, no. 5, pp. 565–572, 2008.

[11] A. Milecki and M. Hauke, "Application of magnetorheological fluid in industrial shock absorbers," *Mechanical Systems and Signal Processing*, vol. 28, pp. 528–541, 2012.

[12] E. Dragasius, V. Grigas, D. Mazeika, and A. Sulginas, "Evaluation of the resistance force of magnetorheological fluid

damper," *Journal of Vibroengineering*, vol. 14, no. 1, pp. 1–6, 2012.

[13] J. Huang, J. M. He, and J. Q. Zhang, "Viscoplastic flow of the MR fluid in a cylindrical valve," *Key Engineering Materials*, vol. 274–276, no. 1, pp. 969–974, 2004.

[14] C. Yu, G. Z. Kang, D. Song, and Q. H. Kan, "Micromechanical constitutive model considering plasticity for super-elastic NiTi shape memory alloy," *Computational Materials Science*, vol. 56, pp. 1–5, 2012.

[15] S. Huang, M. Leary, T. Ataalla, K. Probst, and A. Subic, "Optimisation of Ni–Ti shape memory alloy response time by transient heat transfer analysis," *Materials and Design*, vol. 35, pp. 655–663, 2012.

[16] A. Hadi, A. Yousefi-Koma, M. Elahinia, M. M. Moghaddam, and A. Ghazavi, "A shape memory alloy spring-based actuator with stiffness and position controllability," *Proceedings of the Institution of Mechanical Engineers Part I*, vol. 225, no. 17, pp. 902–917, 2011.

[17] S. Langbein and A. Czechowicz, "Adaptive resetting of SMA actuators," *Journal of Intelligent Material Systems and Structures*, vol. 23, no. 2, pp. 127–134, 2012.

[18] T. Georges, V. Brailovski, and P. Terriault, "Characterization and design of antagonistic shape memory alloy actuators," *Smart Materials and Structures*, vol. 21, no. 3, Article ID 035010, 2012.

[19] J. Huang, G. H. Deng, Y. Q. Wei, and J. Q. Zhang, "Application of magnetorheological fluids to variable speed transmission," in *Proceedings of the International Conference on Mechanical Transmissions (ICMT'01)*, pp. 296–298, April 2001.

[20] J. D. Jiang, X. C. Liang, and B. Zhang, "Research of self-pressurized MR continuously variable transmission," in *Proceedings of the Smart Materials for Engineering and Biomedical Applications*, pp. 386–390, 2004.

[21] J. Z. Ma, G. C. Wang, and D. Zuo, "Geometric analysis in an MR fan clutch," *Advanced Materials Research*, vol. 239-242, pp. 1731–1734, 2011.

[22] W. H. Herschel and R. Bulkley, "Consistency measurements of rubber-benzol solutions," *Kolloid-Zeitschrift*, vol. 39, no. 4, pp. 291–300, 1926.

15

Severe Accident Simulation of the Laguna Verde Nuclear Power Plant

Gilberto Espinosa-Paredes,[1] Raúl Camargo-Camargo,[2] and Alejandro Nuñez-Carrera[1]

[1] *Área de Ingeniería en Recursos Energéticos, Universidad Autónoma Metropolitana-Iztapalapa,*
Avenida San Rafael Atlixco 186 Col. Vicentina, 09340 Mèxico City, DF, Mexico
[2] *Nuclear Safety Division, Comisión Nacional de Seguridad Nuclear y Salvaguardias, Doctor Barragán 779, Col. Narvarte,*
03020 Mèxico City, DF, Mexico

Correspondence should be addressed to Alejandro Nuñez-Carrera, anunezc@cnsns.gob.mx

Academic Editor: Jun Sugimoto

The loss-of-coolant accident (LOCA) simulation in the boiling water reactor (BWR) of Laguna Verde Nuclear Power Plant (LVNPP) at 105% of rated power is analyzed in this work. The LVNPP model was developed using RELAP/SCDAPSIM code. The lack of cooling water after the LOCA gets to the LVNPP to melting of the core that exceeds the design basis of the nuclear power plant (NPP) sufficiently to cause failure of structures, materials, and systems that are needed to ensure proper cooling of the reactor core by normal means. Faced with a severe accident, the first response is to maintain the reactor core cooling by any means available, but in order to carry out such an attempt is necessary to understand fully the progression of core damage, since such action has effects that may be decisive in accident progression. The simulation considers a LOCA in the recirculation loop of the reactor with and without cooling water injection. During the progression of core damage, we analyze the cooling water injection at different times and the results show that there are significant differences in the level of core damage and hydrogen production, among other variables analyzed such as maximum surface temperature, fission products released, and debris bed height.

1. Introduction

Currently Laguna Verde Nuclear Power Plant (LVNPP) uses, for decision-making in emergency case, the Emergency Operating Procedures Guides (EOPG) in order to ensure safe operation and prevent serious consequence in case of possible accident. However, the EOPG does not include the stage of core damage and currently, for the specific case of the LVNPP, there is not a clear definition about the develop of specific guidelines for the management of the severe accident.

LVNPP has two units and is located on the coast of the Gulf of Mexico in the municipality of Alto Lucero in the state of Veracruz. Both units of this plant have a boiling water reactor nuclear steam supply system as designed and supplied by the General Electric Company and designated as BWR 5.

The primary containment is part of the overall containment system, which provides the capability to reliably limit the release of radioactive materials to the environs subsequent to the occurrence of the postulated Loss-of-coolant accident (LOCA) so that offsite doses are below the *reference values* stated in Title 10 of the United States Code of Federal Regulations, Part 50 [1]. The design employs the drywell/pressure-suppression features of the BWR/Mark II containment concept [2].

The Unit 1 started operation in 1990 and the Unit 2 in 1995 with rated power levels of 1931 MWt each. The Unit 1 has been operated with a capacity factor of 80.73% and the Unit 2 with 85.34%, and this NPP contributes with the 4% of the national electricity production. This NPP utilizes a single-cycle forced circulation BWR provided by General Electric (GE). LVNPP was originally designed to operate at a gross electrical power output of approximately 695 MWe and a net electrical power output of approximately 674 MWe.

The thermal power was uprated by 5% (from 1931 MWt to 2027 MWt) during cycle 7 for unit 1 and cycle 4 for unit 2. In December of 1999, both units were authorized to operate to power uprate conditions. In July of 2008, the Comisión Federal de Electricidad (CFE, Mexican Electric Power Company) submitted the applications to the Mexican Regulatory Authority (CNSNS) for an operating license at power level of 2317 MWt. This corresponds to 120% of the original licensed thermal power (OLTP). This approach is referred as constant pressure power uprate (CPPU) because there are no changes in reactor dome pressure for this extended power uprate (EPU).

The BWR are designed with structures, systems, and components (SSCs) to accommodate steam flow rate at least 5% above the original rating. Safety analysis using better computer codes, methodologies, and operation experience allow the increase the thermal power by 5% without any hardware modification in the nuclear steam supply system (NSSS). However, the power increases up to 20% involve major changes in the SSC. Changes in the main condenser, turbine blades, main generator, steam reheater, and booster pump with more capacity are some of the most important changes performed in LVNPP.

In nuclear safety, defense in depth concept should consider accident conditions beyond design basis (severe accidents), although these are highly unlikely, and such conditions have not been explicitly addressed in the original design current nuclear power plants.

An important part of mitigating the damage to the core is the cooling of the debris produced during the melting of the core, but so far not known with certainty at what point should be the cooling by any of the emergency systems and whether it is appropriate to do so. Then, the aim of this work is to present numerical experiments in transient conditions to analyze the behavior of progression of core damage, and consequences such as hydrogen generation and fission products released: (1) without coolant injection in the core region and (2) three different times of the coolant injection with the high-pressure core spray (HPCS) during scenario of a LOCA. The LVNPP model was developed using RELAP/SCDAPSIM code [3–5].

At least until 2003, RELAP/SCDAPSIM [3–5], MELCOR [6], and MAAP4 [7] have been considered as three representative U.S. computer codes that are being widely used for the integral analysis of the core melt accident progression [8]. In this work, we used the RELAP/SCDAPSIM computer code for the analysis of the core and lower plenum phenomena in the simulation of the postulated accident LOCA with and without cooling in LVNPP. RELAP/SCDAPSIM considers the core and vessel with two-dimensional model both axial and radial directions.

2. System Description

In order to understand the phenomena of the thermo-hydraulics process during a hypothetical LOCA with core damage, we presented a brief description of a BWR, where Figure 1 shows the schematic diagram [9].

The reactor water recirculation system (circulates the required coolant flow through the reactor core) consists of two external loops to the reactor vessel. The jet pumps located within the reactor vessel provide a continuous internal circulation path for a major portion of the core coolant flow. The recirculation pumps take the coolant suction from the downward flow in the annulus between the core shroud and the vessel wall. This flow is discharged into the lower core plenum from jet pumps. The coolant water passes along the individual fuel rods inside the fuel channel where it boils and becomes a two-phase steam/water mixture. In the core, the two-phase fluid generates upward flows through the axial steam separators while the steam continues through the dryers and flows directly out through the steam lines into the turbine generator.

The LOCA is postulated as rupture in the suction of the recirculation pipe as is illustrated in Figure 1, in this figure can be observed that the emergency cooling systems: high-pressure core spray system (HPCS), low pressure spray system (LPCS), and low-pressure coolant injection system (LPCI) that inject in the upper plenum.

3. Preliminaries

A severe accident is one that exceeds the design basis of the plant sufficiently to cause failure of structures, materials, systems, and so forth, without which it can ensure proper cooling of the reactor core by normal means [10]. Faced with a severe accident the first response is to try to maintain the core cooling by any means available, but in order to carry out such an effort, the response is understand necessary the progression of core damage, because such action has effects that may be determinants in the progression of the accident.

Due to that the severe accident phenomenology is very broad and not all aspects can be studied with a computer code of analysis of the core damage [8]. Then, the information presented in this paper is focused on the study of the progression of core damage phenomena called "In-Vessel". These phenomena are dominated primarily by the temperature, which increases significantly in the absence of coolant flow. Figure 2 shows the progression of core damage for a BWR reactor as a function of temperature [11].

The overheating in the fuel due to lack of cooling, the decay heat, and the chemical reaction between the zirconium and steam, in about an hour, start to discover the fuel, causing the temperature in the center of the core can reach high values, such as 2000 at 2600°C, so the core begins to melt. With the LOCA scenario, these processes are observed and analyzed in this work.

The molten material will flow slowly down by gravity effects into the lower regions, and colder of the core region, where it resolidifies and blocks the channels between the fuel rods. By this mechanism, a block of solidified material (basin-shaped) of the core (corium) is formed, which collects the molten material. When a sufficient amount of molten material is collected, it will flow to the bottom of the reactor vessel. After a certain time (greater than 30 minutes), the amount of molten corium has flowed to the bottom of the vessel which may correspond to about two thirds of the fuel material from the core region. The rest of the combustible material overheats more slowly and may take several hours to melt. Finally, if the molten material of the core cannot cool due to the loss of geometric and failure of the cooling

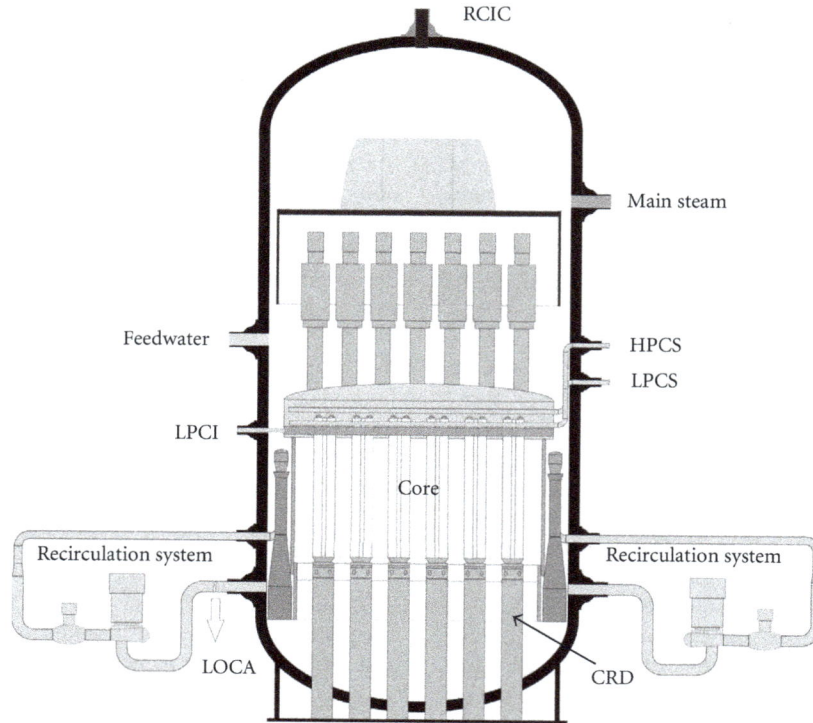

FIGURE 1: Schematic diagram of the boiling water reactor (BWR) [9].

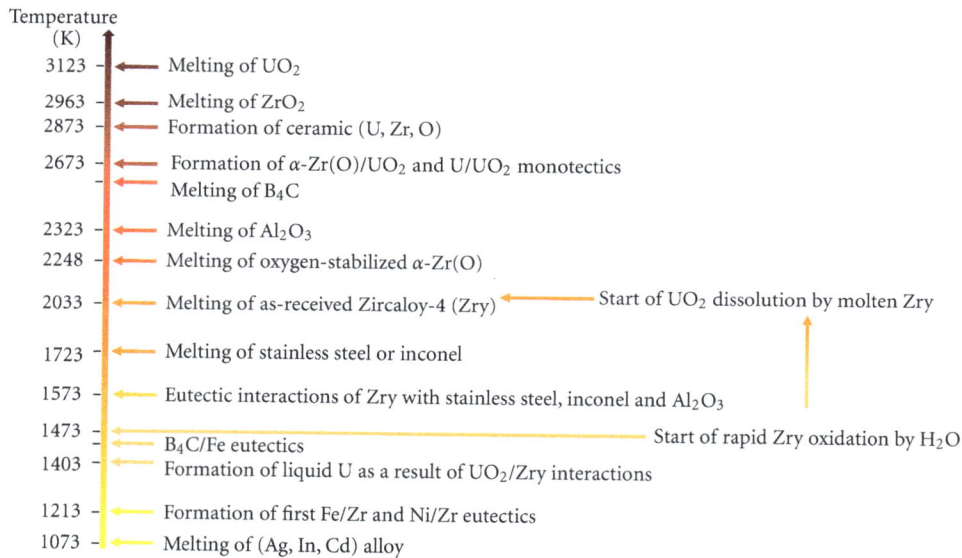

FIGURE 2: Progression of core damage.

systems, the wall of the reactor vessel will fail. If at that time the pressure in the reactor coolant system is low, the corium will flow down the cavity of the container. If the pressure is high, the corium is violently eject and dispersed. The impacts on the containment due to these two scenarios are completely different, but in both cases, the containment can fail early or late or even may remain intact.

3.1. Core Damage Progression. As a result of heat produced by radioactive decay of the active elements, known as the decay

energy, the fuel temperature begins to rise, even when the reactor is off after a scram, if the core cooling is inadequate [12].

In the case of severe accidents, where it is anticipated that the safety engineering systems (IS) functions incorrectly, to the extent that the core loses its cooling, the core will be damaged by overheating and release radioactive elements, mainly the primary system reactor.

At temperatures of 1073 K, molten occurs of the alloy Ag-In-Cd of the control rod for pressurized water reactors

(PWRs, Figure 2), the molten alloy is thermodynamically stable within the fuel element due to pressure vapor of the alloy, especially due to cadmium. For BWR, the molten of the control rods occurs around 1,500 K, and according with CORA 17 experiment, the release of energy and hydrogen production is higher than a PWR as the steam reaction with the remnant B$_4$C, absorber of the BWR control rods, is more exothermic per gram of material than Zircaloy during the quench phenomenon [13].

The guide tube failure will cause the attack of the alloy Zircaloy of the fuel element. The liquid mixture may fall down by gravity effects and move into the core regions, causing localized damage in the fuel cladding. Eventually is resolidified and may produce blockage in cold areas of the lower region of the core region.

At temperatures around 1173 K, the zirconium of the fuel cladding begins to react chemically with steam, which produces hydrogen and reaction heat.

At temperatures above 1500 K, the reaction becomes very large, accelerating overheating of the fuel, the oxidation of zircaloy (Zr) of the fuel cladding by steam effects, it becomes important. This reaction is exothermic and the temperature increases the rate of oxidation increases, the energy release is large as well as the generation of hydrogen (H$_2$).

The hydrogen production rate depends on the temperature of zirconium in the core and the amount of steam available. When the temperature rises about 1500 K, the following reaction can be carried out:

$$Zr + 2H_2O \longrightarrow ZrO_2 + 2H_2 + 586\,kJ/mol. \quad (1)$$

On the other hand, the heating of the fuel elements results in loss of mechanical properties with consequent ballooning that can lead to breakage. The deformation occurs due to the difference in pressure between the inside of the fuel element and reactor pressure. The internal pressure of the fuel element is a function of fuel temperature on the thermodynamic process of heating at constant volume. Regarding the reactor pressure will depend on the type of accident. Large breaks in the primary cooling system have much higher risk of ballooning when the pressure of the system mentioned above is almost atmospheric.

The effects of ballooning in the fuel elements are mainly to (1) reduce the passage section of coolant flow in the core and (2) hinder the cooling of localized areas of the fuel. Both phenomena lead a further increase in temperature and higher deformation. In addition to the deformation by pressure difference, it also can occurs due to thermal expansion. Another possible mechanism is the breaking strain differential expansion between the oxide layer and the inner layer of metal. This latter mechanism may induce breakage of oxidized fuel element even without differential pressure. The break of the fuel rods produces that the reaction between steam and cladding increases twice.

Experiments have shown that the rupture of the fuel does not block the flow of steam but whether it deviates by reducing the natural flow in the core. The fuel temperature increases and most of the hydrogen generated in the accident occur during this early stage.

At intermediate temperatures (2033 and 2273 K), if the thickness of the oxide layer is sufficiently large, the oxide will retain the Zr molten metal, preventing the fall down by gravity and staying in touch with the pellets of dioxide uranium (UO$_2$), which will remain until the outer layer of oxide loses their resistance due to its dissolution by molten Zr or due to mechanical failure or because it reaches the melting temperature (2963 K) of ZrO$_2$.

If the fuel element has ballooned previously, the ability to contact the pellets will be reduced, thereby reducing the ability of UO$_2$ dissolution. In the analysis of severe accidents, the phenomenology is very important because it is capable of producing nuclear fuel liquefaction temperatures (1000 K) below the melting point of the UO$_2$ (3123 K), where the interaction process is complex.

At temperatures above 2873 K will occur the melting of ceramic materials UO$_2$, ZrO$_2$ and mixed oxide (U, Zr) O$_2$, and total loss of the geometry of the affected zone.

4. Severe Accident Scenarios for the LVNPP

The severe accidents simulations consider the occurrence of the postulated LOCA in the recirculation loop of the reactor (Figure 1), with water coolant injection with HPCS at different times. Also, another scenario considered the postulated LOCA without water coolant injection. For this study, we used as a reference the data presented in Figure 2, which is information that Innovative Systems Software (ISS) has published [11] and that is generally similar the reference [14].

5. Numerical Model of the LVNPP

The nuclear steam supply system (NSSS) whose function is to carry the steam from the reactor vessel to the main turbine, and then drive the feed-water flow from the condenser to the vessel, holding constant all parameters in the reactor. The NSSS model representing the LVNPP using SCADAP/RELAP5 is illustrated in Figure 3. The operating conditions correspond to power uprate (PU) LVNPP, that is, 2027 MWt of the thermal power rate. This model includes the following main elements: (1) feed-water system, (2) reactor vessel and internals, (3) recirculation loops, (4) the reactor core, (5) main steam line, (6) bottom of the reactor vessel. The bottom model of the vessel is a fundamental element for the analysis in accident scenarios, for which it is included as a separate element that interacts with the other models.

6. Numerical Experiments

In order to evaluate the effect of time on the core cooling, we considered various cases, starting from a steady-state reactor of the LVNPP at to power uprate conditions (2027 MWt). In particular, we present the results and discussions of three conditions at different times of injection of HPCS, for reference conditions Which were determined without the performance of security systems, subsequently with damage in the core region was injected coolant flow with the HPCS

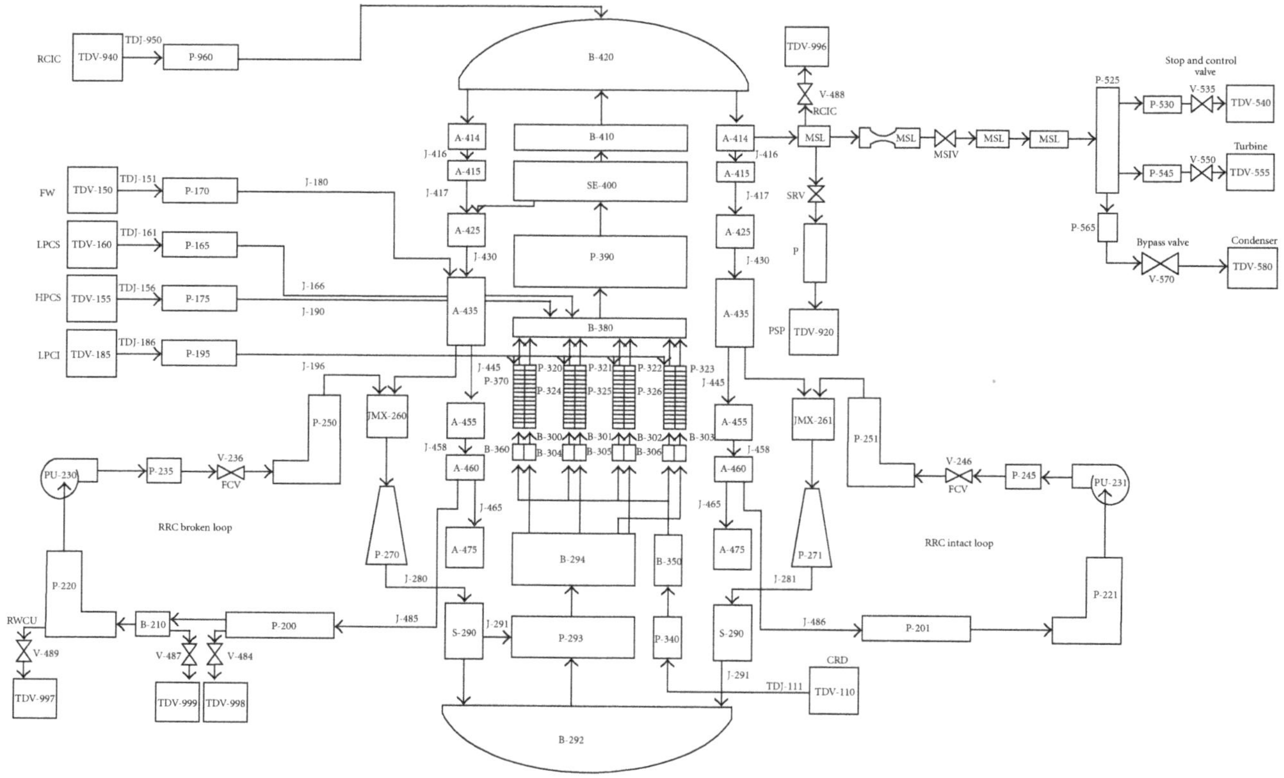

FIGURE 3: SCADAP/RELAP5 Laguna Verde Nuclear Power Plant model [22].

at 700, 900, 1200 s; therefore, four numerical experiments are analyzed in this work.

For each numerical experiment the following parameters were analyzed:

(i) collapsed level,

(ii) pressure in the dome of the vessel,

(iii) maximum surface temperature in the core region,

(iv) noncondensable fission products released,

(v) soluble fission products released.

(vi) hydrogen generation,

(vii) core damage,

(viii) control rod damage,

(ix) debris bed height in lower plenum.

7. Simulations and Discussions

A comprehensive analysis of the nuclear reactor during the transient evolution for the LOCA event with and without cooling is presented in this section.

The studied scenario (base case) is a loss of coolant accident (LOCA) due to rupture of a pipe in a recirculation circuit, imposing no coolant flow injection in the core region. This causes the signal to activate isolation of main steam (MS) and activates the coolant injection systems. Due to this, recirculation pumps trips and reactor scram occure.

Subsequently, we expected decrease of the thermal power, level, pressure in the dome, steam flow, feed-water flow, and mass flow rate in the core, the close of the main steam insolation valves (MISVs) and trips the feed-water turbine pumps. The reactor collapsed level drops dramatically activating the emergency core cooling systems but these do not provide coolant flow, which causes damage in the core region.

The sequence of events is presented in summary form in Table 1, for the case without cooling effects. In this table, it is observed that the hypothetical accident due to LOCA and no available emergency systems to cooling core, where at 324 s the hydrogen generation starts. Oxidation of the cladding, rods, and other components in the core constructed in zirconium base alloy, by steam, is a critical issue in LWR accident producing severe core damage. During a severe accident in a nuclear reactor, several works are focusing on core degradation by metal core components oxidation by air or steam (e.g., [15–18]), on the other hand, Royl et al. [19] have made a hydrogen risk analysis during severe accident using computational fluid dynamics (CFD) codes, to obtain localized detailed information and supplement the results of lumped parameter codes, which focus on global or average effect.

In this table, the maximum surface temperature in the core at different times match the event which is related with the progression of the core damage as is illustrated in the schematic diagram of the Figure 2.

The numerical results of nuclear reactor during the transient evolution for the LOCA event with and without cooling are presented in Figures 4–10.

TABLE 1: Main sequence of events in a LOCA without cooling effects.

Time (seconds)	Event
0.0	The reactor operates to power uprate conditions (2027 MWt).
30.0	Loss of external power, loss of coolant in a recirculation loop, closing of the turbine control valves, and trip of the reactor and recirculation pumps.
34.3	Low level in the reactor vessel (L2), before high pressure occurs in the drywell. The emergency core cooling systems (HPCS* and RCIC*) are active without injecting. Starts closing insolation valves (MSIV*).
37.4	Low-low level (L1) is reached.
40.0	Level in the BAF*.
200.0	Level below the core support plate.
324.0	Hydrogen generation starts.
370.0	Maximum temperature at the core surface of 1073 K.
458.0	Core damage stage starts with release of fission products. Maximum Temperature at the core surface of 1185 K.
936.0	Control rod damage starts Maximum temperature at control rod of 2136 K.
1800.0	Core damage de 19%.
3600.0	Core damage of 78%.

*HPCS: high pressure core spray system; RCIC: reactor core isolation cooling; MISV: main steam insolation valves; BAF: bottom of active fuel.

The core melt accident progression can be divided into two stages.

(1) An early phase up to the partial melting of core material, where involves core uncover, heat up, and partial melting in the reactor core region.

(2) A late phase results in the significant melting of the core material with relocation, and redistribution in the lower plenum. The governing phenomena involves porous debris bed, molten pool, and formation cavity, where the process heat, momentum, and mass are complex due to very high temperatures, multicomponent and multiphase materials, melting and freezing process, and geometrical configurations.

In general terms, if not cooled in the core region, the molten core materials (known as corium that is a conglomerated mixture of various core materials; mainly but not exclusively as oxide component-UO_2, and ZrO_2-, metallic components-U, Zr, Fe, and stainless steel), accumulated on the core support plate, would be eventually relocated into de lower plenum region and the thermal attack on the lower plenum could occur. In the presence of water in the lower plenum, some portions of the relocating molten core material could be fragmented into small solid particles in the lower head, and the remain material will maintain its original

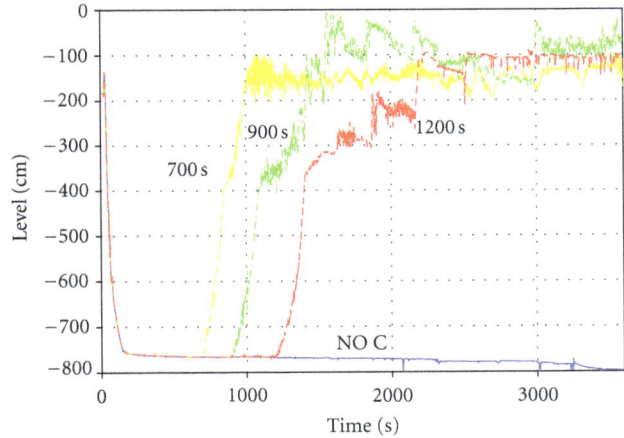

FIGURE 4: Level respect to top of active fuel (TAF) or fuel zone. HPCS injection to 700 s, 900 s, and 1200 s; NO C indicates without HPCS injection.

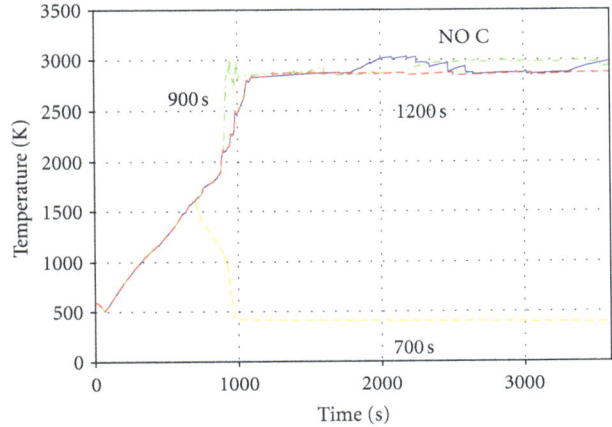

FIGURE 5: Maximum surface temperature in the core. HPCS injection to 700 s, 900 s, and 1200 s; NO C indicates without HPCS injection.

liquid phase [8]. The main mechanism of fragmentation (debris jet) is a hydrodynamics process, where rapid heat transfer from the debris jet to the lower plenum water accompanies the hydrodynamic fragmentation process and debris oxidation, results in steam and hydrogen production and an abrupt increase of the pressure. The heat removal from the lower head debris bed is determined by amount of the heat generated in the corium accumulated in the lower plenum, heat transfer inside the core material, heat transfer from the corium to the reactor vessel, and heat transfer to outer vessel.

7.1. Level Behavior. The transient behavior comparison of the level in the vessel respect to TAF (Top of Active Fuel) with and without cooling is shown in Figure 4. This figure shows that core is discovered for all cases analyzed in this work. The core discovers that time is crucial for the progression of the core damage, due to the temperature of the core region increase, even be lost cooling capacity. In all cases

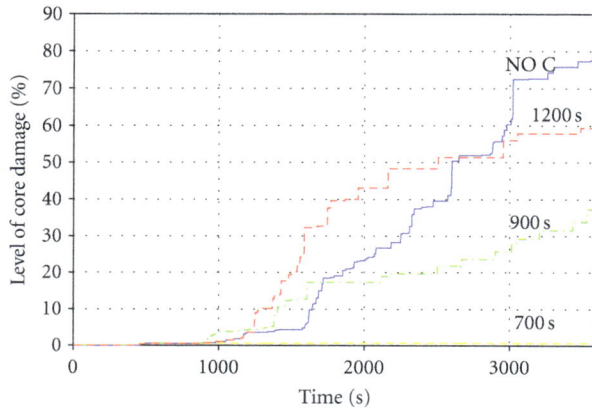

FIGURE 6: Level of core damage. HPCS injection to 700 s, 900 s, and 1200 s; NO C indicates without HPCS injection.

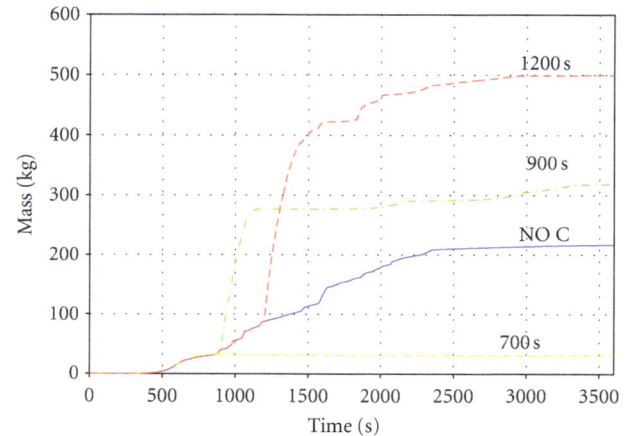

FIGURE 8: Core total hydrogen generation (kg). HPCS injection to 700 s, 900 s, and 1200 s; NO C indicates without HPCS injection.

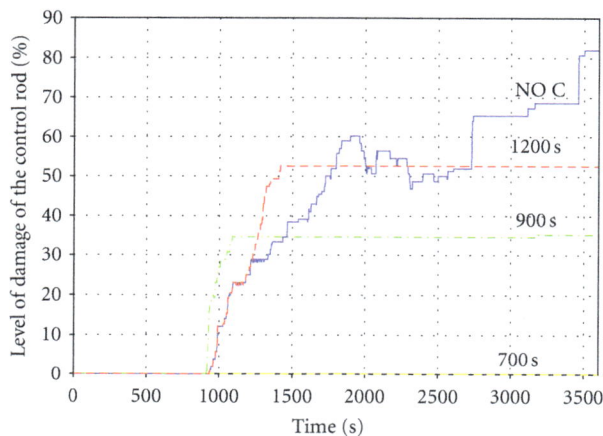

FIGURE 7: Level of damage of the control rod. HPCS injection to 700 s, 900 s, and 1200 s; NO C indicates without HPCS injection.

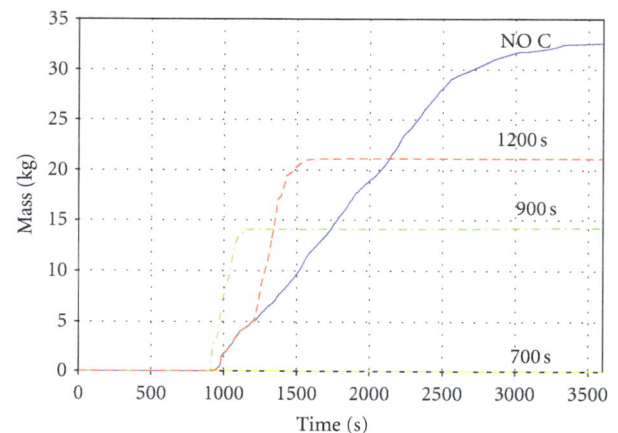

FIGURE 9: Noncondensable fission product released. HPCS injection to 700 s, 900 s, and 1200 s; NO C indicates without HPCS injection.

in which the coolant injected, recovering refrigerant level in the reactor vessel is according to the time of action of HPCS (Figure 4).

According with these results when the HPCS started, the water level in the vessel is partially recovered to ensure the reactor core cooling. Approximately in -400 cm, the increased level presents a change in trend accompanied (or mounted) by oscillations apparently of high frequencies, and these are very strong when the core is practically covered (-150 cm of the level) especially for the case of 700 s of HPCS injection, which is presented between 1000 to 1500 s of elapsed time in the simulation. These effects are produced due to convective heat transfer process between core region and the cooling where the rapid change of liquid phase to gas phase occurs, which causes variations of the pressure in the reactor vessel, these being the main phenomena. Regarding the 900 s of HPCS injection, it can be seen that these effects occur immediately after the coolant interacts with the core region which occurs at -400 cm of water level, while that for the case to 700 s occurs at -150 cm of water level. These differences between two cases (700 and 900 s HPCS injection)

are the energy accumulated in the core region being higher for 900 s. Then, removing energy (900 s) with the same water flow of the HPCS is more complex respect to 700 s, which can be inferred due to the greater instabilities (respect to 700 s) along the simulation time (see 900 s in the Figure 4).

In the case of the 1200 s of HPCS injection, the water level in the reactor vessel is recovered but with significant lower stability, that is, apparently the steam by convective effects and heat transfer mechanism by radiation remove heat generated and accumulated in the core region at this time (1200 s) before the injection of HPCS; also can be observed some agglutinated oscillations of the level in approximately -300 cm and -200 cm characteristics the thermodynamics effects due to phase changes or due to pressure changes. For the case without HPCS injection, the progression of the core damage is expected and the consequences were widely discussed previously.

7.2. Maximum Surface Temperature in the Core. The behavior of the maximum surface temperature in the core is

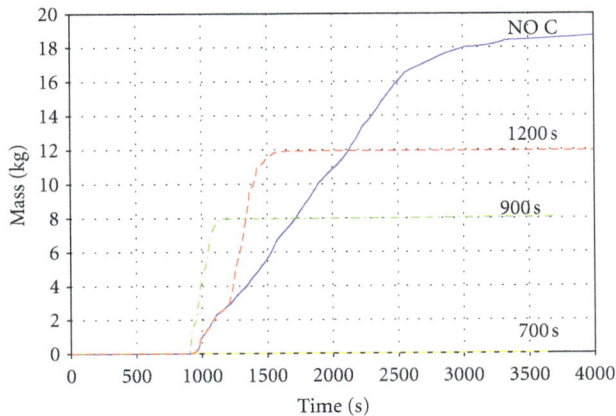

FIGURE 10: Soluble fission products released. HPCS injection to 700 s, 900 s, and 1200 s; NO C indicates without HPCS injection.

presented in Figure 5. For the case of 700 s of HPCS injection, the temperature increases linearly to a temperature of 1600 K at 700 s the elapsed time, under this conditions some fuel elements fail. When acts the HPCS at 700 s; the temperature decreases to 401 K at 900 s of elapsed time, due to presence of coolant removing the heat in the core region primarily by convective mechanism.

For the rest of the numerical experiments, that is, with cooling (HPCS injection to 900 s, 1200 s), and without cooling, it does not cool the core region, as can be observed in Figure 5. However, there are some peculiarities that it should be noted.

(1) At 900 s of HPCS injection, the temperature is approximately of 2126 K that represents conditions of significant core damage, in this case the presence of the coolant flow apparently increases the core damage due to presence the two temperature peaks of approximately of 3000 K. The interpretation physical of this peaks is that the chemicals reactions with materials of the core region (control rod, uranium oxide, Zr, etc.) in average are exothermic, this occurs in time lapse of 900 s to 1000 s the elapsed time, after this time, the temperature decreases to approximately to 2750 K, which is maintained until 2200 s of elapsed time, and after of this time, the temperature increases again to 3000 K, indicating the presence of complex phenomena such as loss geometry in the core region, therefore, some loss of cooling capacity due to coolant flow blockage.

(2) For the case of 1200 s of the HPCS injection and without cooling (NO C in Figure 5), the temperature behavior follows the same trends until 1800 s the elapse time, that is, the behavior is practically the same with and without cooling. It is important to note for cooling case at 1200 s, that the temperature is increased until of 2750 K and remains around this value from the 1150 s to complete simulation, that is, the presence of coolant flow is not responsible that for the temperature does not increase further, due to

that the inflexion point occurs 150 s before injection of HPCS.

(3) For the case without cooling of the core region as was previously mentioned, the temperature behavior follows the same trends of the HPCS injection until 1800 s, in this time, a temperature increase can be observed slightly more than 3000 K, and at 2600 s of the elapse time, the temperature decreases to 2750 K, then afterward, therecan be observed some increases.

The preliminary conclusions according with the behavior of maximum surface temperature in the core, with and without cooling in the core region are as follows: (1) The injection of the HPCS at 700 s, cooling is achieved properly of the core region and the consequence of core damage can be stopped. (2) The injection of the HPCS at 900 s can be not recommendable due to exothermal process that cause large increase of the temperature (temperatures peaks) in the core region. (3) The injection of the HPCS at 1200 s has not significant effect and the behavior is similar to without cooling, apparently due to the loss of cooling capacity due to loss of geometry of the core region.

The energy accumulated and generated by different process such as chemical reactions is a indicative that the time the injection is crucial as after the core region, it is not possible to cool (with HPCS), at least at short times, the order of the hours. Also as is showed by results, the injection at 900 s generates higher thermal shock, which may make it harder to mitigate core damage.

Figure 5 is obtained using the variable BGMCT of RELAP/SCDAPSIM that represent the maximum temperature in the surface of the core, this case is observed that the maximum temperature in the surface of one axial node remains high even after the injection of the HPCS, this is because of the loss of configuration of the core, the hottest axial node is not cooled.

7.3. *Level of the Core Damage.* The behavior of the level of the core damage is presented in Figure 6. This result provides sufficient information to establish the effect of time of injection of cooling water to the core region. In this figure, it can be see that the simulation start with little core damage before of 500 s for four analyzed cases, that is, with injection (HPCS to 700, 900, and 1200 s) and without cooling the core region. The core damage that is represented in the Y axis (Figure 7) refers to 0% when all the axial nodes of the core fuel bundles are intact, and 100% refers when the whole axial nodes of the fuel bundles are in an intermediate stage as ballooning, rubble, and cohesive debris or molten. This core damage is a consequence of the deficiency or lack of cooling the reactor for a prolonged period of time; in this case, the core damage is presented at approximately 750 s. According to Figure 6, the level of core damage is about 1% for the case where the HPCS injection acts at 700 s.

Now, in the following analysis, we take as a reference the behavior without HPCS injection (NO C) to emphasize the importance of time of injection of cooling water in the core region. In Figure 6, it can be seen that without HPCS injection the core damage remains unchanged at a minimum value (1%) until about 1000 s, and after this time the core damage increases. When the injection is at 900 s, the level of the core damage is greater than without cooling until 1750 s where the core damage rate increases slightly. For the case HPCS injection at 1200, this behavior is observed, that is, the level the core damage with and without cooling is practically the same until the 1250 s, after the presence of the cooling water accelerates the core damage whose increase is greater than without cooling case, after 1700 s, this behavior is reversed as can be observed in Figure 6.

In summary form, these results indicate that possible negative effect of cooling can be observed in cases of HPCS injecting at 900 to 1200 s, where a sudden increase (at early times of the injection) to the core damage is fully associated with the entry of coolant into the reactor, being greater the core damage than without coolant injection. However, in Figure 6, it can be observed that the rates of core damage decrease due to the phenomena associated with the heat removal.

7.4. Average Level of Control Rod Damage. The behavior of the level of the core damage is presented in Figure 7. According to these results in general terms, the negative effect of cooling can be observed in cases of HPCS injected at 900 to 1200 s, where a sudden increase to the control rod damage is fully associated with the entry of coolant flow into the core region, being greater than without injection for time elapse of 1400 s and 1800 s, respectively. We must notice that the level of damage in the control rods is a function of the cooling rate; therefore, the damage is greater when the cooling starts late or is missing. The control rods damage that is represented in the Y axis (Figure 7) refers to 0% when all the axial nodes of the control rods are intact, and 100% refers when the whole axial nodes of the control rods are not intact. In Figure 6, it is observed that the level of damage of the control rod decreased between 2250 s and 2600 s is an abnormal behavior and is attributable to a mistake in the version 3.2 of RELAP/SCDAPSIM.

7.5. Hydrogen Generation. The behavior of the hydrogen generation in the core region is presented in Figure 8. The hydrogen generation is coming mainly from the Zircaloy cladding oxidation and partly from the boron carbide oxidation of the control rod. In this figure, it can be observed that the level more low of hydrogen generation is when the HPCS injection acts at 700 s, and, therefore, the heat of reaction is less about the other cases studied. Now, when the HPCS injection acts at 900 s and 1200 s, the masses of the hydrogen generation are practically 11 (which is generated in 150 s) and 17 (is generated in 250 s) times greater than the case of the HPCS injection at 700 s (31 kg of hydrogen generated), respectively. Finally, in 3500 s of the elapsed time, the mass generated for the case of 900 s (HPCS injection) is 320 kg and for the case of 1200 s (HPCS injection) is 500 kg. In both cases, the rate of the heat due to oxidation reaction is very high with respect to the case of 700 s of HPCS injection. For the case without cooling with the HPCS, the mass of the

generation hydrogen was 220 kg, being 7 times higher than the case of 700 s (HPCS injection).

From the point of view of hydrogen generation, it can be observed of the simulations results in Figure 8, that the following hold:

(1) Hydrogen production is maintained at a low level if the coolant flow is appropriate at the right time (in this work correspond to 700 s of the HPCS injection).

(2) If the coolant flow is not appropriate at the right time (cases 900 and 1200 s of HPCS injection), the hydrogen generation increases to more than one order of magnitude, due to coolant flow entering the core region, including the lower the hydrogen generation is smaller than without cooling as is shows in Figure 8.

(3) The consequences are the following: for the cases of the 900 s and 1200 s, the HPCS injection produces high rates of hydrogen generation due to oxidation and therefore high rates of heat of reaction because it is an exothermic reaction (clearly predominates in this numerical experiment this mechanism), which produces high temperatures at the core surface near 3000 K as is observed in Figure 5. Then, the reaction heat is crucial in the cooling of the core region, and the HPCS injection under these conditions is insufficient to cool the core region.

(4) The core and control rod damage (Figures 6 and 7, resp.) has two components that predominate; the first is the time without cooling flow, and reaction heat due to hydrogen generation.

7.6. Fission Products Released from Core Region. The prediction of the behavior of the released fission products from the core are presented in Figure 9 for noncondensable fission products and in Figure 10 for soluble fission products. The released fission products are a function of the level of the core and control rod damage (as was discussed previously in Figures 6 and 7), which is affected by the injection time of the HPCS. According to the progression of the core damage, the mass of fission products (noncondensable and soluble) is higher for the case without cooling (NO C in Figures 9 and 10), which presents higher level of core and control rod damage, with respect to the others cases, that is, 900 s and 1200 s (of HPCS injection).

For the case of 700 s of the HPCS injection, apparently no fission products are released. However, the core damage for this case is less than 1% (Figure 5) and the control rod damage is null (Figure 7). Obviously, one should expect the minimum amount of fission products released if there is failure the reactor core as shown if Figure 7 (see behavior at 700 s).

It is important to note that the release of the fission products contributes to the energy generated and accumulated in the core region; therefore, this fission heat (to call it in some way) the decay heat, reaction heat due to generation of the hydrogen, in other phenomena as melting process causes that the temperature at core region is too high (3000 K) as can be observed in Figure 5 for the case of the 900 s and 1200 s (of

TABLE 2: Severe accident (LOCA) with and without cooling.

Parameter	Without cooling	HPCS injection at 700 s	HPCS injection at 900 s	HPCS injection at 1200 s
Level	Below the core support plate	Below the core support plate. After is recovers	Below the core support plate. After is recovers	Below the core support plate. After is recovers
Pressure	Low by decompression	Low by decompression	Low by decompression	Low by decompression
Maximum surface temperature in the core (K)	3028.00	1640.00 (proper cooling with HPCS)	2960.00 (cooling is not insured with HPCS)	3000.00 (cooling is not insured with HPCS)
Non-condensable fission product (kg)	33.00	0.00	19.00	21.00
Soluble fission product (kg)	18.70	0.00	8.00	11.90
Hydrogen generation (kg)	218.70	30.70	319.50	499.50
Debris bed height (m)	0.36	0.00	0.23	0.21
Core damage (%)	77.70	0.60	37.30	59.40
Control rod damage (%)	82.10	0.00	35.30	52.60

HPCS injection). At this point in the analysis and discussion of the results, we see that the injection of water for the cases of 900 and 1200, generating greater amounts of hydrogen (Figure 8), but smaller amounts of fission products released, with respect to the case without cooling (Figures 9 and 10). Figure 9 refers to noncondensable fission product released (krypton, xenon and Iodine), and Figure 10 is the soluble fission products released (Cesium Iodine (CsI) and Cesium hydroxide (CsOH)) during the LOCA [20, 21].

The comparisons of the four cases studied in this work are summarized in Table 2. In this table, are presented other variables such as control rod damage, hydrogen generation, and debris bed height, which are included in this analysis. But before addressing these variables, it can be observed in this table that the exposition time without cooling is determined for the progression of core damage, for 900 s and 1200 s the cooling cannot assured with HPCS injection, at least in 4000 s of the simulation. The maximum surface temperature in the core indicates that the melting of ZrO_2 and near the melting UO_2 for the case without cooling, 900 s and 1200 s of HPCS injection. The fission products released for the case without cooling are greater than the other cases but not the hydrogen generation being higher for the cases of 900 s and 1200 s of HPCS injection (under this condition, the water accelerated the hydrogen generation, when the temperature of the core region is of the order of 1470 K). The core and control rod damage is proportional to the injection time of the HPCS, being maximum for the case without cooling and minimum for 700 s of HPCS injection.

Now, respect to the debris bed height also can be observed that is proportional to core and control rod damage, being maximum for the case without cooling and null for 700 s of HPCS injection. The core region molten core materials accumulated on the core support plate would be eventually relocated into de lower plenum region and the thermal attack on the lower head vessel could occur. From phenomenological point of the view, the physical and

chemical processes of the core debris are very complex, due to configuration, temperature, and composition. The heat removal from the debris bed in the lower plenum is determined by amount of the heat generated in the molten core materials, heat transfer inside the core material, and heat transfer from the core material to the reactor vessel. In this work, the thermal attack on the lower plenum is not presented, which requires a broad and specific study.

8. Conclusions

The severe accident simulation in the BWR of the Laguna Verde Nuclear Power Plant (LVNPP) to 105% of rated power was analyzed in this work using RELAP/SCDAPSIM code. The severe accidents start with a loss-of-coolant accident (LOCA) in loop of the recirculation of the reactor with and without water cooling injection. The variables analyzed were level, pressure, maximum surface temperature in the core region, maximum temperature of the control rod, fission products released (noncondensable and soluble), hydrogen generation, core and control rod damage, and debris bed height in lower plenum. The numerical experiments consider a little damage in the core with three different times of HPCS injection at 700 s, 900 s, and 1200 s. The following results were obtained.

(i) In the case of the 1200 s of HPCS injection, the level is recovered but with significant lower stability (Figure 4), indicative of the low activity of heat transfer processes, losing the cooling capacity for severe core damage.

(ii) For the case of HPCS injection at 900 s, two temperature peaks in the surface of the core of approximately of 3000 K were observed when the water cooling was injected (Figure 5).

(iii) For the case of 1200 s of the HPCS injection and without cooling, the behavior of the surface temperature in the core is practically until 1800 s (Figure 5).

(iv) The energy accumulated and generated by different process such as reactions chemical is an indicative that the time the injection is crucial due to that after the core region is not possible to cool, at least a short time the order of the hours (Figure 5).

(v) The negative effect of cooling can be observed in cases of HPCS injecting at 900 s and 1200 s, where a sudden increase to the core damage is fully associated with the entry of coolant into the core region, being greater than the core damage than without injection (Figure 6).

(vi) The hydrogen generation is higher with coolant injection for the cases of 900 s and 1200 s (of HPCS injection) that without coolant injection (Figure 8).

(vii) The fission products released for the case without coolant injection are greater than with coolant injection (Figures 9 and 10).

These results are crucial and can be applied to establishing strategies for LVNPP and development of severe accident guide.

References

[1] NRC, "Nuclear Regulatory Commission: 10 CRF 100 "Reactor Site Criteria" and 10 CFR 50 Appendix A, Criterion 19 "Control Room"," 1999.

[2] CFE, Comisión Federal de Electricidad (CFE): Final Safety Analysis Report (FSAR), México, 1979.

[3] Innovative Systems Software, LLC (ISS), "Software, System thermal hydraulics", 2012, http://www.relap.com/sdtp/software.php/.

[4] C. M. Allison and J. K. Hohorst, "Role of RELAP/SCDAPSIM in nuclear safety," Science and Technology of Nuclear Installations, vol. 2010, Article ID 425658, 2010.

[5] RELAP5/MOD3.3 Code Manual Volume I: Code Structure, System Models, and Solution Methods, Nuclear Safety Analysis Division, NUREG/CR-5535/Rev 1-Vol I.

[6] R. Gaunt, R. Cole, C. Erickson et al., "MELCOR Computer code manuals," NUREG/CR-6119, Rev. 1, National Laboratory, USA, 1998.

[7] MAAP4, "MAAP4: Modular accident analysis program for LWR plants, code manual vols. 1–4," Fauske & Associates Inc., Burr Ridge, Ill, USA, 1994.

[8] K. I. Ahn and D. H. Kim, "A state-of-the-art review of the reactor lower head models employed in three representative U.S. severe accident codes," Progress in Nuclear Energy, vol. 42, no. 3, pp. 361–382, 2003.

[9] BWR/6, BWR/6 General Description of a Boiling Water Reactor, Nuclear Energy Division, General Electric Company, 1975.

[10] Guía de Seguridad No. 1.10, Consejo de Seguridad Nuclear, Madrid, España, 1995.

[11] C. Allison, "Phenomenology of Severe Accident," IAEA sponsored training workshops on Analysis of Severe Accident.

[12] B. De Boeck and D. Gryffroy, "Introduction to severe accidents especially the containment behaviour," AVN-97/013, AIB VINÇOTTE, NUCLEAR, 1997.

[13] S. Hagen, P. Hofmann, V. Noack, L. Sepold, G. Schanz, and G. Schumacher, Comparison of the Quench Experiments CORA-12, CORA-13, CORA-17, Forschungszentrum Karlsruhe, 1996.

[14] NUREG/CR-6042, Rev. 2, Perspectives on Reactor Safety, U. S. Nuclear Regulatory Commission, Washington, DC, USA, 2002.

[15] G. Schanz, B. Adroguer, and A. Volchek, "Advanced treatment of zircaloy cladding high-temperature oxidation in severe accident code calculations Part I. Experimental database and basic modeling," Nuclear Engineering and Design, vol. 232, no. 1, pp. 75–84, 2004.

[16] C. Duriez, M. Steinbrück, D. Ohai, T. Meleg, J. Birchley, and T. Haste, "Separate-effect tests on zirconium cladding degradation in air ingress situations," Nuclear Engineering and Design, vol. 239, no. 2, pp. 244–253, 2009.

[17] M. Steinbrück, "Prototypical experiments relating to air oxidation of Zircaloy-4 at high temperatures," Journal of Nuclear Materials, vol. 392, no. 3, pp. 531–544, 2009.

[18] E. Beuzet, J. S. Lamy, A. Bretault, and E. Simoni, "Modelling of Zry-4 cladding oxidation by air, under severe accident conditions using the MAAP4 code," Nuclear Engineering and Design, vol. 241, no. 4, pp. 1217–1224, 2011.

[19] P. Royl, H. Rochholz, W. Breitung, J. R. Travis, and G. Necker, "Analysis of steam and hydrogen distributions with PAR mitigation in NPP containments," Nuclear Engineering and Design, vol. 202, no. 2-3, pp. 231–248, 2000.

[20] SCDAP/RELAP5/MOD3.2 Code Manual, Volume II: Damage Progression Model Theory, NUREG/CR-6150, INEL-96/0422, Revision 1, October 1997.

[21] SCDAP/RELAP5/MOD3.2 Code Manual Volume IIII: User's Guide and Input Manual, NUREG/CR-6150, INEL-96/0422, Revision 1, November 1997.

[22] G. Espinosa-Paredes and A. Nuñez-Carrera, "SBWR model for steady state and transient analysis," Science and Technology of Nuclear Installations, vol. 2008, Article ID 428168, 18 pages, 2008.

16

Image-Processing-Based Study of the Interfacial Behavior of the Countercurrent Gas-Liquid Two-Phase Flow in a Hot Leg of a PWR

Gustavo A. Montoya,[1] Deendarlianto,[2] Dirk Lucas,[3] Thomas Höhne,[3] and Christophe Vallée[3]

[1] *Chemical Engineering Department, Simon Bolivar University, Valle de Sartenejas, Baruta, 1080A Caracas, Venezuela*
[2] *Department of Mechanical and Industrial Engineering, Faculty of Engineering, Gadjah Mada University, Jalan Grafika No. 2, Yogyakarta 55281, Indonesia*
[3] *Helmholtz-Zentrum Dresden-Rossendorf e.V., Institute of Safety Research, P.O. Box 510 119, 01314 Dresden, Germany*

Correspondence should be addressed to Thomas Höhne, t.hoehne@hzdr.de

Academic Editor: Michio Murase

The interfacial behavior during countercurrent two-phase flow of air-water and steam-water in a model of a PWR hot leg was studied quantitatively using digital image processing of a subsequent recorded video images of the experimental series obtained from the TOPFLOW facility, *Helmholtz-Zentrum Dresden-Rossendorf e.V.* (HZDR), Dresden, Germany. The developed image processing technique provides the transient data of water level inside the hot leg channel up to flooding condition. In this technique, the filters such as median and Gaussian were used to eliminate the drops and the bubbles from the interface and the wall of the test section. A Statistical treatment (average, standard deviation, and probability distribution function (PDF)) of the obtained water level data was carried out also to identify the flow behaviors. The obtained data are characterized by a high resolution in space and time, which makes them suitable for the development and validation of CFD-grade closure models, for example, for two-fluid model. This information is essential also for the development of mechanistic modeling on the relating phenomenon. It was clarified that the local water level at the crest of the hydraulic jump is strongly affected by the liquid properties.

1. Introduction

One hypothetical accident scenario in which two-phase countercurrent flow may occur in a PWR hot leg is a loss-of-coolant accident (LOCA), which is caused by the leakage at any location in the primary circuit. During this scenario it is considered that the reactor will be depressurized and vaporization will take place, thereby creating steam in the PWR primary side. Should this lead to "reflux condensation," which may be a favorable event progression, the generated steam will flow to the steam generator through the hot leg. This steam will condense in the steam generator, and the condensate will flow back through the hot leg to the reactor, resulting in countercurrent steam/water flow. In some scenarios, the success of core cooling depends on the behavior of this countercurrent flow.

The stratified countercurrent flow of steam and condensate is only stable for a certain ranges of steam and water mass flow rates. For a given condensate flow rate, if the steam mass flow rate increases to a certain value, a portion of the condensate will exhibit a partial flow reversal and will be entrained by the steam in the opposite flow direction towards the steam generator. This phenomenon is known as countercurrent flow limitation (CCFL) or the onset of "flooding." In case of an additional increase of the steam flow, the condensate is completely blocked and the reflux cooling mode ends. In this situation the cooling of the reactor core from the hot leg is impossible but may be continued by coolant drained through the cold leg to the downcomer. Figure 1 illustrates the countercurrent flow in the hot leg under reflux condensation conditions.

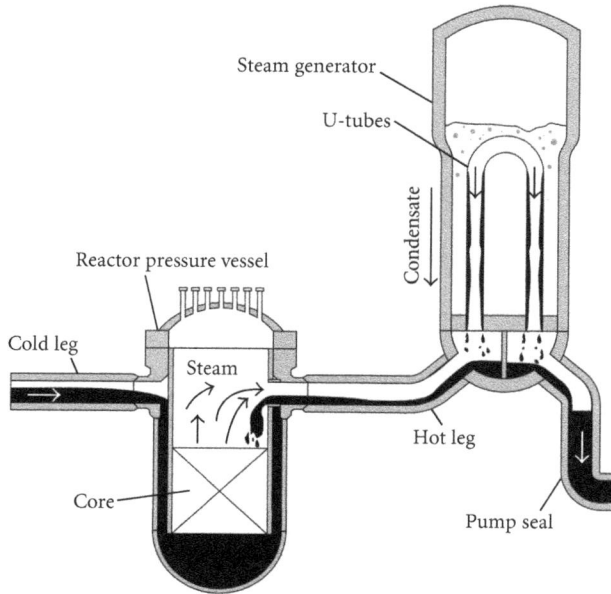

FIGURE 1: Konvoi German PWR piping configuration and reflux condensation flow paths [1].

In order to understand safety-related issues in nuclear power plants, analytical simulations using computational fluids dynamics (CFD) tools have been done, expecting to enhance the accuracy of the simulation predictions compared to the established one-dimensional thermal hydraulic analyses. Compare to the traditionals thermal hydraulic codes, CFD would supply a more reliable scale up to the reactor scale, as well as a more flexible behavior in terms of transferability of models to changes in geometry and thermodynamic boundary conditions. This due to its ability to reveal the interactions between the phases, which are determined by interfacial transfers, and the ability to substitute geometry-dependent empirical closure relations with more physically justified closure laws formulated at the scale of the structures of the interface.

To support the theoretical model development and to validate the CFD codes, a horizontal rectangular channel connected to an inclined riser was constructed, as a model of a PWR hot leg, where air/water countercurrent two-phase flow experiments were performed. This equipment was installed in the pressure chamber of the TOPFLOW test facility (transient two-phase flow) of *Helmholtz-Zentrum Dresden-Rossendorf* (HZDR), Dresden, Germany. This model allows the investigation of co- and counter-current flows under reactor typical boundary conditions (steam/water at pressures up to 5.0 MPa and saturation temperature). This has become the major experimental facility of the German CFD-network, initiated by the GRS (*Gesellschaft für Anlagen und Reaktorsicherheit mbH*).

For the CFD validation, it is of great importance to ensure a good access for measurements of distributed flow parameters, more than to create an exact geometrical similarity with the original equipment. Also, there is no need to quantify the critical mass flow rate since it was done

in the past (e.g., *UPTF, Germany, in UPTF-Fachtagung IV* [2]). In previous experiments, the recognition of bubbles and droplets as detailed structures was not possible mainly because the optical access was limited, and so the observation of the flow was mainly used to support the interpretation of results.

Since investigations in the past, performed in pipes [3–5], were limited by the three-dimensional shape of the interfacial structure, the new test section has been optimized for the application of optical observation using a flat channel model of a PWR hot leg. In order to accelerate the CFD code validation program, the high-resolution pictures were analyzed by own developed image-processing algorithms. In the present paper, experimental procedure and developed image-processing technique will be presented firstly. Next, the time variation of water level inside the hot leg channel around the CCFL will be given. By using this data, the development of the waves around the CCFL can be clarified. Next, the statistical treatment of the data will be presented on the basis of the average parameters. Finally, the effect of fundamental parameters such as liquid mass flow rate, system pressure, and liquid properties will be discussed.

2. Experimental System and Image Processing Technique

2.1. Experimental Setup. A schematic view of the hot leg model test section is shown in Figure 2. It consists of the test section, the reactor pressure vessel (RPV) simulator located at the lower end of the horizontal channel, and the steam generator (SG) separator connected to the inlet chamber. This test section is a reproduction of the diameter of PWR hot leg of a German *Konvoi* type at a 1 : 3 scale. In order to provide better viewing conditions, the test section is not composed of pipes as in the original plant, but rather is a channel of about 50 mm (wide) which represents a vertical cut through the midplane of the hot leg and the inlet chamber of the steam generator. By using a flat section the error produced by the three-dimensional interface in high-speed videos is eliminated, allowing a closer look of the interface, including dispersed structures (drops and bubbles) that cannot be observed in a common pipe. Consequently, the test section is composed of a horizontal rectangular channel, a bend that connects it to an upward inclined and expanded channel, and a quarter of a circle representing the steam generator inlet chamber. The horizontal part of test section is 2.12 m long and has a rectangular cross-section of 0.05 m × 0.25 m. The SG and RPV simulators are identical vessels with 0.8 m × 0.5 m × 1.55 m (D × W × H) cubic shape.

Unfortunately, due to the overall dimensions of the hot leg model (3 m long), it was not possible to visualize the complete test section. Therefore, a region of observation had to be chosen. Previous investigations (e.g., [3]) indicate that the most agitated flow region is located near the bend and that a recirculation zone is formed there. Consequently, it was chosen to observe the bended region of the hot leg and the steam generator inlet chamber as shown in Figure 2. The

Image-Processing-Based Study of the Interfacial Behavior of the Countercurrent Gas-Liquid Two-Phase Flow
in a Hot Leg of a PWR

159

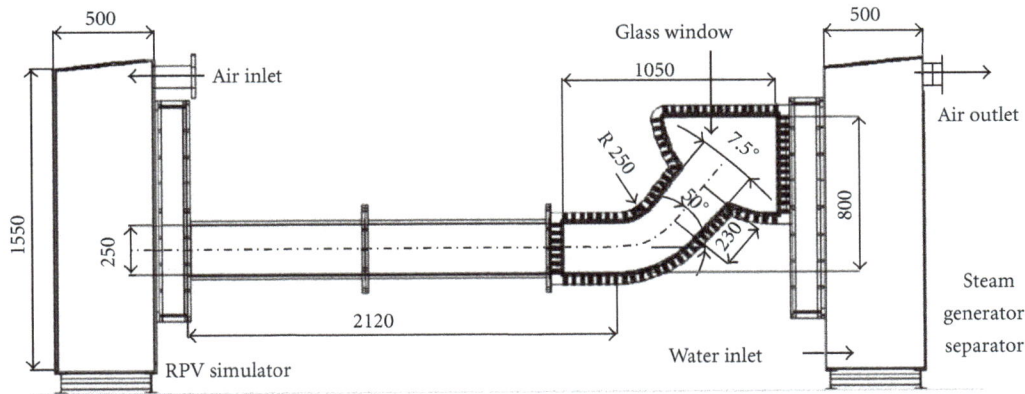

FIGURE 2: Schematic view of the hot leg model test section (dimension in mm).

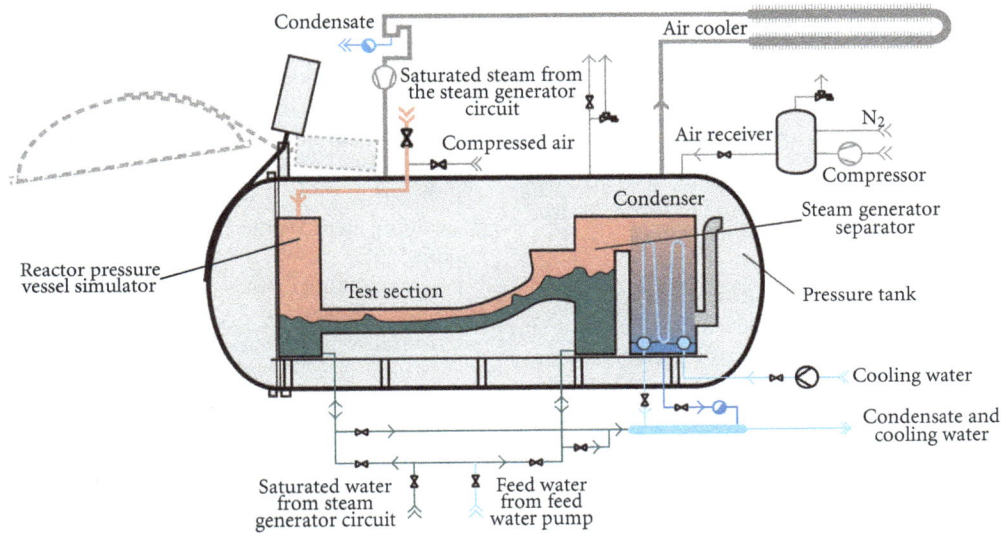

FIGURE 3: Schematic diagram of the experimental apparatus.

flow behavior was recorded by a high-speed video camera at frequencies of 60 to 100 Hz and a shutter speed of 1/1000 s.

The test section was placed in a pressure chamber, in which it was operated in pressure equilibrium with the inner atmosphere of the tank in TOPFLOW (transient two-phase flow) facility as shown in Figure 3. A special heat exchanger condenses the exhaust vapor from the test section directly in the pressure vessel, when steam/water experiments are made. As shown in Figure 3, the cold end of this condenser and the inside atmosphere of the vessel are permanently connected, in order to guarantee full pressure equilibrium at all times [6]. The vessel can be pressurized up to 5 MPa either with air for cold experiments or with nitrogen for steam experiments. Using this method allowed to design the equipment with thin materials.

The injected water mass flow rate was measured by a vortex meter. The injected air mass flow rate was measured and controlled using thermal mass flow meters. The steam flow rate over the pressure drop was measured through a venturi's tube. The temperatures of the fluids were measured

by thermocouples at various positions in the facility. The water levels in both tanks were determined by the measurement of the differential pressure between the top and the bottom of the vessels with differential pressure transducers. The pressure drop over the test section was measured by a differential pressure transducer placed between the SG simulator and the RPV separator. These global parameters were measured via a data acquisition system running at 1 Hz, which was synchronized with the high-speed video camera.

In this experiment, the air was injected in the RPV simulator and flowed through the test section to the SG separator, from which it was released to the inner atmosphere of the pressure tank. The water from the feed water pump was injected in the SG separator, from where it can flow in countercurrent to the air flow through the test section to the RPV simulator. The onset of flooding was obtained by a stepwise increase of the gas mass flow rate with a small increment (9–35 g/s), under a constant water mass flow rate. Due to the internal buffer of the high-speed camera that was limited to 8 GB, a compromise had to be found in each run

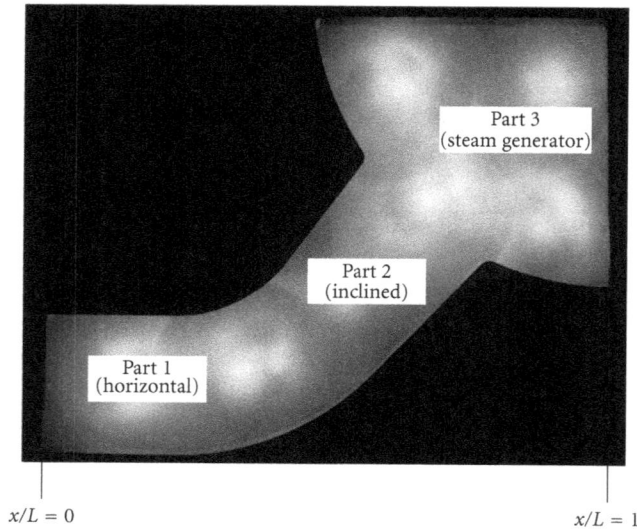

FIGURE 4: Submatrix segmentations in the image processing.

TABLE 1: Analyzed experimental data.

Gas	Pressure (MPa)	Experimental running			
Steam	5.00		06-15		
Steam	3.00		15-07	15-10	
Steam	2.36			11-07	
Steam	1.50		11-01	11-04	
Air	0.30	18-09	30-05		
Air	0.15	19-02	30-09		
		0.15	0.3	0.6	Water (kg/s)

between the number and duration of the plateau of air flow rate. The number of plateau realized during one run was varied between 4 and 8, with a duration of each between 15 and 35 s [7].

The onset of flooding was defined as the limiting point of stability of the countercurrent flow, indicated by the maximum air mass flow rate at which the down-flowing water mass flow rate is equal to the inlet water mass flow rate. This is a method used previously by Zabaras and Dukler [8] and Deendarlianto et al. [9]. The experiments were carried out until the point of zero liquid penetration occurs, when the down-flowing water mass flow rate was equal to zero [7].

During the onset of flooding, the water level reached in the horizontal segment of the hot leg, obstructs the passage of steam enough to generate waves at the interface, which are eventually transform into slugs. Consequently, the pressure difference between the separators increase and become unstable. Then, the gas that impedes the water to flow into the RPV accumulates in the SG separator. The average pressure drop through the test section increases with the level of water in the steam generator, and the slug increases until the zero liquid penetration point occurs.

2.2. Image Process Techniques. For the analysis of the high-speed video data obtained from the TOPFLOW experiments, two programs based on image processing techniques were used. These programs were written in MATLAB code along with simulink files. First the program will cut a sub-matrix from the original data, which becomes separate in a horizontal, inclined, and steam generator segments of the image as shown in Figure 4. An example of one image before processing can be seen in Figure 5(a).

A pre-processing stage is begun in the horizontal segment of the image in order to improve the interface in this part. Next step is to eliminate the drops and bubbles in the interface and the water that is stick in the walls of the test

section. Here the bubbles and drops were considered as noise and possible reason of error. In order to correct this, the smooth filters were used to eliminate any salt and pepper noise.

In order to improve the uneven light conditions found in the data before processing, a background was created using the first image obtained from the Gaussian filter. After this, the background image is subtracted from the one before the background development. Next, an image enhancement function was implemented to improve the contrast. Finally, the image quality can be improved. An example of the processing steps of the horizontal segment can be found in Figure 5(b).

When analyzing the inclined segment, the original image is rotated about −50 degrees in order to make it become a horizontal approximation as in Figure 5(c). For the steam generator, the raw image is cut into a submatrix leaving only the part of interest. When the preprocessed of the steam generator segment ends, then all the obtained data is send from the workspace to a simulink file. In this file, the raw image is binarized in order to detect the borders of the hot leg channel. Finally, the water level can be calculated. An example of the processing steps for the steam generator segment can be found in the Figure 5(d).

To obtain the local information, the picture was divided into twelve points for the horizontal segment, nine points for inclined segment, and ten points for the steam generator segment. The program for analyzing the steam-water data works on the same philosophy as the air-water data. Finally, an example of the detected water level for the steam-water case is shown in Figures 6(a) and 6(b).

Ten experimental data series of HZDR in total were studied. Six of them were steam-water, and four of them were air-water. These sets of data were chosen in order to obtain a wide variety of parameters' combinations for pressure and mass flow rate, where XY-WZ represents the name of the specific experiment based on the date in which the experiment was made. The selected series are shown in Table 1.

Extensive checks were made in order to analyze the quality of the algorithm, but the most useful method in order to verify the reliance of the code is an automated one inside the same algorithm. When the water level's values are found in pixels, the algorithm creates markers that are drawn in the exact place in the original gray-level images where the water level point was calculated. This allows verifying each

Image-Processing-Based Study of the Interfacial Behavior of the Countercurrent Gas-Liquid Two-Phase Flow in a Hot Leg of a PWR

161

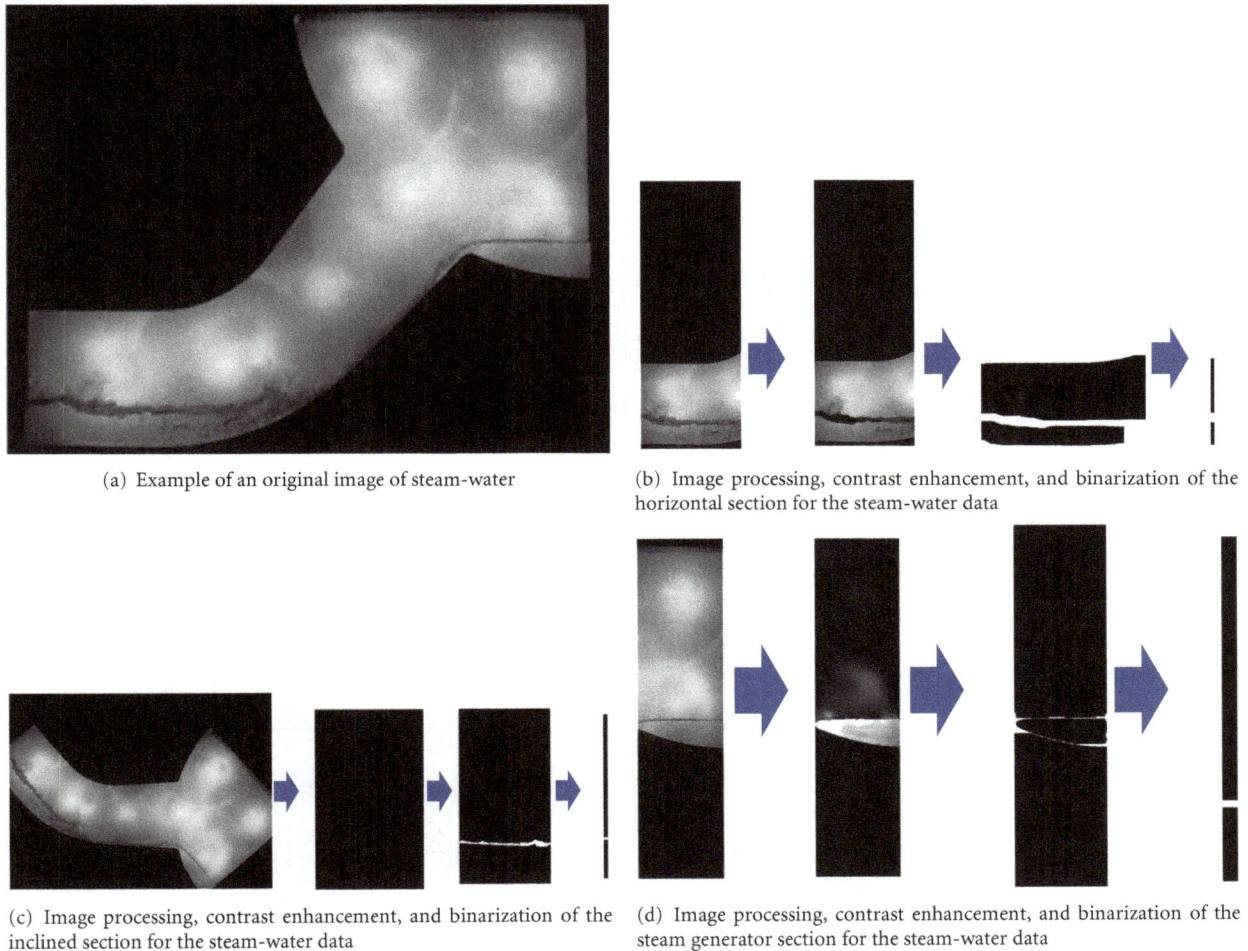

(a) Example of an original image of steam-water

(b) Image processing, contrast enhancement, and binarization of the horizontal section for the steam-water data

(c) Image processing, contrast enhancement, and binarization of the inclined section for the steam-water data

(d) Image processing, contrast enhancement, and binarization of the steam generator section for the steam-water data

FIGURE 5: Detected water level and boundaries of the hot leg channel (steam-water).

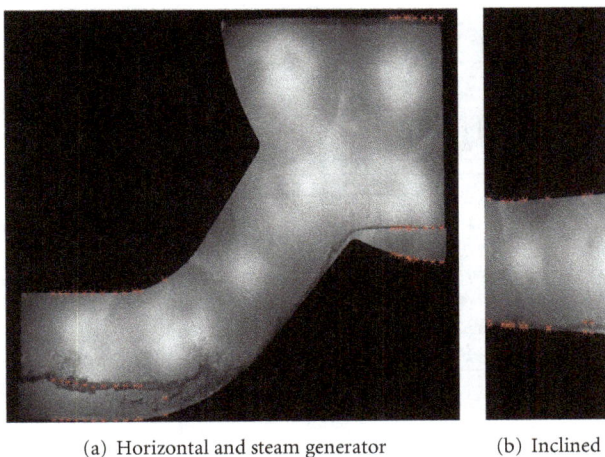

(a) Horizontal and steam generator

(b) Inclined

FIGURE 6: Detected water level and boundaries of the hot leg channel (steam-water).

exact value that is calculated for each water level in every image. This is made automated, and the result images with the markers are saved in a separate folder, which allow the possibility of verifying the results during any point after the processing of a set of images.

Analyzing a representative set of data of 5000 pictures after using the image processing algorithm on them, allowed to find that 233 points out of 155000 were deviated from the actual interphase, which represents a discrepancy of 0.15%.

3. Results and Discussions

Figures 7(a) and 7(b) show the time variation of the water level during the CCFL of air-water and steam-water, respectively. The injected liquid mass flow rate was 0.3 kg/s. The temperature of the fluids for the experiments with air/water was 18–24°C (room temperature) and for stem/water was saturation temperature. As shown in the figure, the water level in the horizontal and inclined parts exhibits notable oscillations, and the shape of the wave varies along the channel. This indicates that the wave patterns during the CCFL are space and time dependent. Next, it can be seen that the successive waves do not follow the direction of liquid injection, and for that matter the CCFL can be identified.

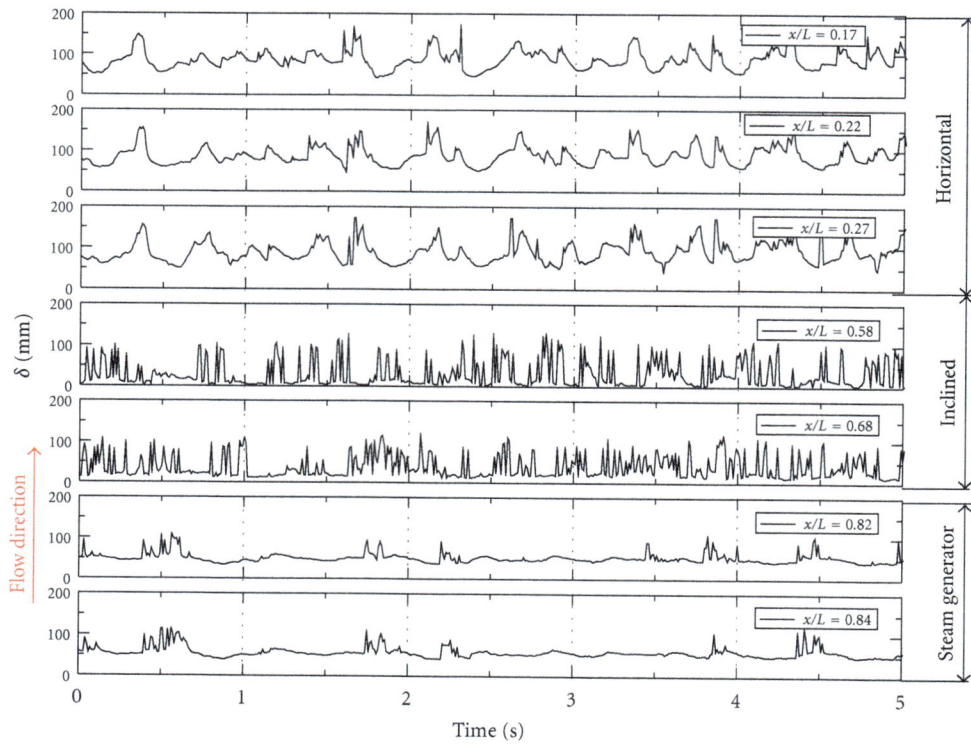

(a) Air-water ($P = 0.15$ MPa, $m_L = 0.30$ kg/s, $m_G = 0.28$ kg/s)

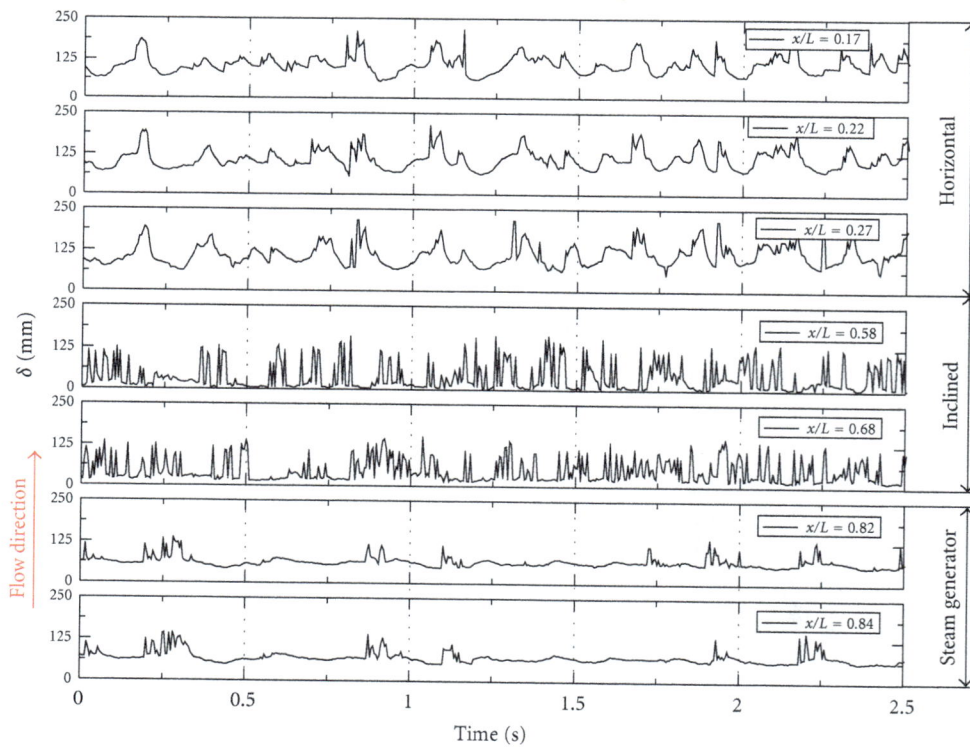

(b) Steam-water ($P = 1.5$ MPa, $m_L = 0.30$ kg/s, $m_G = 0.53$ kg/s)

FIGURE 7: Time variation of water level inside the hot leg channel during CCFL.

Image-Processing-Based Study of the Interfacial Behavior of the Countercurrent Gas-Liquid Two-Phase Flow in a Hot Leg of a PWR

163

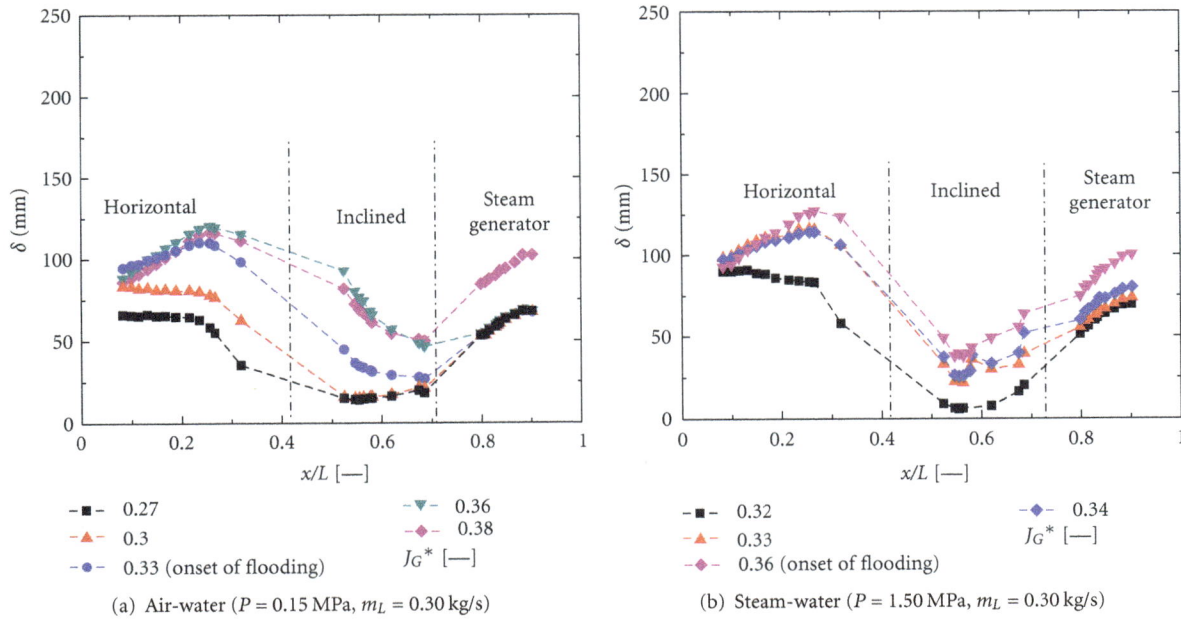

(a) Air-water ($P = 0.15\,\text{MPa}$, $m_L = 0.30\,\text{kg/s}$)

(b) Steam-water ($P = 1.50\,\text{MPa}$, $m_L = 0.30\,\text{kg/s}$)

FIGURE 8: Time average of the water level distribution inside the hot leg channel.

In addition, the maximum of the water level was lower than 250 mm. This indicates that there is no total blockage of the water inside the channel during the CCFL.

3.1. Average Parameters. Figure 8 illustrates the distribution of the water level inside the hot leg channel. Figures 8(a) and 8(b) correspond to the cases of air-water and steam-water, respectively. The injected liquid mass flow rate was 0.3 kg/s of each case. For a meaningful comparison, a nondimensional gas superficial velocity $J_G{}^*$, namely, as gas Wallis parameter, is used. Here the gas Wallis parameter in Figure 8 is defined as follows:

$$J_k{}^* = J_G \sqrt{\frac{1}{gH} \cdot \frac{\rho_G}{(\rho_L - \rho_G)}}, \qquad (1)$$

where J_G indicates the superficial gas velocity, ρ the density, g the acceleration of gravity, and H the height of the channel. Close inspection of Figure 8 reveals that before the onset of flooding the water levels in horizontal and inclined parts increase with the increase of the gas Wallis parameter. Meanwhile it is almost constant in the steam generator until the onset of flooding. Next, it shows also that the water level in the horizontal part decreases when x/L closes to the inclined part. The decrease of water level at the beginning of the inclined part indicates the change of the liquid film profile from the subcritical to supercritical condition as described by Deendarlianto et al. [7]. Meanwhile this profile changes dramatically when the gas mass flow rate approaches the onset flooding. This means that the change of wave direction is begun here. Next, it is noticed the observed phenomenon of air-water is the same as of the steam-water.

Figure 9 illustrates the standard deviation of water level inside the hot leg channel. The flow condition was similar to

that of Figure 8. The figure shows that the standard deviation in horizontal and inclined parts increases with the increase of gas Wallis parameter. This means that wave fluctuations increase also with the increase of the gas superficial velocity. On the other hand, the wave in the steam generator begins fluctuating near the onset of flooding.

A small algorithm to calculate the probability distribution function was created. The probability distribution of a random variable can be defined as a function that assigns the probability that an event occurs to each defined variable. The program computed the *normpdf* (X, mu, sigma). This function determines the PDF for each value of *X*, using the data, the average, and standard deviation. The mathematical function that defined this is

$$y = f(x \mid \mu, \sigma) = \frac{1}{\sigma\sqrt{2\pi}} e^{-(x-\mu)^2/2\sigma^2}. \qquad (2)$$

After the calculation of the PDF, the obtained values are ordered from highest to lowest to get the PDF chart with the values placed in a corresponding manner, without requiring any subsequent operation. This operation is performed for the 31 data points (x/L).

At the end of the program, the results for the 31 points of the PDF are stored in two vectors to be used or transferred to another cell's working program in the future (e.g., Excel).

Figure 10 shows the example of the probability distribution function (PDF) of the water level at the horizontal segment. This figure reveals that the PDF begins to spread when the gas velocity approaches the onset of flooding. From this result, it is possible to notice that the spread of PDF in the horizontal position under this flow condition can be used as a good indicator to detect the occurrence of flooding.

When $J_G{}^*$ is small compared with the other values studied in this case, the probability function is represented

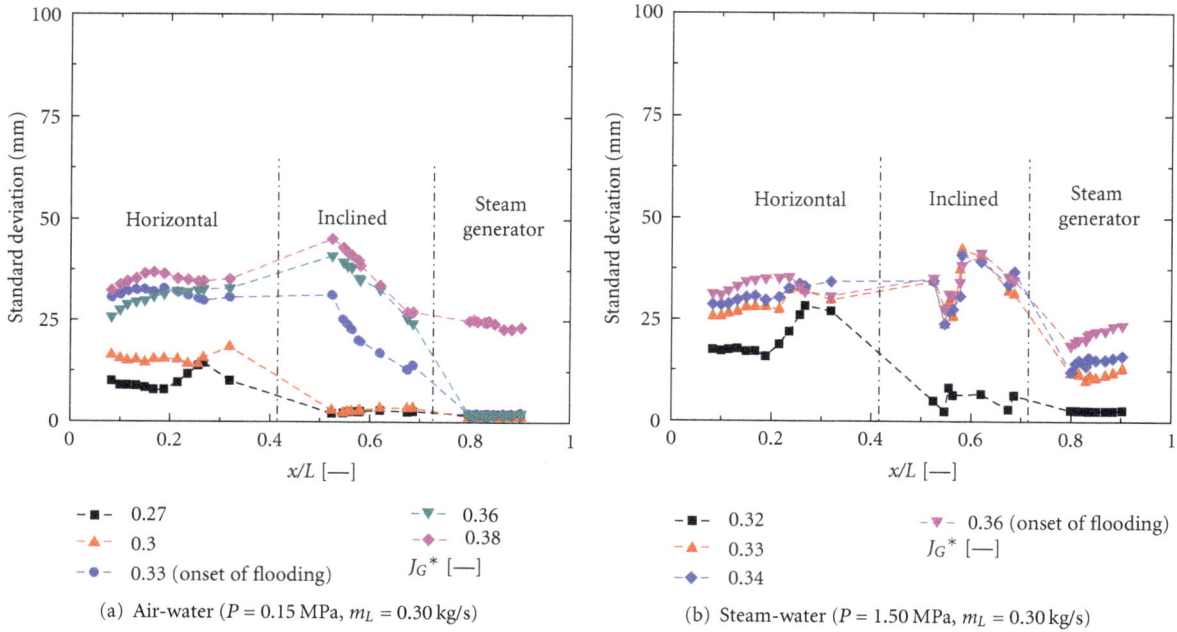

(a) Air-water ($P = 0.15\,\text{MPa}$, $m_L = 0.30\,\text{kg/s}$)

(b) Steam-water ($P = 1.50\,\text{MPa}$, $m_L = 0.30\,\text{kg/s}$)

FIGURE 9: Distribution of the standard deviation of the water level inside the hot leg channel.

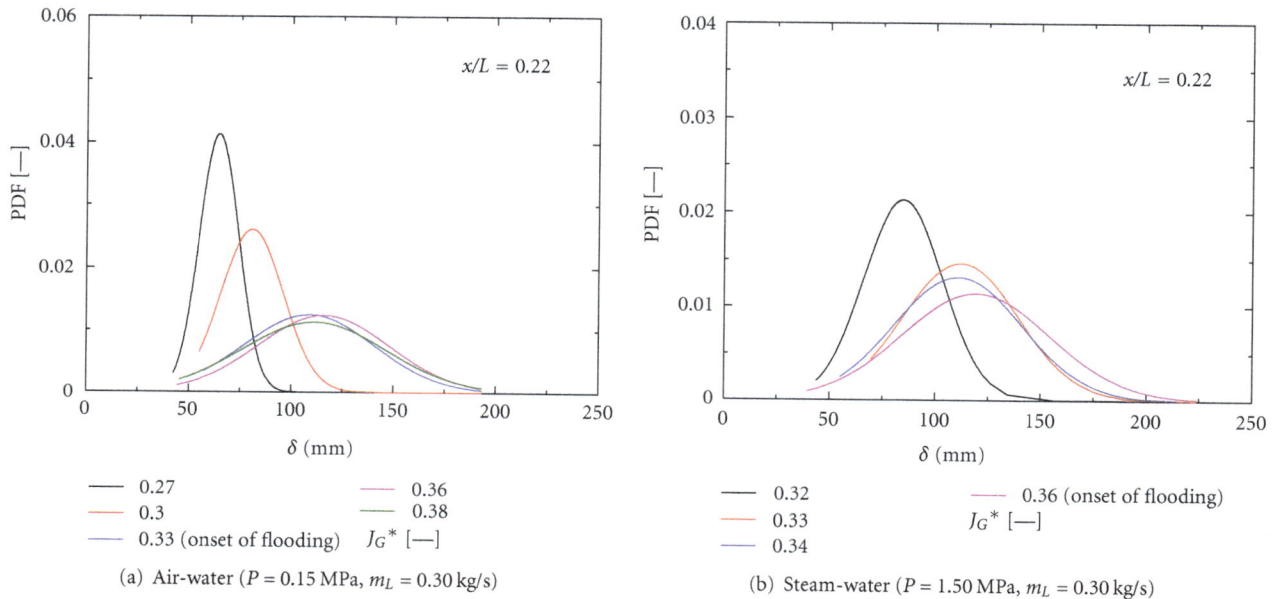

(a) Air-water ($P = 0.15\,\text{MPa}$, $m_L = 0.30\,\text{kg/s}$)

(b) Steam-water ($P = 1.50\,\text{MPa}$, $m_L = 0.30\,\text{kg/s}$)

FIGURE 10: Probability distribution function ($x/L = 0.22$, $m_L = 0.3\,\text{kg/s}$).

by a curve with the smallest dispersion of water levels in comparison with the rest. This curve is equivalent to a stratified and almost laminar flow. As J_G^* increases, the curve opens and the dispersion of probabilities on the water level rises. This means that as J_G^* increases, the flow becomes more turbulent and less stratified, which is why it can be seen an opening of the probability to multiple values of water level instead of tending to one.

In the case of Figure 10(a), before reaching the CCFL ($J_G^* < 0.33$), the values of the liquid phase level in which the

probabilities are greater are displaced to the left, indicating a lower water level at this point. Then, when the CCFL occurs ($J_G^* = 0.33$), the values became more scattered (less stratified flow), and the level of the phase is most likely to become greater in the region near the elbow. It can be seen that, once the onset of flooding is reached, the curves for higher J_G^* are maintaind around the same values.

While in Figure 10(b) could be difficult to find the point for the onset of flooding, it can be noticed that, based on the same principle, the values keep changing, becoming

Image-Processing-Based Study of the Interfacial Behavior of the Countercurrent Gas-Liquid Two-Phase Flow in a Hot Leg of a PWR

165

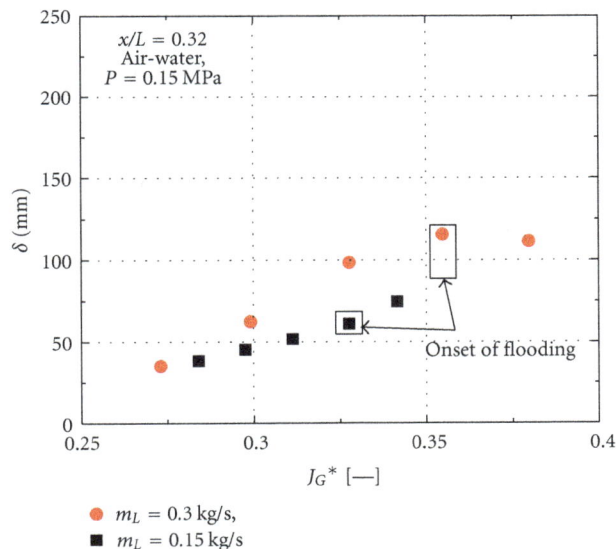

FIGURE 11: Effect of injected liquid mass flow rate on the water level.

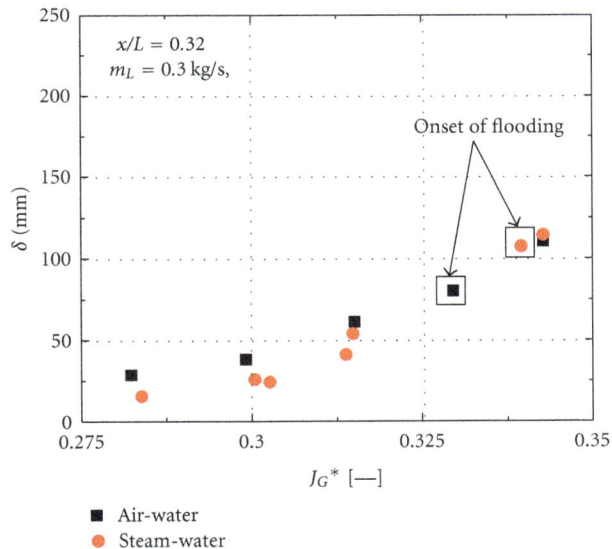

FIGURE 13: Effect of physical properties on the water level.

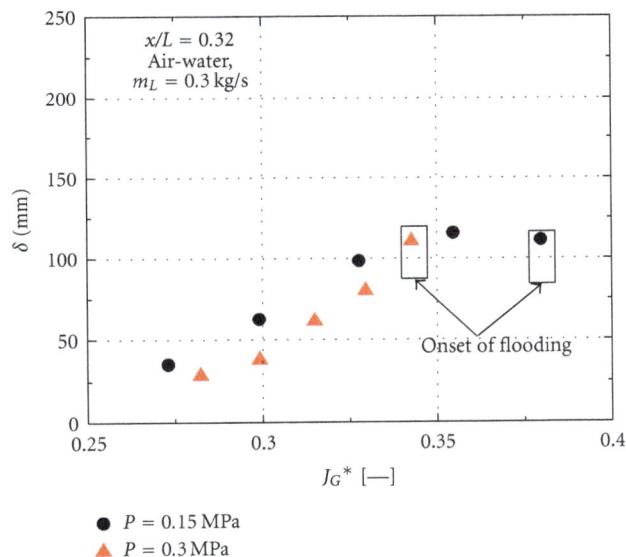

FIGURE 12: Effect of system pressure on the water level.

increasingly dispersed until $J_G^* = 0.36$. This same pattern repeats for all the other experimental conditions. One of the main differences between air/water and steam/water experiments when analyzing the PDF is that because the steam/water interphase is not completely laminar since low values J_G^*, then the change to the onset of flooding seems more drastic in the air/water experiments.

Figures 11 to 13 show the effects of the fundamental parameters such as the injected liquid mass flow rate, system pressure, and liquid properties on the water level inside the hot leg channel, respectively. The shown data is the water level at the horizontal part of $x/L = 0.32$, which is the closest location to the elbow. It was taken due to the consideration

that the flooding coincides to the slugging inception in the lower leg of the elbow close to the bend [10, 11].

Figure 11 shows that for the conditions presented, when every parameter but the liquid mass flow rate is fixed, the water level will depend of this last value. When $m_L = 0.30$ kg/s, the water level will be approximately 100 mm greater than that when a liquid mass flow rate of 0.15 kg/s is used.

In Figure 12, the effect of system pressure of water level is shown. It can be seen that, when every parameter is fixed but the system pressure, the water phase is maintained almost at the same level. Finally, Figure 13 shows the effect of physical properties over the water level, where it can be seen that when air-water are used as the fluids for the experiments, the water level in the onset of flooding increases in around 80 mm when compared to the experiment in which the steam-water are the working fluids. The same pattern repeats when other experiments are analyzed. This clearly indicates that the liquid mass flow rate and the physical properties greatly affect the onset of flooding in the experiments, while the pressure is a parameter that would not affect the water level in a drastic way when changed.

Then, it is shown that before the onset of flooding the water level increases with the increase of gas superficial velocity. Meanwhile, at the onset of flooding, it is dependent on the injected liquid mass flow rate and the physical properties. Those effects were not considered by previous researches; therefore, the proposed flooding correlations in their literature were not successful.

4. Conclusions

An image-processing technique was developed to analyse the sequence video images of the countercurrent gas-liquid two-phase flow of air-water and steam-water in a model of PWR hot leg. The data images were taken from the experimental series of TOPFLOW facility, HZDR, Dresden, Germany. This technique allows us to access the quantitative local flow information; therefore, a detailed CFD-grade data for the water level evolution inside the hot leg channel including its statistic characteristics were set up. Extensive checks were also made with regard to the quality of the method. The data can be used to validate the CFD code, and physics behind the observed phenomenon were clarified by being able to understand more in deep the behavior of the system and the particular properties that affected the water level during the onset of flooding, as well as the effect of the $J_G{}^*$ in the water level and the time-dependent interfacial behavior of the flow.

Acknowledgments

This work is carried out within the frame work of a current research project funded by the German Federal Ministry of Economics and Technology, Project number 1501329. The authors would like to thank also the TOPFLOW team for their work on the test facility and the preparation of the experiments. Dr. Deendarlianto is an Alexander von Humboldt Fellow in the Institute of Safety Research, Helmholtz-Zentrum Dresden-Rossendorf e.V., Germany. The present research is also supported by the Alexander von Humboldt Foundation in Germany.

References

[1] T. Seidel, C. Vallée, D. Lucas, M. Beyer, and Deendarlianto, "Two-Phase Flow Experiments in a Model of the Hot Leg of a Pressurised Water Reactor," Wissenschaftlich-Technische Berichte/Forschungszentrum Dresden-Rossendorf; FZD-531, 2010.

[2] UPTF-Fachtagung IV, Versuchsergebnisse, Analysen, Mannheim 25. Marz , Siemens AG, KWU, KWU R 11/93/005, 1993.

[3] A. Ohnuki, H. Adachi, and Y. Murao, "Scale effects on countercurrent gas-liquid flow in a horizontal tube connected to an inclined riser," Nuclear Engineering and Design, vol. 107, no. 3, pp. 283–294, 1988.

[4] S. Wongwises, "Experimental investigation of two-phase countercurrent flow limitation in a bend between horizontal and inclined pipes," Experimental Thermal and Fluid Science, vol. 8, no. 3, pp. 245–259, 1994.

[5] M. A. Navarro, "Study of countercurrent flow limitation in a horizontal pipe connected to an inclined one," Nuclear Engineering and Design, vol. 235, no. 10–12, pp. 1139–1148, 2005.

[6] H. M. Prasser, M. Beyer, H. Carl et al., "The multipurpose thermalhydraulic test facility TOPFLOW: an overview on experimental capabilities, instrumentation and results," Kerntechnik, vol. 71, no. 4, pp. 163–173, 2006.

[7] Deendarlianto, C. Vallée, D. Lucas, M. Beyer, H. Pietruske, and H. Carl, "Experimental study on the air/water counter-current flow limitation in a model of the hot leg of a pressurized water reactor," Nuclear Engineering and Design, vol. 238, no. 12, pp. 3389–3402, 2008.

[8] G. J. Zabaras and A. E. Dukler, "Counter-current gas-liquid annular flow, including the flooding state," AIChE Journal, vol. 34, no. 3, pp. 389–396, 1988.

[9] Deendarlianto, A. Ousaka, A. Kariyasaki, and T. Fukano, "Investigation of liquid film behavior at the onset of flooding during adiabatic counter-current air-water two-phase flow in an inclined pipe," Nuclear Engineering and Design, vol. 235, no. 21, pp. 2281–2294, 2005.

[10] K. H. Ardron and S. Banerjee, "Flooding in an elbow between a vertical and a horizontal or near-horizontal pipe. Part II: theory," International Journal of Multiphase Flow, vol. 12, no. 4, pp. 543–558, 1986.

[11] S. Wongwises, "Two-phase countercurrent flow in a model of a pressurized water reactor hot leg," Nuclear Engineering and Design, vol. 166, no. 2, pp. 121–133, 1996.

17

Design and Simulation of a New Model for Treatment by NCT

Seyed Alireza Mousavi Shirazi[1] and Dariush Sardari[2]

[1] Department of Physics, Sciences College, Islamic Azad University, South Tehran Branch, Tehran 1581819411, Iran
[2] Department of Nuclear Engineering, Technical College, Islamic Azad University, Science and Research Branch, Tehran, Iran

Correspondence should be addressed to Seyed Alireza Mousavi Shirazi, a_moosavi@azad.ac.ir

Academic Editor: Xing Chen

In this investigation, neutron capture therapy (NCT) through high energy neutrons using Monte Carlo method has been studied. In this study a new method of NCT for a sample liver phantom has been defined, and interaction of 12 MeV neutrons with a multilayer spherical phantom is considered. In order to reach the desirable energy range of neutrons in accord with required energy in absence of eligible clinical neutron source for NCT, this model of phantom might be utilized. The neutron flux and the deposited dose in the all components and different layers of the mentioned phantom are computed by Monte Carlo simulation. The results of Monte Carlo method are compared with analytical method results so that by using a computer program in Turbo-Pascal programming, the deposited dose in the liver phantom has been computed.

1. Introduction

Neutron capture therapy (NCT) has been one of the most important methods for treatment of cancers in recent years. This method of radiation therapy is applicable in treatment of liver cancer. During clinical practice, it is always essential to stop absorption of additional dose by normal tissue. On the other hand, measurement and assessment of the absorbed dose and its calibration is an important matter [1, 2]. Thus computation and modeling of the deposited dose by Monte Carlo method before practical treatment is recommended. An appropriate software tool for this purpose is MCNP4C code. It is a particular engineering solution when BNCT facilities such as low energy neutron source are not available.

In this paper for simulation by MCNP4C code, a phantom is considered so that it has been encased by polyethylene sphere with 20 cm radius. This sphere is covered with a layer of cadmium which has 100 μm thickness [3, 4]. The cadmium layer has high absorption cross-section for thermal neutrons and helps to reentrance the scattered neutrons from surface of the sphere to the phantom [5]. The polyethylene sphere is surrounded by a graphite shell which has 25 cm radius and 5 mm thickness (according to moderation ratio:

$\xi(\Sigma_s/\Sigma_a)$). This layer serves as a reflector to reduce escaping the fast neutrons [6].

In the present work, neutrons are emitted from an external source, and after passing through polyethylene and slowing down, their deposited energy in the phantom's materials is computed by the MCNP4C code. The F6 tally in the MCNP4C code is applied. The absorbed energy in the liver is computed through analytical computations as well. It includes generation of random numbers along with using the neutron diffusion equation [7]. The outcomes of two methods are finally compared.

The results of computation provide assurance that whether the absorbed dose in cancerous and healthy tissues is in accord with requirements [8, 9]. Then neutron therapy would be implemented on the patient. In neutron capture therapy it is suggested to use monoenergetic neutrons. For this purpose, one choice is D-Be source which produces 14 MeV neutrons [10].

As Am-Be source produces a wide spectrum of neutron energies, thus, is not suitable for this purpose [11]. Liver tissue includes substances such as water, glycogen, and heavy molecules like protein. Figure 1 shows interaction of incident neutron and depositing the energy during neutron irradiation.

TABLE 1: The components and the structural materials of a liver tissue (relating to male sex).

Mass percent	Material
69.69%	Water
0.35%	Glycogen ($C_{24}H_{42}O_{21}$)
29.9%	Protein and Glucose ($C_{44189}H_{71252}N_{12428}O_{14007}S_{321}$ and $C_6H_{12}O_6$)

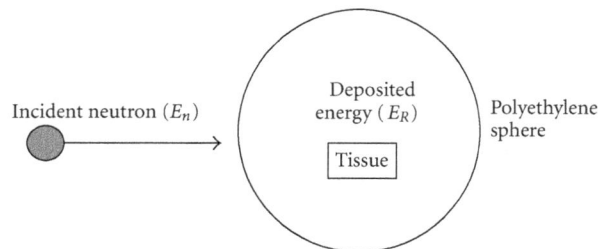

FIGURE 1: The emitted neutron from source and entering into sphere including tissue.

2. Monte Carlo Calculations

To simulate neutron therapy and related dosimetry, a phantom is considered as shown in Figure 2.

Figure 3 depicts the geometry which has been defined by MCNP4C code, and Figure 4 is a schematic view of the physical shape of the phantom.

Laboratory experiments have already rendered the precise molecular composition of liver tissue according to Table 1 [12].

The composition and geometrical data belonging to liver tissue have been given to MCNP4C as input data [13, 14]. In the simulation, neutron slowing down has been taken into account as well [15]. The input data to MCNP4C code are the radius of polyethylene sphere (20 cm), the thickness of cadmium layer (100 μm), radius and thickness of graphite reflector (25 cm and 5 mm, resp.), and the weight of liver tissue that is 190 g.

3. Analytical Calculations

It is essential to obtain data on neutron scattering cross-section and its angular distribution because of computing the neutron penetration in liver tissue (or in the corresponding equivalent material) and obtaining the absorbed energy in it. Since neutrons pass through the polyethylene sphere, the knowledge about neutron angular distribution after scattering is vital. Thus, using the random sampling techniques might help to compute the probability of those neutrons which might be scattered in a definite angle [16]. Since it might be comprehended, the atomic composition of the soft tissue is approximated as well with hydrocarbon materials.

In the process of neutron transport in matter, it is subjected to three major kinds of interactions with carbon and hydrogen nuclei. These are elastic scattering, inelastic scattering, and radioactive capture. Each kind of interactions

takes place with a specific probability with a magnitude proportional to related interaction cross-section according to following equations (1)–(4):

$$P_1 = \frac{\sum_{H(\text{Elastic})}}{\sum_{\text{tot}}},$$

$$P_2 = \frac{\sum_{C(\text{Elastic})}}{\sum_{\text{tot}}}. \tag{1}$$

The probability of inelastic scattering in carbon (for the first excitation level in: 4.43 MeV):

$$P_3 = \frac{\sum_C}{\sum_{\text{tot}}}. \tag{2}$$

The probability of inelastic scattering in carbon (for the second excitation level in: 7.65 MeV):

$$P_4 = \frac{\sum_C}{\sum_{\text{tot}}}. \tag{3}$$

The probability of neutron absorption in hydrogen and carbon:

$$P_5 = \frac{\left[\sum_C + \sum_{H(\text{absorption})}\right]}{\sum_{\text{tot}}}. \tag{4}$$

Collision of neutrons on carbon and hydrogen nuclei is caused to transfer some energy from neutrons to the target nucleus. The recoiled nucleus moves a short distance through the matter and deposits energy along its path. The problem is to compute the energy of recoiled nucleus. By using the Monte Carlo method, collision history of each neutron might be tracked down to the energy where it is either absorbed or escapes from the volume of material.

The lost energy of neutron in each elastic scattering is described by (5):

$$E_1 - E_2(\cos\theta)P(\cos\theta)d(\cos\theta). \tag{5}$$

E_2 is also obtained by the following equation (6):

$$E_2 = E_1 \cdot \frac{A^2 + 2A\cos\theta + 1}{(A+1)^2}. \tag{6}$$

Thus the average energy of recoiled nucleus through assuming the isotropic scattering is

$$\overline{E_R} = \frac{2AE_1}{(A+1)^2}. \tag{7}$$

The inelastic scattering of neutron in polyethylene is important. There are two excited statuses in carbon nucleus, namely, 4.43 MeV and 7.65 MeV. Thus remarkable value of energy is absorbed in the recoiled nucleus, and E_R is less than the energy of inelastic scattered neutrons [17].

The transferred energy to a recoiled nucleus with mass number A due to collision of a neutron with energy E_n is computed with (8) [18]:

$$E_R = \frac{2A}{(A+1)^2}E_n\cos^2\psi. \tag{8}$$

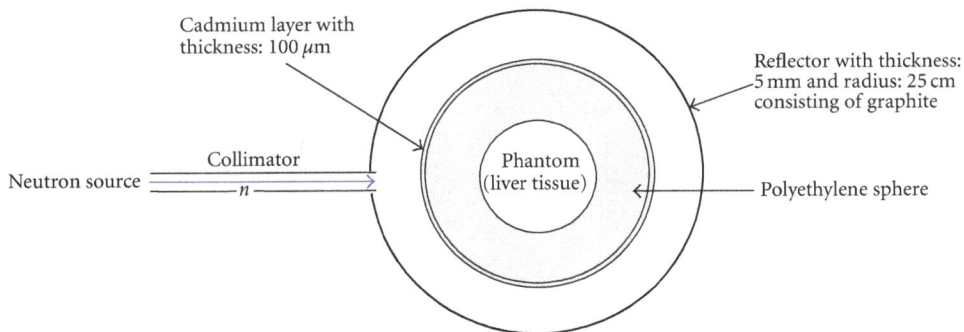

FIGURE 2: The sphere including the liver tissue for the neutron capture therapy.

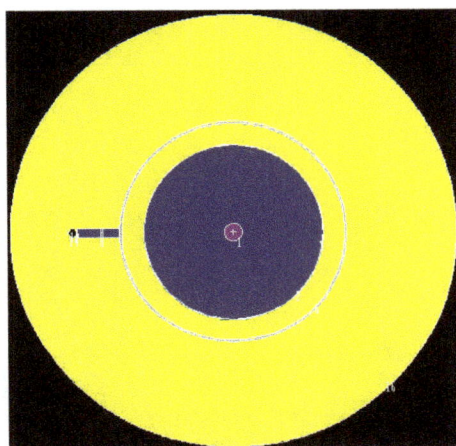

FIGURE 3: The geometry of the phantom defined by MCNP4C code.

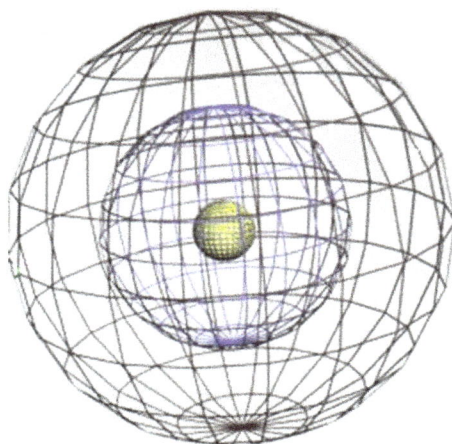

FIGURE 4: The schematic view (3D) of the related geometry.

At 14 MeV the contribution of recoiled proton (hydrogen nucleus) is related to energy deposition phenomenon, that is, 66% of total, and the rest is due to α particle plus other heavy nuclei [20, 21].

In fact, the penetration of neutron through polyethylene sphere and its interaction in liver tissue require precise knowledge about interaction cross-sections and angular distribution of the scattered neutron.

For this purpose by using the Turbo-Pascal programming and considering several values for E_n in the interval 0.001–12 MeV and providing the mass number (A) of the components of the phantom as the input to the program, the absorbed dose in the liver tissue (E_R) and other components such as polyethylene sphere, air, and collimator are computed [Gy]. The number of neutron collisions is calculated according to (9):

$$n = \frac{\ln(E_n/E_R)}{\xi}. \tag{9}$$

In order to determine absorbed dose in the phantom or other components such as polyethylene, the numeric range of E_n (energy of incident neutrons) and E_R (deposited energy) into the number of neutron collisions (n) is divided, and then its value is added to early energy (10):

$$E_{R(\text{new})} = E_R + \frac{E_n - E_R}{n}. \tag{10}$$

The Turbo-Pascal programming which has been developed in the present work computes the energy transferred from incident neutrons to the tissue based on given neutron energy, scattering angle (ψ), and mass number of the target nuclei.

Figure 5 shows a schematic view of the phantom which is comprised by four layers. A hypothetical track and slowing down of neutron in consecutive collisions are shown as well.

Since remarkable percentage of the soft tissue is hydrogen, the interaction of neutron with hydrogen must be studied as well. Approximately 85%–95% of neutron energy transferred to soft tissue is attributed to its interaction with hydrogen [19]. For instance, for $E_n > 8$ MeV the (α, n) reaction makes the significant fraction of absorbed dose in the tissue.

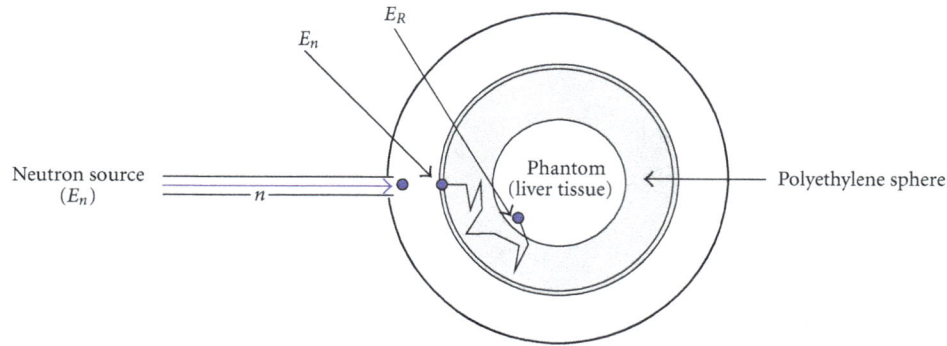

FIGURE 5: The collision of emitted neutron from the external source to the polyethylene shell.

Some useful equations in analyzing the energy transfer of neutron from high energy to the medium are according to (11)-(12) [22]:

$$E_R = E_n e^{-n\xi}, \tag{11}$$

$$L_{sl}^2 = \frac{n(\lambda_{tr})^2}{3}, \tag{12}$$

$$n = \frac{3L_{sl}^2}{\lambda_{tr}^2} \approx \frac{\Sigma_{tr}}{\Sigma_{sl}} = \frac{\Sigma_a + \Sigma_s(1 - (2/3A))}{\Sigma_{sl}}.$$

Thus,

$$n \approx \frac{\Sigma_a + \Sigma_s(1 - (2/3A))}{\Sigma_{sl}}. \tag{13}$$

According to (11), (14) is obtained as follows:

$$E_R = E_n e^{-n\xi} = E_n e^{-((\Sigma_a + \Sigma_s(1-(2/3A)))/\Sigma_{sl}) \times (2/(A+(2/3)))}. \tag{14}$$

The variables of A, E_n, and Σ_{sl} are as the inputs for Turbo-Pascal programming so that E_n is inputted: 12 MeV. Therefore according to (14) the values of absorbed dose in the phantom and other components using Turbo-Pascal programming are obtained.

4. Results

The derived graphs for absorbed dose in the liver tissue and other components of mentioned system (per emitted neutrons) by both Monte Carlo simulation with nps: 10^6 and analytical method (Turbo-Pascal programming) are as Figures 6, 7, 8, 9, 10, 11, and 12.

In addition the neutron fluence [n/cm^2] or on the other hand the number of neutrons which reach the liver tissue and other components is obtained as Figure 13. Meantime the X-axis means the neutron energies reached the phantom's components that are low energies [eV].

5. Conclusion

The absorbed dose by liver tissue in the course of neutron capture therapy has been simulated as function of irradiation time and neutron physical data. This simulation might be

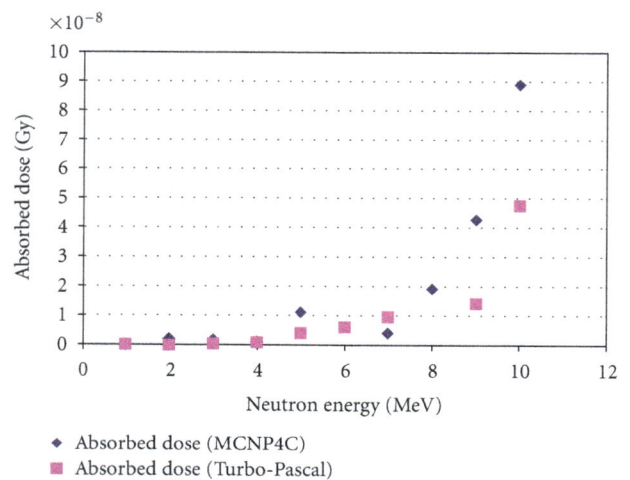

FIGURE 6: The absorbed dose in the liver tissue.

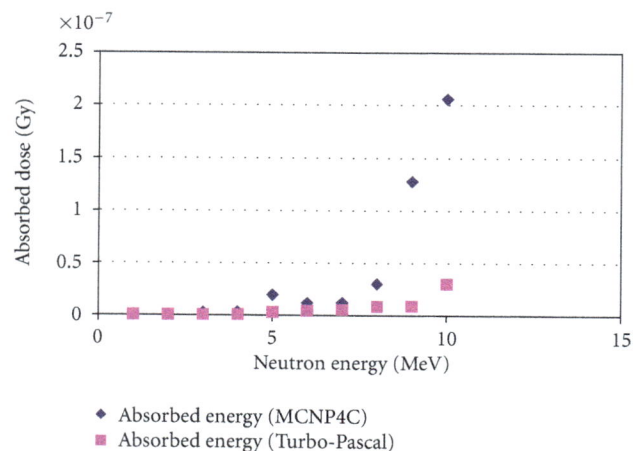

FIGURE 7: The absorbed dose in the polyethylene sphere.

carried out for various shapes and size of liver, and the results must be saved in a data bank. For each patient, in accord with given required dose for treatment by the clinic and also knowing the specifications of neutron source, the eligible irradiation time according to both the neutron fluence

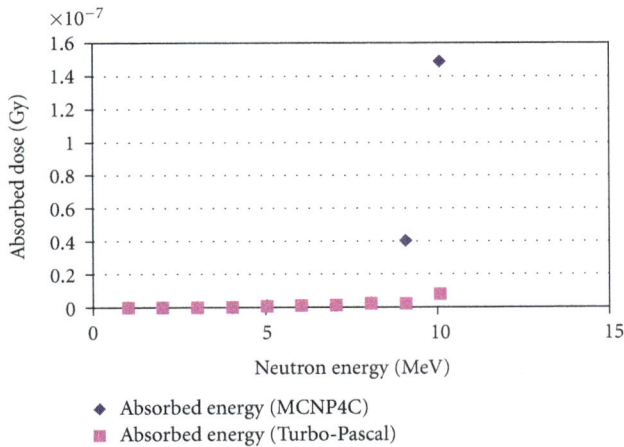

FIGURE 8: The absorbed dose in the cadmium layer.

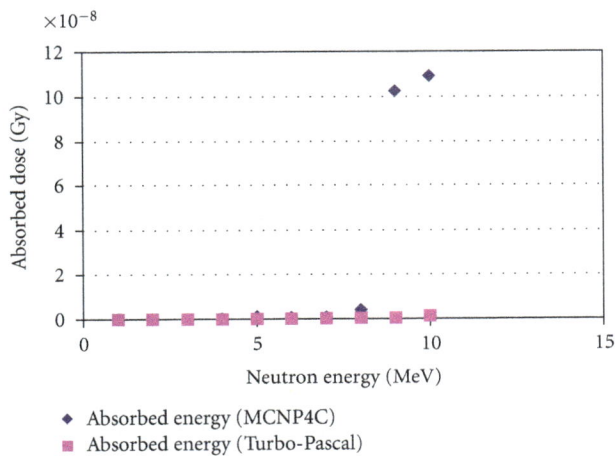

FIGURE 9: The absorbed dose in the air of sphere.

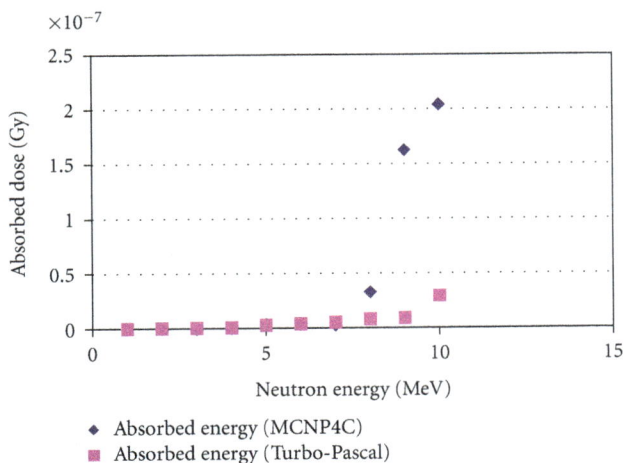

FIGURE 10: The absorbed dose in the collimator.

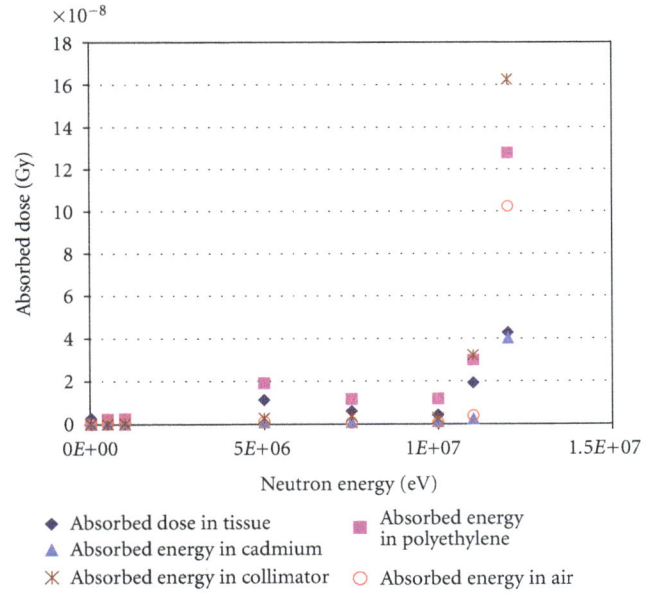

FIGURE 11: The absorbed doses in the liver tissue and other components obtained by MCNP4C.

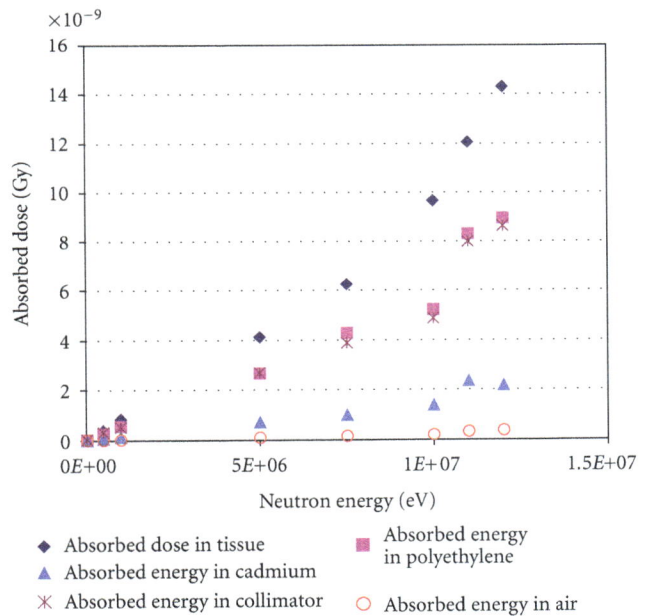

FIGURE 12: The absorbed doses in the liver tissue and other components obtained by Turbo-Pascal programming.

and energy of existing clinical neutron source through the present simulation is obtained. Meantime the neutron energy reached the eligible interval of energy on the tissue computed by (14).

Therefore in case there is no eligible clinical neutron source for NCT, then this model of phantom might be utilized to reach the desirable energy range of neutrons according to the required energy in NCT.

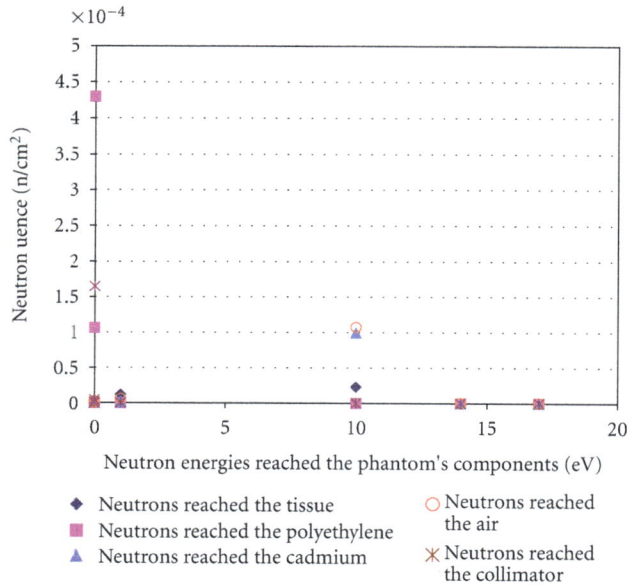

FIGURE 13: The neutron fluence in the phantom's components obtained by MCNP4C.

From Figures 6–10 it is observed that within neutron energy range of 0.001 eV–8 MeV the computed dose by Turbo-Pascal programming is approximately similar to obtained results by MCNP code, and the derived graphs of both methods agree together as well for neutron energy below 8 MeV. Actually, through increasing incident neutron energy, the absorbed dose is increased. For neutron energy above 8 MeV, the results of these two approaches produce significant error. This is because of computation based on MCNP code that computation bases of this code are transport equation and neutron tracking. The equations in Turbo-Pascal programming include diffusion approximations. In addition for neutron energy greater than 8 MeV, (n, α) reactions become more probable to occur. In such nonelastic reactions, the angular distribution of incident neutron and the recoiled nucleus are complicated more than lower energies. This is observed in Figure 7 for polyethylene as well. The absorbed dose in polyethylene phantom is increased through neutron energy so that excess doses may be delivered to the liver tissue. Meantime the absorbed doses because of thermal and epithermal neutron irradiation are negligible.

Nomenclature

P_1, P_2, P_3, P_4, P_5:	Probability of events
E_1:	Initial energy of neutron
E_2:	Final energy of neutron after interaction
ψ:	The angle between the course of projectile neutron and recoiled nucleus in the laboratory system
n:	The number of collisions to reach the eligible energy
A:	Mass number of the components of the phantom
E_n:	Energy of incident neutron
$\overline{E_R}$:	Deposited energy
E_R:	The energy of thermal neutrons which have reached the eligible energy
ξ:	Lethargy
L_{sl}:	Slowing down length
λ_{tr}:	Mean free distance of transport
Σ_{tr}:	Macroscopic transport cross-section
Σ_{sl}:	Macroscopic slowing down cross section
Σ_a:	Macroscopic absorption cross section
Σ_s:	Macroscopic scattering cross section
Σ_{tot}:	Macroscopic total cross section
Σ_C:	Macroscopic absorption cross section for carbon
Σ_H:	Macroscopic absorption cross section for hydrogen.

Acknowledgment

This paper is related to a research project entitled: "Design and Theoretical Simulation of a New System for Neutron Capture Therapy (NCT) of a Sample Tissue for Determination of Requirement Duration and Determination of Absorbed Dose and Preparing the Favorite Energy of Clinical Source" that by sponsorship and financial supporting the "Islamic Azad University-South Tehran Branch" has been carried out.

References

[1] E. C. C. Pozzi, S. Thorp, J. Brockman, M. Miller, D. W. Nigg, and M. Frederick Hawthorne, "Intercalibration of physical neutron dosimetry for the RA-3 and MURR thermal neutron sources for BNCT small-animal research," *Applied Radiation and Isotopes*, vol. 69, no. 12, pp. 1921–1923, 2011.

[2] R. L. Moss, O. Aizawa, D. Beynon et al., "The requirements and development of neutron beams for neutron capture therapy of brain cancer," *Journal of Neuro-Oncology*, vol. 33, no. 1-2, pp. 27–40, 1997.

[3] M. Reginatto, "What can we learn about the spectrum of high-energy stray neutron fields from Bonner sphere measurements?" *Radiation Measurements*, vol. 44, no. 7-8, pp. 692–699, 2009.

[4] M. L. Andrieux, B. Dinkespiler, J. Lundquist, O. Martin, and M. Pearce, "Neutron and gamma irradiation studies of packaged VCSEL emitters for the optical read-out of the ATLAS electromagnetic calorimeter," *Nuclear Instruments and Methods in Physics Research*, vol. 426, no. 2, pp. 332–338, 1999.

[5] M. P. Dhairyawan, P. S. Nagarajan, and G. Venkataraman, "Response functions of spherically moderated neutron detectors," *Nuclear Instruments and Methods*, vol. 169, no. 1, pp. 115–120, 1980.

[6] A. Bolewski, M. Ciechanowski, A. Dydejczyk, and A. Kreft, "On the optimization of the isotopic neutron source method for measuring the thermal neutron absorption cross section: advantages and disadvantages of BF3 and 3He counters," *Applied Radiation and Isotopes*, vol. 66, no. 4, pp. 457–462, 2008.

[7] H. R. Vega-Carrillo, V. Hernandez-Davila, E. Manzanares-Acuña et al., "Neutron spectrometry using artificial neural networks," *Radiation Measurements*, vol. 41, no. 4, pp. 425–431, 2006.

[8] F. Trompier, P. Battaglini, D. Tikunov, and I. Clairand, "Dosimetric response of human bone tissue to photons and fission neutrons," *Radiation Measurements*, vol. 43, no. 2–6, pp. 837–840, 2008.

[9] G. Bartesaghi, J. Burian, G. Gambarini, M. Marek, A. Negri, and L. Viererbl, "Evaluation of all dose components in the LVR-15 reactor epithermal neutron beam using Fricke gel dosimeter layers," *Applied Radiation and Isotopes*, vol. 67, no. 7-8, pp. S199–S201, 2009.

[10] D. Zhou, E. Semones, R. Gaza et al., "Radiation measured during ISS-Expedition 13 with different dosimeters," *Advances in Space Research*, vol. 43, no. 8, pp. 1212–1219, 2009.

[11] S. J. González, M. R. Bonomi, G. A. S. Santa Cruz et al., "First BNCT treatment of a skin melanoma in Argentina: dosimetric analysis and clinical outcome," *Applied Radiation and Isotopes*, vol. 61, no. 5, pp. 1101–1105, 2004.

[12] J. McBride, M. Mason, and E. Scott, "The storage of the major liver components," *The Journal of Biological Chemistry*, vol. 1, pp. 943–952, 1941.

[13] J. T. Goorley, W. S. Kiger, and R. G. Zamenhof, "Reference dosimetry calculations for Neutron Capture Therapy with comparison of analytical and voxel models," *Medical Physics*, vol. 29, no. 2, pp. 145–156, 2002.

[14] T. Tagami and S. Nishimura, "Intercalibration of thermal neutron dosimeter glasses NBS-SRM612 and corning 1 in some irradiation facilities: a comparison," *International Journal of Radiation Applications and Instrumentation. Part*, vol. 16, no. 1, pp. 11–14, 1989.

[15] D. Rochman, R. C. Haight, S. A. Wender et al., "First measurements with a lead slowing-down spectrometer at LANSCE," in *Proceedings of the International Conference on Nuclear Data for Science and Technology (AIP '04)*, pp. 736–739, October 2004.

[16] T. Matsumoto, H. Harano, Y. Ito, A. Uritani, K. Emi, and K. Kudo, "Development of a fast neutron spectrometer composed of silicon-SSD and position-sensitive proportional counters," *Radiation Protection Dosimetry*, vol. 110, no. 1–4, pp. 223–226, 2004.

[17] J. Chuncheng, G. H. R. Kegel, J. J. Egan et al., "Measurement of U-235 fission neutron spectra using a multiple gamma coincidence technique," in *Proceedings of the International Conference on Nuclear Data for Science and Technology (AIP '04)*, vol. 769, pp. 1051–1053, October 2004.

[18] H. Kahn, "Application of Monte Carlo," USAEC Report AECU-3259, Rand Corporation, Santa Monica, Calif, USA, 1954.

[19] S. Y. Hohara, M. Imamura, T. Kin et al., "Development of gas proportional scintillation counter for light heavy-ion detection," in *Proceedings of the International Conference on Nuclear Data for Science and Technology(AIP '04)*, pp. 773–775, October 2004.

[20] S. H. Shinde and T. Mukherjee, "Sensitization of glycine (spectrophotometric read-out) dosimetric system using sorbitol," *Radiation Measurements*, vol. 44, no. 4, pp. 378–383, 2009.

[21] T. Taosheng, L. Dong, and H. Li, "A Monte Carlo design of a neutron dose-equivalent survey meter based on a set of 3He proportional counters," *Radiation Measurements*, vol. 42, no. 1, pp. 49–54, 2007.

[22] M. Stacy, *Nuclear Reactor Physics (SE)*, Chapter 5, John Wiley Publishing Company, New York, NY, USA, 2007.

18

Aqueous Nanofluid as a Two-Phase Coolant for PWR

Pavel N. Alekseev, Yury M. Semchenkov, and Alexander L. Shimkevich

NRC "Kurchatov Institute", 1 Kurchatov Square, Moscow 123182, Russia

Correspondence should be addressed to Alexander L. Shimkevich, shall@dhtp.kiae.ru

Academic Editor: Boštjan Končar

Density fluctuations in liquid water consist of two topological kinds of instant molecular clusters. The dense ones have helical hydrogen bonds and the nondense ones are tetrahedral clusters with ice-like hydrogen bonds of water molecules. Helical ordering of protons in the dense water clusters can participate in coherent vibrations. The ramified interface of such incompatible structural elements induces clustering impurities in any aqueous solution. These additives can enhance a heat transfer of water as a two-phase coolant for PWR due to natural forming of nanoparticles with a thermal conductivity higher than water. The aqueous nanofluid as a new condensed matter has a great potential for cooling applications. It is a mixture of liquid water and dispersed phase of extremely fine quasi-solid particles usually less than 50 nm in size with the high thermal conductivity. An alternative approach is the formation of gaseous (oxygen or hydrogen) nanoparticles in density fluctuations of water. It is possible to obtain stable nanobubbles that can considerably exceed the molecular solubility of oxygen (hydrogen) in water. Such a nanofluid can convert the liquid water in the nonstoichiometric state and change its reduction-oxidation (RedOx) potential similarly to adding oxidants (or antioxidants) for applying 2D water chemistry to aqueous coolant.

1. Introduction

It is well known [1] that the microstructure of liquid water is not understood, and its dynamic hydrogen-bonds (HB) structure has been the subject of intense debate for decades. Ice, whose HB structure was long ago well established, forms a tight "tetrahedral" lattice of molecules each binding to four others. The prevailing model of liquid water holds that as ice melts, the molecules loosen their grip but remain generally arranged in the same tetrahedral groups. This hydrogen-bonding pattern has been assumed to account for water properties.

However the majority of molecules were found in higher density regions with an asymmetric disordered structure where some islands of tetrahedral order were floated [2]. The greater density of liquid water in these regions implies that the molecules are more closely packed there than the simple tetrahedrons seen in ice [3, 4].

The conclusion [2, 5] that a dominant fraction of the molecules in liquid water are very asymmetrically hydrogen bonded with only two well-defined H-bonds (one donating and one accepting) is in strong contrast to the accepted picture as being near tetrahedral, H bonded. From small-angle X-ray scattering studies, they furthermore find evidence for density nonhomogeneity on a length-scale of 1 nm indicating that the two components are spatially separated on the time scale of the experiment. The recent controversy about the structure of liquid water pits new models involving water molecules in relatively stable rings-and-chains structures against the standard model that posits water molecules in slightly distorted tetrahedral coordination. The current study is giving new life to familiar Rontgen's "two-structure" model of liquid water [3, 4].

A topological structure of density fluctuations in condensed matter has been studied by molecular-dynamics (MD) simulation as ramified clusters of almost regular Delaunay's simplexes (tetrahedrons) built on the fours of densely packed atoms and connected in pairs by faces as tetrahedral Bernal's chains [6, 7]. The review of publications on this subject is presented in the monograph [8] and a topological criterion [9–11] is offered for finding these simplexes exactly. Such a criterion allows making the selection of the dense-part simplexes by fixing the overall length of their edges in a point of maximum number of obtained clusters in the MD cell.

At the same time, any nanofluid as a new coolant is a suspension of nanoparticles with sizes less than 100 nm and volume fractions typically less than 4% [12]. Such a coolant has shown the ability of enhancement (up to 40%) in thermal conductivity compared with the base liquid [13] and a significant increase of critical heat flux [14]. Oxides (Al_2O_3, CuO, TiO_2), nitrides (AlN, SiN), carbides (SiC, TiC), and metals (Ag, Au, Cu, Fe) can be used in the nanofluid as nanoparticle materials [15].

Presently, nanofluids are produced by two techniques [15]. A two-step technique starts with nanoparticles produced by one of physical or chemical synthesis techniques as a dry powder and then dispersed into the base liquid. This method may result in a large degree of nanoparticle agglomeration. The single-step (evaporation) one simultaneously makes and disperses the nanoparticles directly into the base liquid. The two-step process produces nanofluids with oxide nanoparticles and the single-step one produces the nanofluids with metal nanoparticles.

They are unlikely to become the mainstay of commercial nanofluid production due to the required low vapour pressure (typically less than 10 Pa) that limits the rate of nanofluid production and makes it expensive. Although nanoparticle agglomeration in this case is minimized as a result of the liquid flowing continuously, the effect of temperature and operation conditions on allocation of nanoparticles may be significant due to changing the electric potential on the surface of colloidal particles as a main factor to provide the stability of nanofluid [16].

Therefore systematic studying of the aqueous nanofluids is needed [17–20] since a key factor in understanding their thermal properties is the clustering effects that provide paths for rapid heat transport and stabilize nanofluid composition in different conditions.

2. Some Aspects of Liquid Water Microstructure

Precise experimental techniques for determining the local structure of liquid water are missing since each water molecule undergoes rapid rearrangement on the order of femtoseconds. The need for a better understanding of water at the microscopic level has forced the development of computational methods that describe the individual and cooperative structure and dynamics of water molecules, and many studies have been carried out using these techniques. These simulations predict locally ordered hydrogen-bonding clusters of water molecules that continually form and break [21].

MD simulations of density fluctuations in liquid metals [9, 10] have shown that their dense tetrahedral clusters are characterized by vertex connections as well as in liquid water the low-density ice crystallites are divided by dense tetrahedral clusters with an asymmetrical structure [1]. However these clusters (see Figure 1(a)) are more complicated due to the effect of hydrogen bonds but the frame of them as a

broken red line connecting the centers of tetrahedrons is also ramified [20].

One can expect that water molecules have enough time for rebuilding hydrogen bonds in a dense part of water density fluctuations due to very fast librations and rotations of them in liquid phase. From this, the model for instant dense clusters [22] is built with helices of hydrogen bonds (see Figure 1(b)). All angles between them are equal to 106.8°. Each water molecule in such clusters is tetrahedrally bonded with three molecules of the same cluster and with one of some ice crystallites in liquid water.

The topology of helical clusters is essentially differed from the one of the crystalline ice. From this and only this point of view, the liquid water is considered as a two-structural fluid by dynamic forming the two topological kinds of clusters in density fluctuations. The mole fraction of helical clusters in liquid water can amount to 0.6 [22].

The dense helical clusters save, in principle, four hydrogen bonds for each molecule of water; three of them are internal and one is external for connecting to ice crystallites. They can stimulate coherent proton vibrations in the coil of helical cluster if these protons are ordered in the helix (see Figure 1(b)). A spectral series, $v(n)$, of coherent proton vibrations in the helical cluster as a function of water-molecules number, n, is [22]

$$v[\text{THz}] \sim 22\frac{\sqrt{n-1}}{n}, \quad \text{at } n \geq 6. \tag{1}$$

A possible generation of coherent proton vibrations by an external electromagnetic impact at these resonance frequencies can selectively amplify each mode and, thus, can strengthen water microheterogeneity far from its thermodynamic equilibrium. Therefore studying the electromagnetic absorption at these frequencies can be interesting both for revealing spectral lines (1), and for creating a possible technique for managing the microstructure of water. Moreover, additives in aqueous solution can be concentrated in the ramified interface of two dynamic microstructures of liquid water that can cause fluctuation-induced clustering of the impurities in nanoparticles and thus, forming a stable two-phase state of aqueous solution [20, 23].

3. Clustering Impurities in Liquid Water

A spatial position of any impurity in liquid solution has dual character due to density fluctuations of the solvent that build instant dense clusters of almost regular tetrahedrons connected in pairs by faces in ramified n-chains from solvent atoms [6, 11]. At a low concentration of impurity, its atoms place on the external faces of the dense solvent clusters. In increasing concentration, the impurity atoms build their own dense tetrahedral clusters in the solution so that it becomes microheterogeneous as a nanofluid [23].

In liquid water, tetrahedral clusters of water density fluctuations (see Figure 1(b)) are more complicated than the ones in simple liquids due to hydrogen bonds but the frame of them as a broken red line (see Figure 1(a)) connecting the centres of tetrahedrons is also ramified [19].

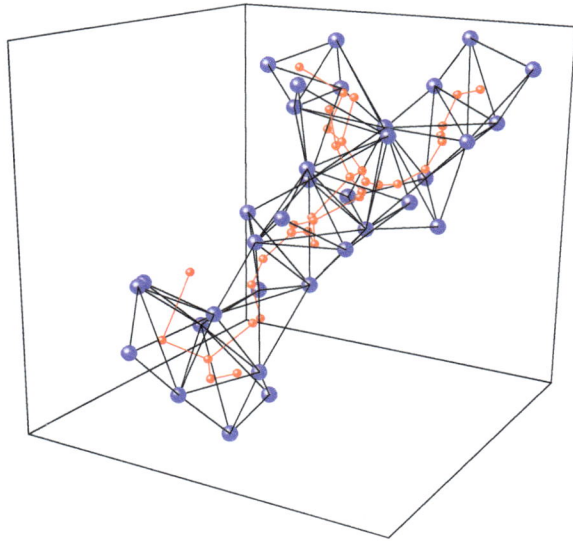

(a)

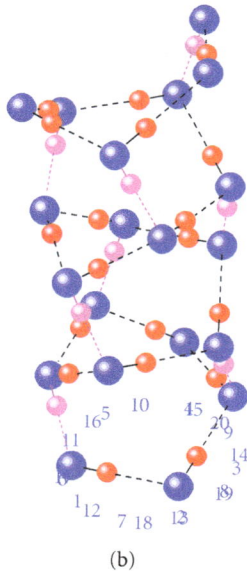

(b)

FIGURE 1: (a) A dense tetrahedral cluster in MD model of liquid water at 300 K and its frame (broken red line); blue points are the molecules and red points are centres of the cluster tetrahedrons; (b) the helical atomic model for this cluster with possible coherent exchange of protons (red balls) between oxygen atoms (blue balls) along the helix of hydrogen bonds (black lines) and hydrogen-bond bridges (pink balls); the projections of oxygen atoms in the plane are denoted by numbers.

Thus, the impurities residence in liquid water is practised in the interface of low- and high-density regions. At some concentration of impurity atoms, they form their own dense tetrahedral clusters in the aqueous solution. Such clusters ("inherent" nanoparticles) will be stable at different conditions of the water-coolant operation [20].

In this connection, it is offered a fractal model for inherent nanoparticles that can appear in the water coolant under some conditions as percolation clusters of solid-like

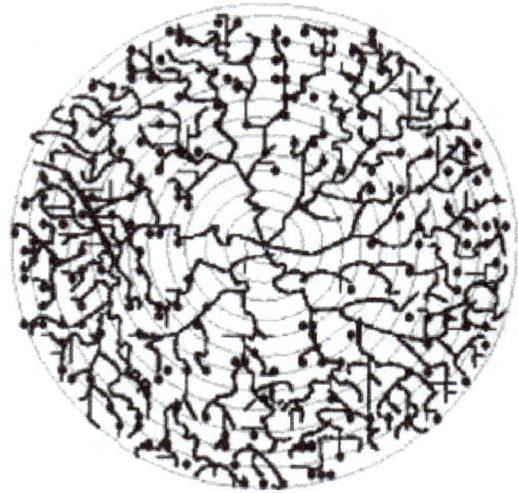

FIGURE 2: The scheme of percolation fractal cluster of solid-like filaments.

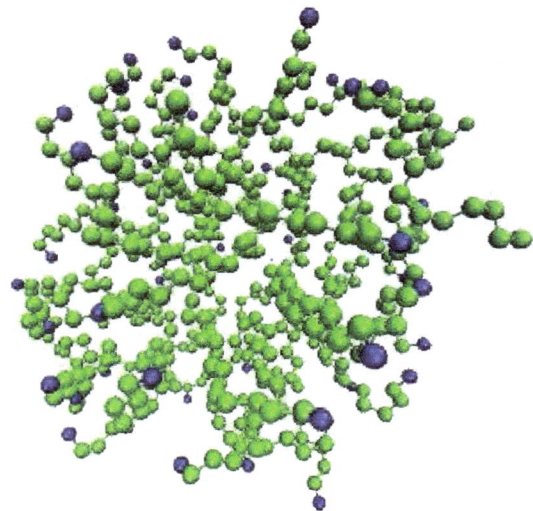

FIGURE 3: The fractal cluster as an inherent micelle in liquid water.

filaments [19] shown in Figure 2. Such a fractal nanoparticle is a micelle presented in Figure 3.

In [24], it is shown that fractal matter, M, enclosed in a sphere of radius, r, satisfies the scaling law

$$M(r) \sim r^D, \qquad (2)$$

where D is Hausdorff's dimension of particle fractal which is equal to ~2.5.

In any real cluster, the fractal structure observed on scales, r, satisfied the condition of $a < r < d/2$, where d is the size of fractal nanoparticle and a is the thickness of fractal filament. In that case, one can easily obtain the volume fraction, φ_f, of particle material in the fractal as [19]

$$\varphi_f = (3/D)(2a/d)^{3-D}. \qquad (3)$$

4. Evaluating Heat Transfer in Liquid Water by Fractal Nanoparticles

Thermal properties of fiber constituents (filaments) are locally anisotropic but the same properties of a clew of filaments are isotropic [19]. Therefore the thermal conductivity, λ_f, of fractal mater as a percolation cluster filled with liquid is the same as the solid nanoparticle with respect to heat flow in the fluid and presented as $\lambda_f \equiv \lambda_m$. Here λ_m is the thermal conductivity of dispersed material.

At the same time, spherical fractal particles have developed interface of the solid/liquid contact (see Figure 2) and its heat-variable resistor is negligible. Then, we can use the potential theory of Maxwell [25] for well-dispersed fractal nanoparticles that gives a simple relationship for the thermal conductivity, λ_n, of nanofluid (with randomly distributed and noninteracting spherical particles) in the reduced form

$$\frac{\lambda_n}{\lambda} \approx 1 + 3\varphi_p \frac{(1-\alpha)}{(1+2\alpha)}, \tag{4}$$

where λ is the thermal conductivity of liquid matrix, φ_p is the volume fraction of nanoparticles in liquid, and $\varphi \equiv \varphi_f \varphi_p$ is the volume fraction of their material in the liquid; $\alpha = \lambda/\lambda_m \ll 1$. Since $\varphi_p \equiv \varphi/\varphi_f = (D/3)(2a/d)^{D-3}\varphi$ in [19], the following equation is obtained:

$$\frac{((\lambda_n/\lambda)-1)}{\varphi} \approx D\left(\frac{2a}{d}\right)^D - \frac{3(1-\alpha)}{1+2\alpha}. \tag{5}$$

The function (5) of three parameters: D, $2a/d$, and α is calculated for estimating the effect of fractal structure of nanoparticles on the thermal conductivity of aqueous nanofluid that is given in Table 1.

5. Discussion of Results

According to [20, 23], the lower limit of the impurity concentration for its clustering in liquid water is $0.1\,C_s$, where C_s is the saturation impurity concentration. It is clear that this range for clustering impurities dissolved in water as "solid-like" nanoparticles is an effective way to stabilize the aqueous nanofluid structure for different conditions (high temperature, flow rate, radiation, etc) of its operation in any power system. In the limit of a small volume fraction of such nanoparticles and their high thermal conductivity, the enhancement of thermal conductivity of aqueous nanofluid can be 18φ for spherical fractal particles of 10–50 nm in size composed of fine filaments of 0.5–1.0 nm in diameter. The significant property of considered fractal structure of disperse phase in nanofluids is an explanation of observed enhancement of their kinetic characteristics obtained in different experiments.

At the same time, it is important to understand that the fractal particles are the product of complex chemical reactions between any dissolved impurity and the aqueous solvent. Therefore, it is necessary to develop a special technology for getting them and stabilizing them in liquid matter.

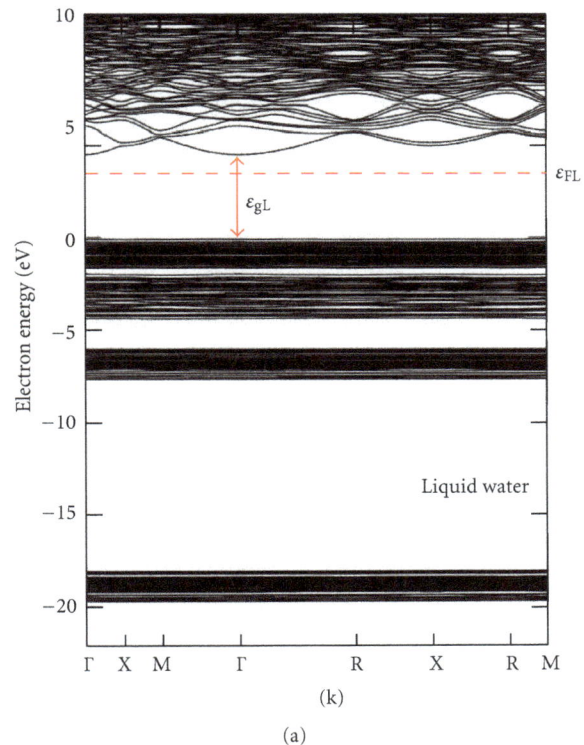

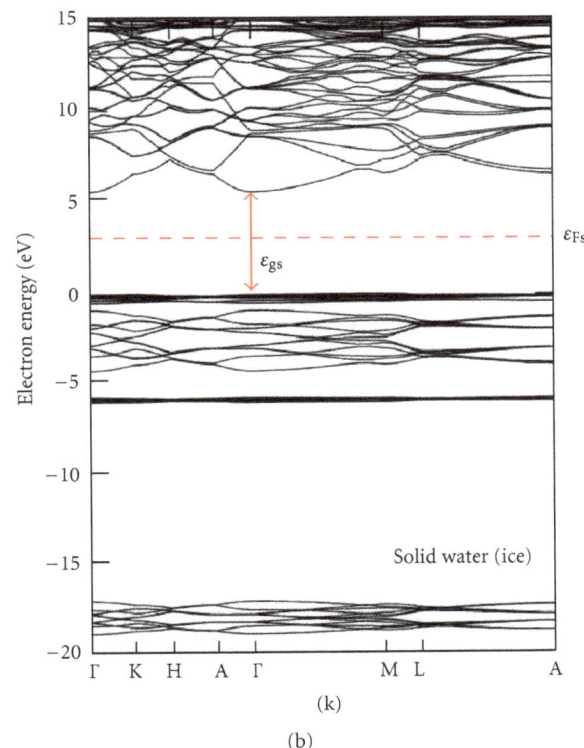

FIGURE 4: Brillouin bands of liquid (a) and solid (b) water [26]; here the zero energy of electrons corresponds to the top of valence band and red dotted lines are Fermi levels, ε_{Fl} and ε_{Fs}, of electrons in liquid and solid water with corresponding band gaps, ε_{gl} and ε_{gs}.

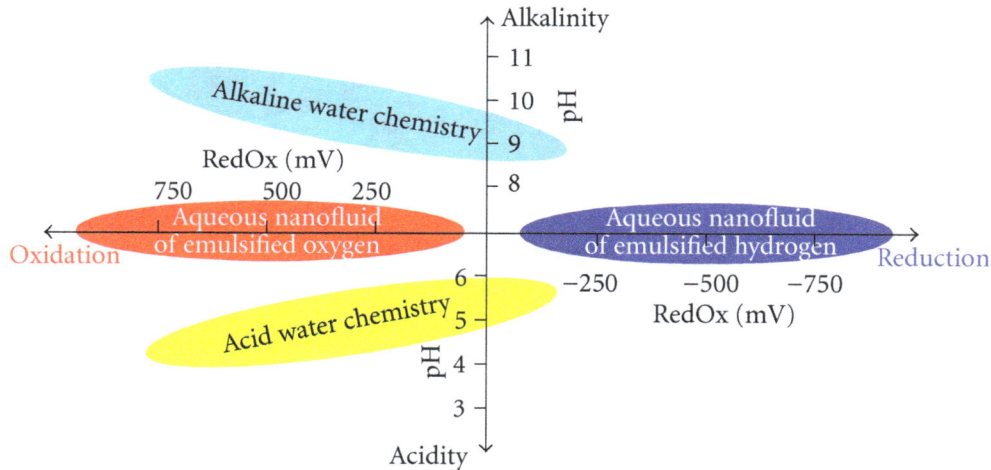

FIGURE 5: Diagram of aqueous chemistry as 2D plot of pH and RedOx parameters.

TABLE 1: The enhancement of water thermal conductivity by fractal nanoparticles.

	Al_2O_3			Al			Cu			Carbon fiber		
D	3.0	2.7	2.5	3.0	2.7	2.5	3.0	2.7	2.5	3.0	2.7	2.5
$2a/d$	—	0.02	0.1	—	0.02	0.1	—	0.02	0.1	—	0.02	0.1
α	0.02	0.02	0.02	0.002	0.002	0.002	0.001	0.001	0.001	0	0	0
$(\lambda_0/(\lambda-1))/\varphi$	2.87	8.35	7.56	2.98	8.68	7.86	2.99	5.37	17.6	3.0	8.73	17.7

The theoretical studies [11] show that the fluctuation-induced clustering of impurities is possible only if the solid-like disperse phase (colloids) in aqueous solution is hydroxides. Therefore, it is necessary to select correctly impurities for water solution in order to get the stable nanosuspension in it [20].

In this connection, a nanofluid with gaseous nanoparticles (bubble stones) is interested especially with oxygen (hydrogen) nanobubbles. These gaseous nanoparticles are gathered in fluctuation-induced defects of tetrahedral water structure mentioned above. Obviously, such formation of gaseous nanoemulsion in aqueous solution allows obtaining comparatively stable nanobubbles that can considerably exceed the molecular solubility of gaseous impurities in liquid water at the given temperature.

Moreover, the aqueous nanofluid of emulsified oxygen (hydrogen) converts the liquid water in the nonstoichiometric state that is illustrated in Figure 4(a) by a shift of Fermi level, ε_{Fl}, in the band gap, ε_{gl}, of liquid water. Here is the hypostoichiometric state of water by emulsifying hydrogen in it. As seen in Figure 4(b), Fermi level, ε_{Fs}, of solid water is in the middle of band gap, ε_{gs}, that is corresponded to the stoichiometric state of ice as a characteristic electron state of solid water.

It is known [27] that the change of Fermi level in the band-gap of any nonmetal liquid is equivalent to changing a Reduction-Oxidation (RedOx) potential of this melt. This can concern the aqueous nanofluid with emulsified oxygen (hydrogen) as it is shown in Figure 5 in the plane of two coordinates, pH and RedOx [28]. The first parameter defines the logarithmic portion of hydrogen cations in liquid water,

that is, alkalinity or acidity of the aqueous matter, and the second one reflects nonstoichiometric state, $H_2O_{1\pm x}$, of liquid water as the condensed matter, that is, the presence in it of dissolved hydrogen or oxygen.

It is visible that the aqueous chemistry accepted in power engineering of pressure-water reactor (PWR) is the oxidative one due to continuous additives of feed water that naturally dissolves oxygen in it. This impurity is not desirable for the aqueous chemistry of the PWR first-loop coolant due to strong oxidizing structural materials providing continuous growth of oxide films on the surface of fuel cladding.

At the same time, the PWR operation experience shows that this film of thickness more than 10–15 microns is fragile and sensitive to thermal cycles. The film is locally breaking and opens the fuel cladding to corrosion by zirconium-steam reaction.

Therefore, it is important to organize an effective technological process for removing oxygen from the feed water. In hydrogen "rinsing" of this water, it is possible to convert RedOx potential of the aqueous coolant in the negative value area (see Figure 5) and obtain the corrosion-passive one that can inhibit the growth of oxide films on the surface of fuel cladding and local breakup of them.

6. Conclusions

The theoretical studies show how one can provide the stable formation of nanoparticles in water solution. It is important to form clusters in water directly from impurities that are dissolved there. Then, ramified fractal clusters, as natural solid-like part of solution, can be stable constituents of the aqueous nanofluid.

It will provide the stability of water nanofluid embedded with such nanoparticles that flow in the first loop of PWR without changing their microstructure.

Acknowledgments

The authors are pleased to acknowledge Dr. A. S. Kolokol for giving some data on molecular-dynamic simulation of water structure and discussing this work that is supported by the Russian Foundation of Basic Researches (Grant no. 10-08-00217).

References

[1] C. Huanga, K. T. Wikfeldtb, T. Tokushima et al., "The inhomogeneous structure of water at ambient conditions," *Proceedings of the National Academy of Sciences of the United States of America*, vol. 106, no. 36, pp. 15214–15218, 2009.

[2] P. Wernet, D. Nordlund, U. Bergmann et al., "The structure of the first coordination shell in liquid water," *Science*, vol. 304, no. 5673, pp. 995–999, 2004.

[3] E. Cartlidge, *The Strangest Liquid*, New Scientist, 2010.

[4] K. Tuttle, *Researchers Rediscover the Structure of Water*, SLAC Today, 2010.

[5] T. Tokushima, Y. Harada, O. Takahashi et al., "High resolution X-ray emission spectroscopy of liquid water: the observation of two structural motifs," *Chemical Physics Letters*, vol. 460, no. 4–6, pp. 387–400, 2008.

[6] J. D. Bernal, "The structure of liquids," *Proceedings of the Royal Society A*, vol. 280, p. 299, 1964.

[7] Y. Waseda, *The Structure of Non-Crystalline Materials, Liquids and Amorphous Solids*, McGraw-Hill, New York, NY, USA, 1980.

[8] N. N. Medvedev, *Voronoy—Delaunay Method in Research of Structure of Non-Crystalline Systems Novosibirsk*, Siberian Branch of the Russian Academy of Science, 2000.

[9] A. S. Kolokol and A. L. Shimkevich, "Topological structure of liquid metals," *Atomic Energy*, vol. 98, no. 3, pp. 187–190, 2005.

[10] A. L. Shimkevich, A. S. Kolokol, and I. Y. Shimkevich, "Two-structure model for simple metals," *Journal of Non-Crystalline Solids*, vol. 353, no. 32–40, pp. 3472–3474, 2007.

[11] A. S. Kolokol and A. L. Shimkevich, "Topological structure of density fluctuations in condensed matter," *Journal of Non-Crystalline Solids*, vol. 356, no. 4-5, pp. 220–223, 2010.

[12] S. Choi, "Enhancing thermal conductivity of fluids with nanoparticles," in *Development and Applications of Non-Newtonian Flows*, D. A. Siginer and H. P. Wang, Eds., p. 99, ASME, New York, NY, USA, 1995.

[13] M. Bahrami et al., "Assessment of relevant physical phenomena controlling thermal performance of nanofluids," in *Proceedings of the ASME International Mechanical Engineering Congress and Expo (IMECE '06)*, Chicago, Ill, USA, November 2006.

[14] S. M. You, J. H. Kim, and K. H. Kim, "Effect of nanoparticles on critical heat flux of water in pool boiling heat transfer," *Applied Physics Letters*, vol. 83, no. 16, pp. 3374–3376, 2003.

[15] W. Yu et al., "Review and assessment of nanofluid technology for transportation and other applications," Report of Argonne National Laboratory ANL/ESD/07-9, 2007.

[16] Nanofluid Datasheet, "Meliorum Technologies," 2008.

[17] C. H. Lo and T. T. Tsung, "Low-than-room temperature effect on the stability of CuO nanofluid," *Reviews on Advanced Materials Science*, vol. 10, no. 1, pp. 64–68, 2005.

[18] P. Keblinski, S. R. Phillpot, S. U. S. Choi, and J. A. Eastman, "Mechanisms of heat flow in suspensions of nano-sized particles (nanofluids)," *International Journal of Heat and Mass Transfer*, vol. 45, no. 4, pp. 855–863, 2001.

[19] A. L. Shimkevich, "On enhancing water heat transfer by nanofluids," in *Proceedings of the 17th International Conference on Nuclear Engineering (ICONE '09)*, pp. 19–22, July 2009.

[20] P. N. Alekseev et al., "On basic principles for modifying water as a coolant of PWR," in *Proceedings of the Transactions of European Nuclear Conference (ENC '10)*, Barcelona, Spain, 2010.

[21] R. Roy, W. A. Tiller, I. Bell, and M. R. Hoover, "The structure of liquid water; novel insights from materials research; potential relevance to homeopathy," *Materials Research Innovations*, vol. 9, no. 4, pp. 98–103, 2005.

[22] A. L. Shimkevich and I. Y. Shimkevich, "On water density fluctuations with helices of hydrogen bonds," *Advances in Condensed Matter Physics*, vol. 2011, Article ID 871231, 5 pages, 2011.

[23] A. L. Shimkevich, *The Composition Principles for Designing Nuclear-Reactor Materials*, Edited by N. N. Ponomarev-Stepnoi, IzdAt, Moscow, Russia, 2008.

[24] V. E. Tarasov, "Fractional hydrodynamic equations for fractal media," *Annals of Physics*, vol. 318, no. 2, pp. 286–307, 2005.

[25] J. C. Maxwell, *A Treatise on Electricity and Magnetism*, Dover, New York, NY, USA, 3rd edition, 1954.

[26] G. Galli, *Electronic Properties of Water*, University of California, http://angstrom.ucdavis.edu/.

[27] P. N. Alekseev and A. L. Shimkevich, "On voltage-sensitive managing the redox-potential of msr fuel composition," in *Proceedings of the 16th International Conference on Nuclear Engineering (ICONE '08)*, pp. 21–29, May 2008.

[28] D. Langmuir, *Aqueous Environmental Chemistry*, Prentice Hall, New Jersey, NJ, USA, 1997.

19

The Effective Convectivity Model for Simulation of Molten Metal Layer Heat Transfer in a Boiling Water Reactor Lower Head

Chi-Thanh Tran and Pavel Kudinov

Division of Nuclear Power Safety, Royal Institute of Technology, Roslagstullsbacken 21, D5, 10691 Stockholm, Sweden

Correspondence should be addressed to Chi-Thanh Tran; thanh@safety.sci.kth.se

Academic Editor: Xu Cheng

This paper is concerned with the development of approaches for assessment of core debris heat transfer and Control Rod Guide Tube (CRGT) cooling effectiveness in case of a Boiling Water Reactor (BWR) severe accident. We consider a hypothetical scenario with stratified (metal layer atop) melt pool in the lower plenum. Effective Convectivity Model (ECM) and Phase-Change ECM (PECM) are developed for the modeling of molten metal layer heat transfer. The PECM model takes into account reduced convection heat transfer in mushy zone and compositional convection that enables simulations of noneutectic binary mixture solidification and melting. The ECM and PECM are (i) validated against relevant experiments for both eutectic and noneutectic mixtures and (ii) benchmarked against CFD-generated data including the local heat transfer characteristics. The PECM is then applied to the analysis of heat transfer in a stratified heterogeneous debris pool taking into account CRGT cooling. The PECM simulation results show apparent efficacy of the CRGT cooling which can be utilized as Severe Accident Management (SAM) measure to protect the vessel wall from focusing effect caused by metallic layer.

1. Introduction

We consider a hypothetical severe accident in a BWR with subsequent core degradation, melt relocation, and debris bed (or cake) formation in the lower plenum filled with water. In case of inadequate cooling the debris bed (cake) will be heated up and remelted and a melt pool(s) will be formed. Prediction of transient melt pool formation, thermomechanical loading on the vessel and subsequent vessel failure modes [1], and timing and melt discharge characteristics is of paramount importance for the ex-vessel melt risk quantification in the Swedish BWR with a deep water-filled cavity under the reactor [2].

The lower plenum of a BWR contains a forest of CRGTs. In normal operation there is a purging water flow into the reactor through the CRGTs. In a severe accident the CRGT purging flow can be used for cooling the core melt materials and thus to become a potentially effective SAM measure for Swedish BWRs. Namely, the CRGT cooling may help to remove effectively the decay heat from a debris bed or melt pool formed in the lower plenum and thus delay or even prevent vessel failure [3] leading in the last case to in-vessel melt retention.

Besides the CRGT cooling, the other factors which may affect in-vessel progression of an accident are the phase changes involved in the melt pool formation process, and pool stratification (with separation of oxidic and metallic layers). There are large aleatory and epistemic uncertainties in the scenario and phenomena of melt pool formation, especially in the presence of core material physicochemical interactions [4]. However, homogeneous and horizontally stratified melt pools are the two most probable configurations which are considered in the present work.

In case of melt pool formation in the lower plenum, direct simulation of flow and heat transfer is a computational challenge, due to large length scale of the reactor lower plenum and high Rayleigh numbers (10^{15}–10^{17} of the oxidic pool and 10^8–10^{11} of the metal layer). The difficulty is augmented with the presence of phase changes, complex 3D geometry of the lower head with a forest of CRGTs, complex flow patterns, and heat transfer induced by the CRGT cooling.

For prediction of decay-heated homogeneous melt pool formation in the Light Water Reactor (LWR) lower plenum, recently the ECM and PECM have been developed [5–7]. However, the stratified melt pool configuration with a metal layer atop which is characterized by Rayleigh-Benard convection or mixed natural convection (i.e., in the presence of side cooling, e.g., CRGT cooling) has not been considered. It is necessary to extend the ECM method to a metal layer. In the present work we develop prediction tools which enable simulations of the melt pool formation and melt-structure-water interaction with taking into account the CRGT cooling, homogeneity and stratification of melt pool configurations, and phase changes during the melt pool formation process.

The technical approach adopted in the present study is as follows. ECM and PECM are further developed to enable simulations of metal layer heat transfer on the base of effective convectivity approach [5, 6]. The CFD study is performed to gain insights into flow physics and examine Rayleigh-Benard convection and mixed natural convection heat transfer. The CFD study on the one hand, supports selection of appropriate correlations to be implemented in the ECM/PECM. On the other hand, CFD method is used to generate spatial distribution data of local heat transfer characteristics which are necessary for validation of the ECM and are not recoverable by average heat transfer correlations. The metal layer ECM and PECM are then validated against heat transfer experiments for both eutectic and non-eutectic binary mixture (melt), as well as against the CFD-generated data. The validated PECM is used to simulate heat transfer of a reactor-scale stratified melt pool whose formation is probable upon a certain BWR severe accident scenario which is also discussed in the paper.

The structure of the paper is as follows. In Section 2, we introduce development of the metal layer ECM and PECM. In Section 3 the CFD study is presented. Validation of the metal layer ECM and PECM is shown in Section 4. Section 5 contains the PECM application and discussions. Section 6 provides a brief summary of the paper.

2. Development of the Metal Layer Effective Convectivity Model

2.1. Review of Available Methods for Prediction of Metal Layer Heat Transfer. Metal layer heat transfer can be predicted using several methods which can be grouped into two classes: the lumped parameter and distributed parameter methods.

The lumped-parameter method is widely used [8–10] for calculation of heat fluxes from a metal layer to surroundings. In the lumped parameter method, the heat fluxes at the metal layer boundaries are calculated using the energy balance equation and empirical heat transfer correlations.

To determine the upward heat transfer coefficient (through the layer), the Globe-Dropkin correlation [11] is used in the work of Theofanous et al. [8], SCDAP/RELAP5-3D [9]. In MELCOR [10] code the Globe-Dropkin correlation is applied for the lower surface and the correlation produced in the ACOPO experiments [12] is applied for the upper surface of the metal layer. Modified Globe-Dropkin correlation

$Nu = 1 + 0.069 \times Ra^{0.333} Pr^{0.084}$ which accounts for conductive heat transfer by adding the unity to the Nusselt number is employed in the MAAP code [13].

For determination of the sideward heat transfer coefficients a wider spectrum of different correlations is used. In the work of Theofanous et al. [8] and SCDAP/RELAP5-3D [9], the sideward heat transfer coefficient along a vertical cooled wall is the average Churchill-Chu correlation [14], while in the MELCOR code [10] it is the ACOPO correlation [12]. In the MAAP code, the sideward heat transfer coefficient is a function of the heat flux transferred at the lower boundary of the oxidic pool to the vessel wall [13]. In a later study which focused on the implemented correlations in MAAP applied to French PWRs [15], the sideward heat transfer model was replaced by the Churchill-Chu correlation which results in higher probability of focusing effect in MAAP calculations for French PWRs.

Clearly, the lumped-parameter method is a simple and effective method for calculations of heat fluxes and energy splitting. Note that the lumped-parameter method established in [8] was validated against the MELAD experiments which were built specially for validation of mixed natural convection heat transfer models. However, the phase change crust dynamics are not considered in the lumped-parameter methods. Thus the effect of latent heat of fusion on the transient heat fluxes at the metal layer boundaries cannot be quantified (we will show the effect of latent heat on heat fluxes in this paper). Furthermore, the local effect of the heat flux along an inclined cooled surface is not considered in the method.

The CFD method such as Direct Numerical Simulation (DNS) and Large Eddy Simulation (LES) is extensively used to examine flow characteristics; low-Prandtl number effects of a molten metal fluid layer heated from below and cooled from the top [16–19]. To understand complex fluid flows, the CFD method is indispensable and helps to reveal separate-effect phenomena. However, the CFD method is computationally expensive and thus difficult to be applied to simulations of melt pool heat transfer. The difficulty is further increased for long (hours) transient scenarios, for the cases of parametric studies, and especially when the phase change is involved.

The Effective Convectivity Conductivity Model (ECCM) can be classified as a distributed parameter method and is a computationally affordable alternative of CFD simulations. Originally the ECCM was developed by Bui and Dinh [20], later on extended to simulations of metal layer heat transfer, and reported in detail in the work of Sehgal et al. [21]. The metal layer ECCM employs the effective convectivity model to represent turbulent heat transport (Rayleigh-Benard natural convection) through a fluid layer using so-called effective velocity U_{RB} which is defined as follows:

$$U_{RB} = 2 \frac{\alpha}{H_{RB}} Nu_{RB}, \qquad (1)$$

where H_{RB} is the thickness of the fluid layer and Nu_{RB} is empirical Rayleigh-Benard convection heat transfer coefficient. To describe turbulent heat transport in the horizontal direction, the effective conductivity model is used

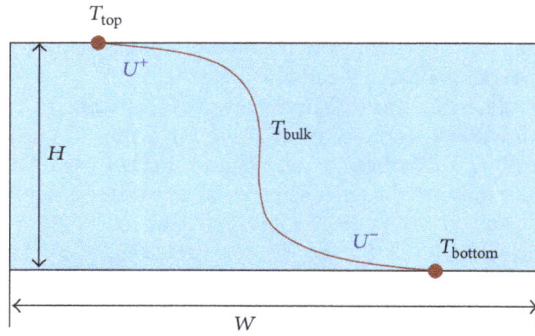

FIGURE 1: Temperature profile of a fluid layer with classical Rayleigh-Benard convection.

($k_{\text{eff}} = k \times \text{Nu}$). In contrast to the lumped-parameter methods, the boundary layer model [22] is applied to describe the local effect of the heat transfer coefficient along a vertical/inclined cooled wall:

$$\text{Nu}_{\text{side}} = 0.508 \text{Pr}^{1/4} \left(\frac{20}{21} + \text{Pr} \right)^{-1/4} \text{Ra}_y^{1/4}. \quad (2)$$

The metal layer ECCM was implemented in MVITA code which was applied to simulations of stratified melt pool heat transfer in a PWR lower head and to examine the focusing effect. However, MVITA is a 2D code, thus it is not applicable to 3D geometry of a BWR lower plenum. Furthermore, in the ECCM non-eutectic mixture phase change and mushy zone heat transfer were not considered.

2.2. Governing Equations and Heat Transfer Characteristic Velocities. Following the concept of effective convectivity [6, 20] the ECM is developed to enable effective heat transfer simulations for a metal layer which may appear atop of an oxidic melt pool in the lower plenum.

The metal layer ECM uses the directional heat transfer characteristic velocities to describe turbulent heat transfer in both the upward and sideward (in the presence of side cooling) directions of a fluid layer. As the heat transfer characteristic velocities $U_{x,y,z}$ (later on denoted as U_{up} and U_{side}) are used instead of instantaneous velocity $u_{x,y,z}$ hence, the following energy conservation equation is solved:

$$\frac{\partial \left(\rho C_p T \right)}{\partial t} + \left(\frac{\partial \left(\rho C_p U_x T \right)}{\partial x} + \frac{\partial \left(\rho C_p U_y T \right)}{\partial y} \right.$$
$$\left. + \frac{\partial \left(\rho C_p U_z T \right)}{\partial z} \right) = \nabla \cdot (k \nabla T). \quad (3)$$

The heat transfer characteristic velocities are determined as follows (see Figure 1).

For a fluid layer heated from below and cooled from the top, using the upward heat transfer characteristic velocity U_{up}, the amount of convective heat transferred from the bottom surface to the bulk fluid (and from the bulk fluid to the top surface) is defined as follows:

$$Q_{\text{conv}} = \rho C_p U_{\text{up}} \left(\frac{\Delta T}{2} \right) \cdot S, \quad (4)$$

where S is the area of the cooled top/heated bottom surface and ΔT is the driving temperature difference between the bottom and top surfaces.

Conduction heat transfer from the bulk fluid to the top cooled surface is

$$Q_{\text{cond}} = \left(\frac{\Delta T}{2} \right) \frac{k \cdot S}{(H/2)}. \quad (5)$$

The heat balance equation is then written as follows:

$$\rho C_p U_{\text{up}} \left(\frac{\Delta T}{2} \right) \cdot S + \left(\frac{\Delta T}{2} \right) \frac{k \cdot S}{(H/2)} = \frac{k \cdot \text{Nu}_{\text{up}} \Delta T}{H} \cdot S. \quad (6)$$

From (6), the upward heat transfer characteristic velocity is derived as follows:

$$U_{\text{up}} = \frac{2\alpha}{H} \left(\text{Nu}_{\text{up}} - 1 \right). \quad (7)$$

The heat transfer coefficient in (7) is defined using external Rayleigh number Ra which is a function of the temperature difference across the layer.

In the case of mixed cooling by the top wall and one side wall (other side wall is either adiabatic or symmetrical, Figure 1), using a formal analogy, the sideward characteristic velocity is derived as follows:

$$U_{\text{side}} = \frac{\alpha}{H} \left(\text{Nu}_{\text{side}} - \frac{H}{W} \right). \quad (8)$$

Note that the heat transfer coefficient Nu_{side} is defined using the sideward Rayleigh number which is a function of the sideward temperature drop, that is, the difference between the bulk temperature and vertical/inclined cooled boundary temperature. Selection of the upward and sideward heat transfer correlations and description of the sideward heat transfer coefficient profile are based on the findings and insights gained from the CFD study which is presented in Section 3.

To implement the ECM, the convective terms with the heat transfer characteristic velocities in (3) are defined as a modified heat source term. In such a way (3) is reduced to a heat conduction equation, can be solved using a commercial CFD code, for example, the *Fluent* code.

For a fluid layer heated from below and cooled from the top (and side) walls, the characteristic velocity is positive on a cooled surface and negative on a heated surface (Figure 1). Apparently the convective terms added to the grid cells, where temperature gradients are not zero, are artificial. In order to keep energy balance of the computational domain, the fluid bulk temperature must be adjusted based on the summarized heat of the artificial convective terms which have been added to the grid cells in the previous time step.

2.3. Phase-Change Effective Convectivity Model. In the phase-change ECM (PECM) for a metal layer, the single enthalpy conservation equation which is common for solid, mushy, and liquid phases is solved:

$$\frac{\partial \left(\rho h \right)}{\partial t} + \nabla \cdot \left(\rho U_{x,y,z} h \right) = \nabla \cdot \left(\frac{k}{C_p} \nabla h \right) - \frac{\partial \left(\rho \Delta H \right)}{\partial t}. \quad (9)$$

Similarly to the PECM for an internally heated volume [7], (9) is solved using the source-based method with a fixed grid [23]. The main approach for the metal layer treatment in PECM is to define the heat transfer characteristic velocities in a mushy zone (U_M) and to employ the compositional convection model.

Heat transfer characteristics of a mushy zone depend on the certain fluid velocity in it. One may assume that the fluid velocity is a function of liquid fraction which in turn is related to temperature of the binary mixture in a mushy zone by either linear, or linear-eutectic, or Scheil, or power-law relationship [23]. The liquid fraction can be determined as follows:

$$F_L = \frac{T - T_{\mathrm{SOL}}}{T_{\mathrm{LIQ}} - T_{\mathrm{SOL}}}. \tag{10}$$

Thus in the PECM, based on (10) different models of the mushy zone characteristic velocities can be realized using different types of liquid fraction dependency, that is, $U_M = f(U_{x,y,z}, F_L)$. Therefore in a mushy zone, reduced characteristic velocities are employed to describe mushy zone heat transfer.

Natural convection in a fluid layer involving non-eutectic mixture phase change is affected not only by the temperature difference but also by the compositional concentration difference of a solute across the liquid layer (double-diffusive convection). In such a case, the effective Rayleigh number which defines the intensity of turbulent natural convection in a fluid layer can be determined as a combination of the thermal Rayleigh number Ra_T and compositional Rayleigh number Ra_C. Thermal Rayleigh number Ra_T is defined as follows [24]:

$$\mathrm{Ra}_T = \frac{g\beta_T \Delta T H^3}{\alpha \nu}, \tag{11}$$

where ΔT is defined as $\Delta T = T_\infty - T_E$. Compositional Rayleigh number Ra_C can be determined as:

$$\mathrm{Ra}_C = \frac{g\beta_C (r_C \times \Delta C) H^3}{\alpha \nu} \tag{12}$$

where ΔC is the concentration difference across the fluid (liquid) layer and r_C is the rejectability coefficient. The presence of the rejectability coefficient in the compositional Rayleigh number expression is explained as follows.

According to the definition, the concentration difference is $\Delta C = C_\infty - C_E$. However, binary mixture solidification experiments [25, 26] show that the value of ΔC remains uncertain due to rejectability of the higher concentration (compared with the bulk fluid) solute from a mushy zone to the bulk fluid. For one binary mixture, the solute with higher concentration may easily be rejected from the mushy zone, while for the other mixture, this higher concentration solute may be stuck in the mushy zone and be solidified. The rejectability of the solute from a mushy zone also depends on the mushy zone structure, the intensity of natural convection in both the mushy zone and bulk fluid, and the fluid properties and the cooling rate. Note that depending on

the hypereutectic or hypoeutectic mixture and the cooling direction, that is, cooling from the upper or lower surface, compositional convection may weaken/strengthen thermal convection. We envision the necessity of examining the effect of mushy zone heat transfer characteristic velocity models and rejectability coefficient on heat transfer of non-eutectic binary mixture solidification/melting.

In the present paper, we show that with appropriate selection of the liquid fraction dependence model for the mushy characteristic velocity, and the rejectability coefficient, the PECM is able to predict the transient behavior of non-eutectic binary mixture solidification observed in different experiments (Section 4).

3. CFD Study

The main purpose of the CFD study is to examine Rayleigh-Benard convection and mixed natural convection heat transfer in order to support selection of appropriate correlations to be implemented in the metal layer ECM. Also CFD data help to quantify the local sideward heat transfer coefficient distribution for validation of the ECM.

The CFD method used in the present study is named Implicit LES (ILES) [27]. The ILES method uses a high-resolution grid to effectively provide LES without explicit Subgrid Scale (SGS) turbulence modeling. To capture the wall-boundary effects, a higher-resolution grid is provided in the near-wall region of the computational domain. The results obtained with ILES for heat transfer coefficients are grid independent.

Prior to presenting CFD simulation results, let us make a brief summary of the past works related to Rayleigh-Benard turbulent convection with respect to low Pr number fluids. It was found that heat transfer coefficient depends on Rayleigh number (external) and the dependency is an exponential law function, where the value of the exponent ranges from 1/4 to 1/3 [28]. In the work of Goldstein et al. [29], it was predicted that the experimental exponent of Ra seems to increase with increasing Pr number. Very weak dependence of the Nusselt number on Pr is observed in the range of $0.7 < \mathrm{Pr} < 21$ [30]. However, for low Pr number fluids Nusselt number becomes a function of not only Ra number but also of Pr number.

In the present study, the CFD predicted results are compared with three correlations which were obtained in the experiments for low Prandtl number fluids. The first is Globe-Dropkin correlation [11], the second is O'Toole-Silveston correlation [31], and the third is Cioni correlation [32]. The Globe-Dropkin correlation is given as follows:

$$\mathrm{Nu} = 0.069 \mathrm{Ra}^{0.333} \mathrm{Pr}^{0.074}$$
$$3 \times 10^5 < \mathrm{Ra} < 7 \times 10^9; \ 0.02 < \mathrm{Pr} < 8750. \tag{13}$$

The O'Toole-Silveston correlation is

$$\mathrm{Nu} = 0.104 \mathrm{Ra}^{0.305} \mathrm{Pr}^{0.084}. \tag{14}$$

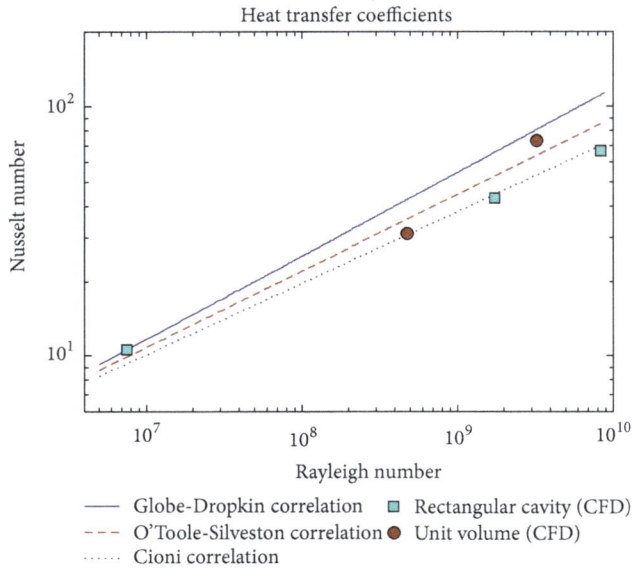

FIGURE 2: Upward heat transfer coefficients (Rayleigh-Benard convection).

The Cioni correlation is expressed as follows:

$$Nu = 0.25(RaPr)^{2/7}$$

$$\text{For } 10^6 < Ra < 10^9; \ Pr < 0.3.$$

(15)

The CFD method (ILES) is applied to simulations of low Pr number fluid layer (liquid metal) in rectangular cavity and Unit Volume (UV) geometries. The UV is a representative rectangular cavity of molten metal surrounding a CRGT. The height of UV is varying depending on the metal layer thickness. To examine the effect of aspect ratio (the height-to-width ratio), the CFD simulations are performed for different heights of rectangular cavity and UV.

Figure 2 shows plots of heat transfer coefficients across the layer heated from below and cooled from the top (Rayleigh-Benard convection), calculated by heat transfer correlations (13), (14), and (15) and the CFD predicted data for different geometries. The predicted Nusselt number is well agreed with the Cioni correlation for the rectangular cavity with the aspect ratio of less than unity (Figure 2). For the UV geometry, two predicted Nusselt numbers shown in the figure are for two different aspect ratios. In the UV with an aspect ratio larger than unity (the upper circle dot), the Nusselt number is consistent with the Globe-Dropkin correlation, while in the other (the lower circle dot with an aspect ratio of less than unity) the Nusselt is slightly lower and consistent with the Cioni correlation. Apparently the aspect ratio does affect the heat transfer; however, the influence can be assumed insignificant.

To examine the effect of side cooling on heat transfer coefficients, a CFD simulation is performed for a fluid layer cooled by both the top and side walls (mixed natural convection). Figure 3 presents temperature profiles across the metal layer of two cooled configurations: the first is Rayleigh-Benard convection and the second is mixed natural

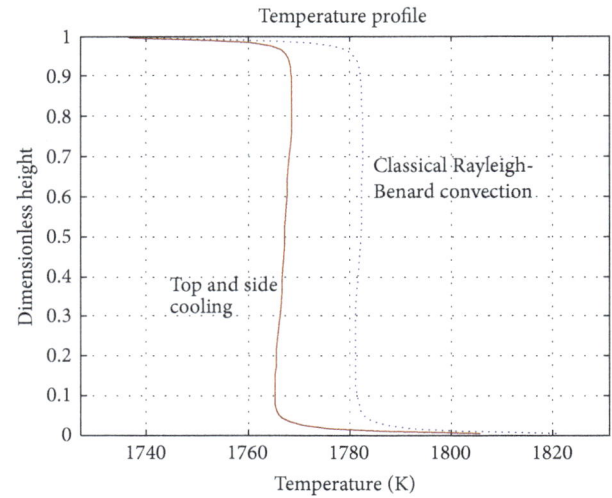

FIGURE 3: Temperature profiles across the fluid layer in two configurations of cooling (with/without side cooling).

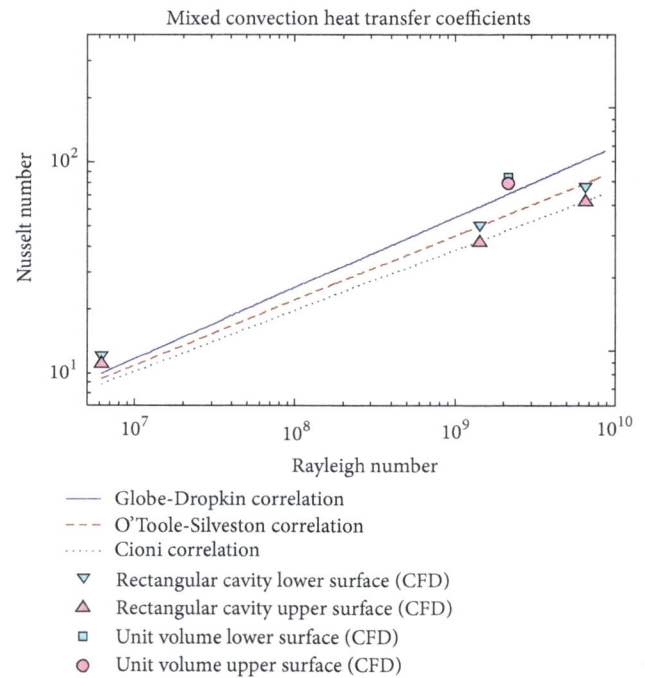

FIGURE 4: Upward heat transfer coefficients (mixed natural convection).

convection (i.e., with side cooling) in the same geometry. Clearly the profiles are qualitatively similar (Figure 2). In case of side cooling, the bulk temperature of the layer is decreased that results in changing the heat transfer coefficients along the top and bottom walls.

Figure 4 presents the CFD predicted heat transfer coefficients of a mixed natural convection fluid layer. As it is shown in the figure, the heat transfer coefficients along the upper surface are lower than those of the lower surface (the heated surface). The CFD predicted Nusselt numbers are

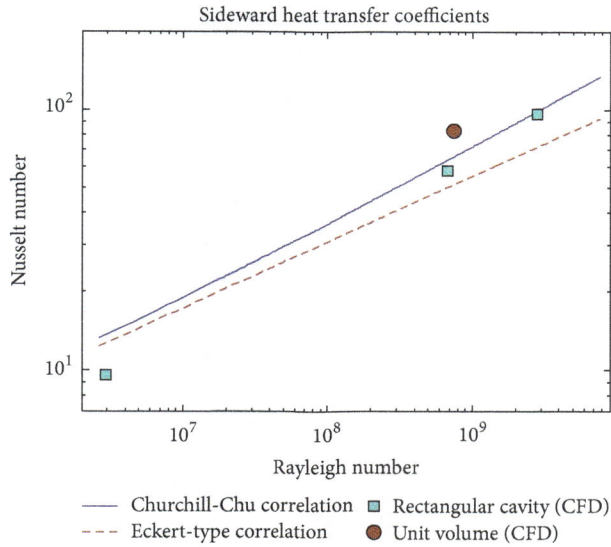

FIGURE 5: Sideward heat transfer coefficients (mixed natural convection).

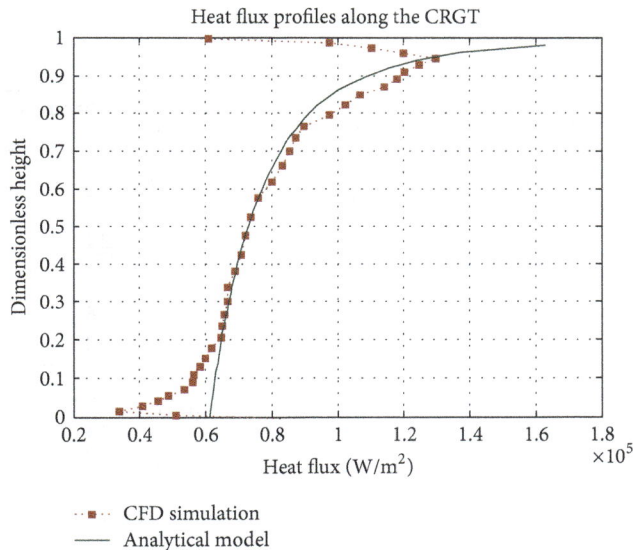

FIGURE 6: Heat flux profiles along the CRGT wall (CFD versus analytical model).

fairly consistent with the experimental correlations. It is seen that the heat transfer coefficients predicted for the UV are higher than those of the rectangular cavity and consistent with the Globe-Dropkin correlations. This may be related to specific geometry of the UV, that is, the presence of a CRGT which changes flow structure and thus results in more intensive heat transfer.

For the mixed natural convection fluid layer, the sideward heat transfer coefficients (along the vertical wall and UV cooled CRGT wall) are presented in Figure 5. The predicted Nusselt numbers are well agreed with the Churchill-Chu correlation.

Figure 6 shows the heat flux profile along the CRGT (mixed natural convection in UV geometry) and the analytical model which is derived from the average Churchill-Chu correlation [14] and the boundary layer model (2). The two profiles are found to be in good agreement along the vertical wall. Although in the uppermost and lowermost regions, the heat flux values are slightly divergent. An explanation of these deviations is as follows. At the upper part of CRGT surface the heat flux is decreased significantly due to the isothermal boundary conditions (with the same temperature) applied for both top (horizontal) and CRGT (vertical) walls. For the lower part of the vertical wall, due to the cold fluid descending from the boundary layer, temperature of the mixed fluid flow is decreased, causing decrease of the heat flux along this part (smaller temperature difference). Note that rapid increase of the heat flux obtained with the analytical model at the top of CRGT is artificial. However, the analytical model presents a conservative estimation of local heat flux.

On the basis of analysis of the CFD simulations results we adopt the Globe-Dropkin correlation for the upward heat transfer coefficient and the Churchill-Chu correlation for the sideward heat transfer coefficient in the ECM and PECM. We select these correlations because they are consistent with the CFD simulation results obtained for the specific BWR geometries under mixed natural convection conditions with low Pr number fluids. Furthermore, these correlations are valid in a wide range of Rayleigh number which is close to the prototypic reactor condition and applicable for low Pr number fluids (e.g., liquid metals). The profile of the sideward heat transfer coefficient can be described by the boundary layer model, (2) which is also confirmed by the CFD study with geometry of interest.

For validation of the developed method against phase change heat transfer we consider simulation of an experiment on solidification and heat transfer in a fluid layer [33]. The experiment was performed in a rectangular cavity of 6.35 cm × 3.81 cm × 8.89 cm; the simulant used was pure gallium (Ga). The cavity is cooled from the top surface till freezing while the bottom surface is heated at constant temperature. Temperature profiles across the layer and crust thickness evolution during the experiment were reported.

The CFD ILES method is applied to simulate this experiment. Figure 7 shows the CFD predicted velocity and liquid fraction contours in 22 min. A Benard's cell is clearly shown in the figure. Temperature profiles across the layer (in 2 min and 8 min) are presented in Figure 8, along with the experimental data. Apparently, the CFD predicted results are well agreed with the experimental data. The CFD simulation data are used for validation of the PECM (see the next section).

4. Validation of the ECM and PECM

Validation of the ECM and PECM covers a wide spectrum of physical phenomena involved in metal layer heat transfer such as Rayleigh-Benard convection mixed natural convection boundary layer development and transient phase change and crust formation. As the dual approach is adopted in the present study, both experimental and CFD-generated (with DNS and ILES methods) data are used for the validation of

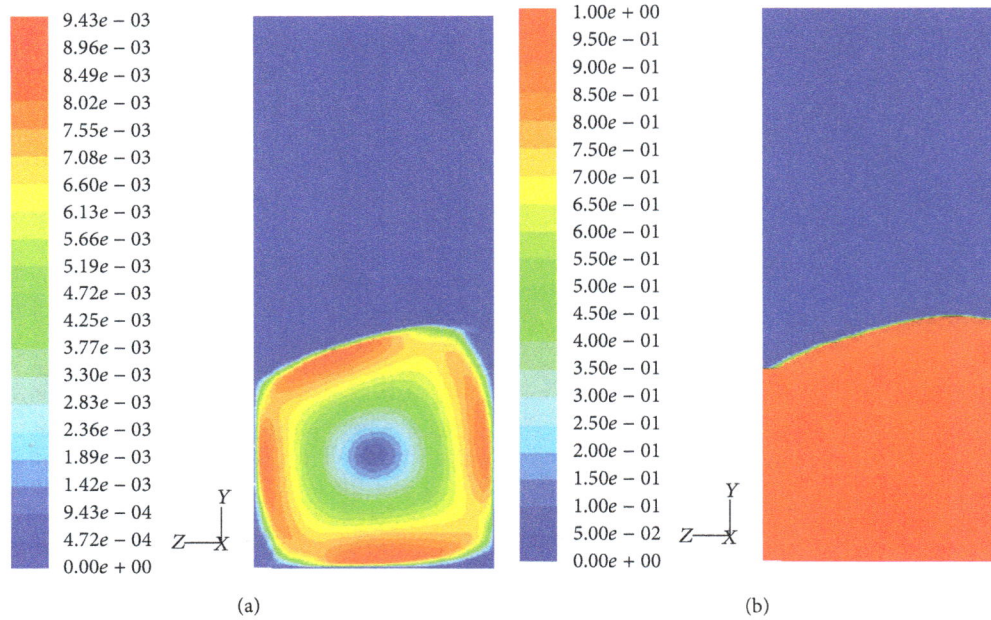

FIGURE 7: CFD simulation velocity (a) and liquid fraction (b) contours at 22 min.

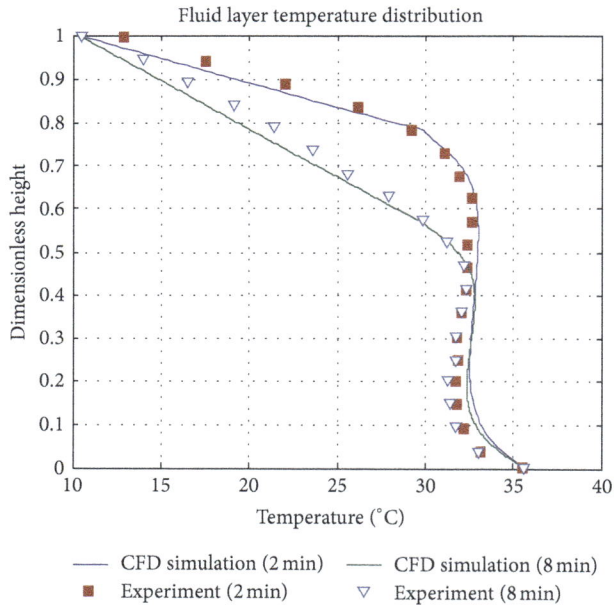

CFD simulation (2 min)   CFD simulation (8 min)
■ Experiment (2 min)   ▽ Experiment (8 min)

FIGURE 8: CFD and experimental temperature profiles across the fluid layer.

TABLE 1: ECM and PECM validation matrix.

Models	Number	Description	Experimental and CFD data
ECM	1	Rayleigh-Benard convection	DNS data [34]
	2	Rayleigh-Benard convection	DNS data [19]
	3	Mixed natural convection	MELAD A1 experiment [8]
	4	Mixed natural convection	CFD data (ILES method)
PECM	1	Solidification from top surface, eutectic mixture	Experimental data, gallium (Ga) fluid [33], and CFD data (ILES)
	2	Solidification from top surface, non-eutectic mixture	Experimental data, aqueous ammonium chloride 5% [35]

the ECM and PECM models. Various heat transfer problems are considered in the validation matrix (Table 1), including classical Rayleigh-Benard convection, mixed natural convection and as well as eutectic and non-eutectic binary mixture solidification.

4.1. ECM Validation. DNS was used in [34] to predict the temperature profile of a fluid layer heated from below and cooled from the top. The fluid Pr number was 0.022 and the layer Rayleigh number reached 2.2×10^7. The metal layer ECM is used to predict the temperature profile across the fluid layer. Figure 9 presents temperature profiles predicted by the ECM and DNS methods. It can be seen in the figure, good agreement of the two predicted results is obtained.

The ECM is also used to predict the Rayleigh-Benard convection temperature profile with a lower Ra number (Ra = 6.3×10^5) and of order of unity Pr number (Pr = 0.71). Figure 10 shows that the ECM predicted the temperature profile agrees well with that predicted by the DNS method [19].

To validate the ECM in prediction of temperature and energy splitting of a mixed natural convection fluid layer,

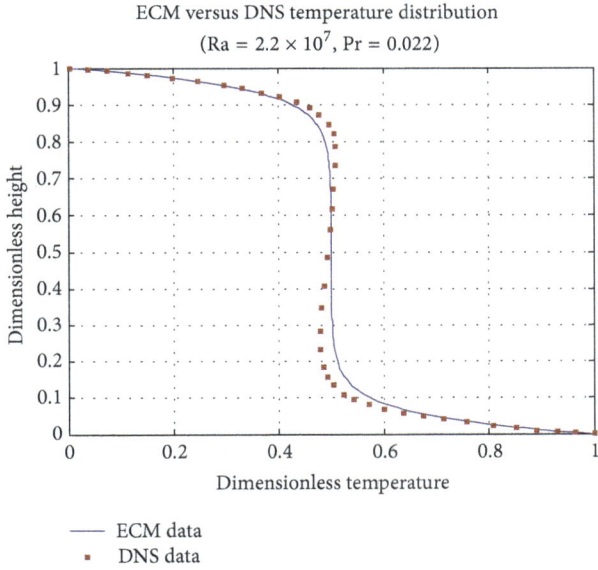

FIGURE 9: ECM predicted versus DNS [34] temperature profiles across the layer.

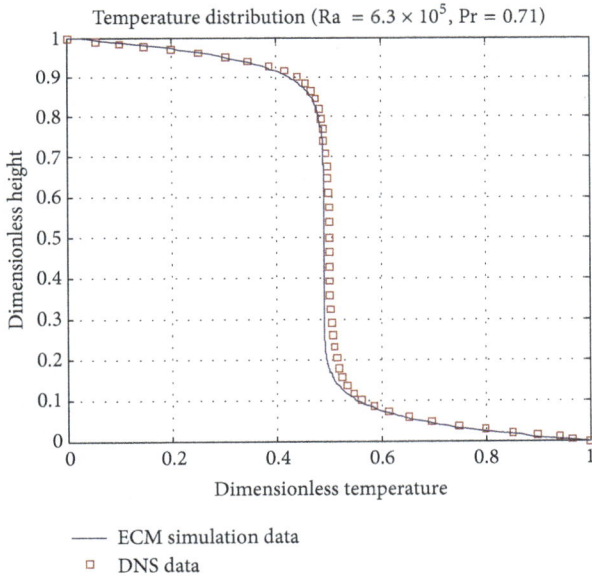

FIGURE 10: ECM predicted versus DNS [19] temperature profiles.

TABLE 2: MELAD A1 experiment [8] versus ECM prediction energy splitting and temperature.

Methods	MELAD A1	ECM
Input power (kW/m^2)	24.7 (+2%)	24.7
Top plate power output (kW/m^2)	21.7	20.6
Side plate power output (kW/m^2)	17.5	19.5
Bottom plate temperature $T_{l,i}$ ($^\circ$C)	59.1	57.2
Average temperature of the top and bottom plates ($^\circ$C)	33.9	32.9
Bulk temperature T_{bulk} ($^\circ$C)	37.5	31.5

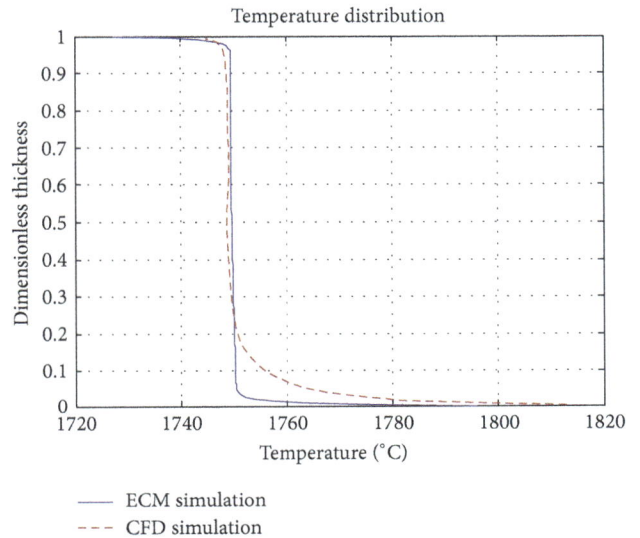

FIGURE 11: ECM and CFD predicted temperature profiles.

the MELAD A1 experiment [8] is used. The ECM predicted results are presented in Table 2. Clearly, the ECM predicted temperature and heat fluxes are very close to the experimental ones. A slight difference in the bulk temperature is observed. This may relate to possible slight thermal stratification in the fluid layer which is not captured by the ECM method.

As the next step, the ECM is benchmarked against CFD simulation data. It is well known that CFD methods can be used to perform reliable "numerical experiments" for classical Rayleigh-Benard convection [16]. In the present study, the CFD method (ILES) is used to simulate heat transfer of a mixed natural convection fluid layer. A UV of

31.26 cm × 31.26 cm × 40 cm is heated from below and cooled from the top wall and from the CRGT wall. The obtained steady state temperature profile and heat flux distribution along the CRGT wall are used for ECM validation.

Figure 11 presents plots of the CFD and ECM predicted temperature profiles across the layer. The two profiles are very close across the most fluid layer. Figure 12 shows the heat flux profiles along the CRGT wall; they are well agreed in the middle part of the CRGT wall. In the upper region, the ECM predicted heat flux value is slightly higher than that obtained with the CFD (less than 10%). In the lower region of the cooled wall, the CFD method is able to predict diminishing heat flux due to the flow stagnation, while the ECM is unable to capture this effect. However, ECM estimation of heat flux can be considered as reasonable and conservative in the sense that the heat flux is not underestimated by ECM model.

4.2. PECM Validation. In this section, validation of the PECM against both eutectic and non-eutectic binary melts is presented. First, the PECM is used to simulate a eutectic melt solidification experiment [33] which was also simulated by the CFD ILES method.

Figure 13 presents crust thickness evolutions predicted by the PECM, CFD, and the experimental data. Excellent

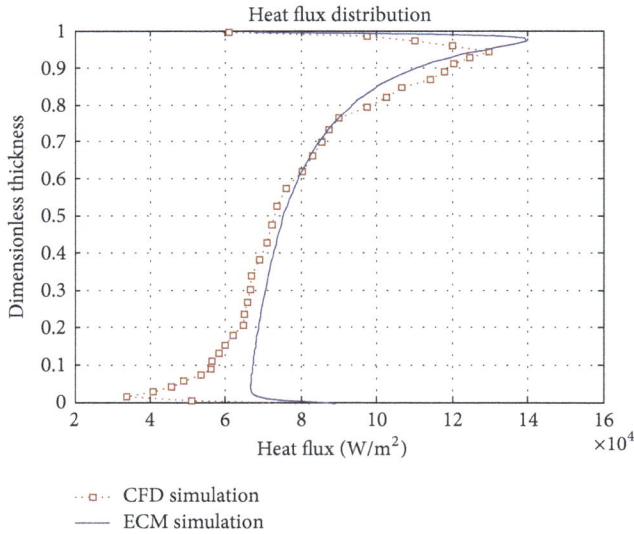

FIGURE 12: ECM and CFD predicted heat flux profiles along the CRGT cooled wall.

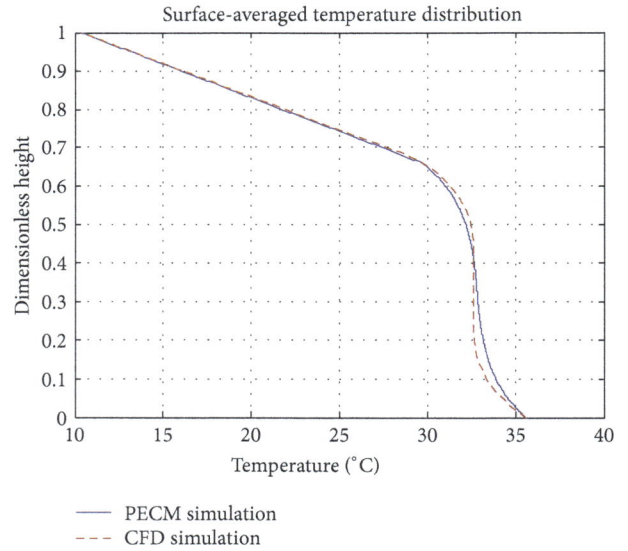

FIGURE 13: Average PECM, CFD predicted, and experimental crust thickness evolutions.

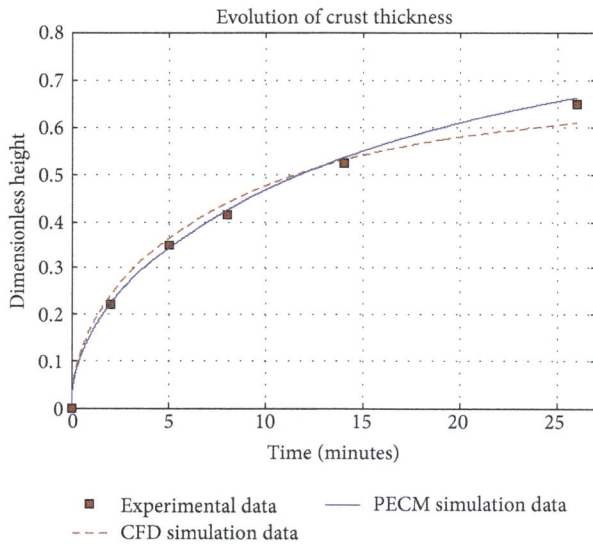

FIGURE 14: PECM and CFD predicted temperature profiles across the layer at 5 min.

agreement between the methods is obtained. In the later time period (after 20 min), the CFD predicted crust is slightly thinner than that of the experiment and the PECM simulation. This is the result of a larger CFD predicted heat transfer coefficient across the convective fluid layer. However, the difference between the CFD and experimental data is insignificant.

The PECM is also benchmarked against the CFD-generated data (Section 3). Figure 14 shows the PECM simulation temperature profile along with that of the CFD simulation. The temperature profiles predicted by the two methods are in good agreement. Note that the CFD simulation of the experiment takes about 5 days of 2.8 GHz CPU time while the PECM simulation lasts only few hours.

To validate the mushy characteristic velocity model and rejectability coefficient implemented in the PECM we use data from one of the experiments of Cao and Poulikakos [35]. In the experiments, non-eutectic binary mixtures of the aqueous ammonium chloride ($NH_4Cl–H_2O$) in a cavity of $48.3\,cm \times 25.4\,cm \times 12.7\,cm$ are cooled from the top surface till freezing. Different compositions of ammonium chloride were used. Evolutions of crust thickness (solid and mushy) were recorded.

The PECM is used to simulate the experiment where the NH_4Cl composition is 5%. PECM simulations show that crust thickness evolution is sensitive to different models of mushy characteristic velocity and to different values of rejectability coefficient. Figure 15 presents the experimental crust thickness evolution and results predicted by the PECM simulation where the mushy characteristic velocity model is described by a third-order power law of liquid fraction and the rejectability coefficient is 0.07. Although there is a slight deviation in prediction of the solid thickness start point (in around 4 h), the PECM well captures the evolutions of the both solid and mushy crusts thicknesses.

Extensive validation of the ECM and PECM against various experiments and CFD-generated data has confirmed the applicability of the ECM and PECM for simulations of turbulent natural convection heat transfer in a fluid layer heated from below, cooled from the top and side walls, as well as for simulations of transient heat transfer which involves the phase change in a BWR lower plenum configuration.

5. Application of the PECM to BWR Melt Pool Simulations

5.1. A Severe Accident Scenario in a BWR. An accident scenario which can lead to formation of stratified melt pool (Figure 16) in the BWR lower head is considered in this section. The scenario is started from state 1, when a debris

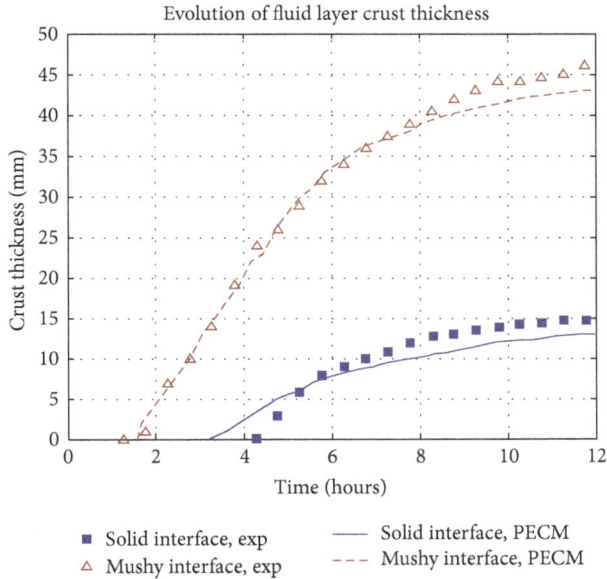

FIGURE 15: Evolution of PECM predicted and experimental crust thickness (solid and mushy), mushy velocity model $U_M = U_{up} \times F_L^3$, combined convection with $r_C = 0.07$.

bed (cake) is formed in the water-filled lower plenum. It is assumed that water ingress is possible only in the upper layer of the debris bed, resulting in formation of a coolable debris layer (bed) atop, while the lower part of the bed is reheated up with time. As temperature of the bed (lower part) exceeds liquidus temperature of metals, metallic components of the bed will be melted and flowing down, filling the pores in the lower part of the debris bed. Due to much higher liquidus temperature the ceramic component will remain in the solid phase (particulates). It can be assumed that displacement of solid oxidic particulates will be possible due to the gradual motion downwards and avalanches. Consequently a heterogeneous debris pool (i.e., oxidic particulates in a molten metal pool) will be formed in the lower plenum (state 2 in Figure 16). Depending on the mass fraction of metallic components in the quenched debris, the debris pool may become stratified; that is, oxidic particulates will be submerged in a liquid metal pool with a metal layer floating atop. In case of insufficient molten metals to fill the pores in the debris bed, a uniform (unstratified) heterogeneous debris pool will be formed in the lower plenum.

In the present study, we apply the metal layer PECM to simulate heat transfer of the stratified heterogeneous debris pool. It is assumed that the IGTs are intact or plugged and do not affect heat transfer of the pool, and chemical interaction is not considered.

5.2. PECM Simulation, Results, and Discussion. The PECM heat transfer simulation of stratified heterogeneous debris pool is presented in this section. The stratified heterogeneous debris pool is comprised of a debris pool which is characterized by a matrix of solid particulates and molten metals and a molten metal layer atop of the pool. The metal layer thickness depends on the fraction of the metallic components presented in the initial debris bed with the porosity of 40% (Figure 17).

As an example, we consider the debris pool depth of 0.8 m and the metal layer thickness of 0.2 m (total $H = 1$ m). The decay heat (about $1\,\mathrm{MW/m^3}$) remains only in the debris pool. The simulation is performed for a representative slice of the BWR lower plenum. The slice is a 3D segment which is filled with core materials (i.e., solid ceramic particulates and liquid metals) and includes 6 cooled CRGTs and a section of the vessel wall from below (Figure 18).

The simulation is started from a hot condition when temperature of the debris pool is higher than liquidus temperature of metals (2000 K is applied). In the presence of CRGT cooling, it is supposed that initial crusts are available on the CRGT wall surfaces. The thickness of the initial crusts is predicted to be of 40 mm, based on the simulation performed for a uniform heterogeneous debris pool. The remained volume of the metal layer is hot and in liquid state. Such a scenario is rather of low probability because formation of metal layer is a process which takes time. However, we consider such a "hot state" as a bounding scenario of instantaneous metal layer formation. Due to low permeability of the porous media, it is supposed that there is no convection in the debris pool; only heat conduction is considered. Conductivity of the debris pool is defined as the effective conductivity of the ceramic oxide and metals ($k_{\mathrm{eff}} = \sum c_i \times k_i$). Natural convection is assumed to take place only in the molten volume of the metal layer where the PECM is applied.

The boundary conditions are as follows. For the CRGT wall internal surfaces the isothermal boundary condition is applied. The top surface of the metal layer is in close contact with the decay heated coolable debris layer (Figure 16), the lower boundary of which may be heated up to high temperature (e.g., liquidus temperature of metals). Therefore the radiation heat transfer boundary condition is used for the metal layer top surface. External radiation temperature is assumed to be liquidus temperature of metals for the sake of conservatism. The external surface of the vessel wall is assumed to be insulated during the accident, thus a small heat flux (about $50\,\mathrm{W/m^2}$) is allowed on this surface. The other boundaries are adiabatic or symmetrical.

Figure 18 presents a transient state of the upper metal layer and the debris pool. There are growing crusts on CRGT surfaces and on the vessel wall submerged in the metal layer. Note that the debris pool state remains unchanged, that is, a matrix of particulates and molten metals, although it is shown in the figure as solid.

The PECM simulation shows that after about 30 min, although temperature of the lower debris pool (core) becomes higher than 2100 K, the upper metal layer is fully solidified. This is because of efficient heat removal from the high thermal conductivity metal layer to the cooled CRGTs. Steady state temperature distribution in the pool is presented in Figure 19 where the metal layer is shown at low temperature. Steady state temperature of the lower debris pool is about 2040 K indicating that the metal components remain in the liquid phase, and ceramic oxidic particulates are not melted.

Figure 20 presents steady state temperature profiles of the vessel wall external surfaces for two cases under the same total

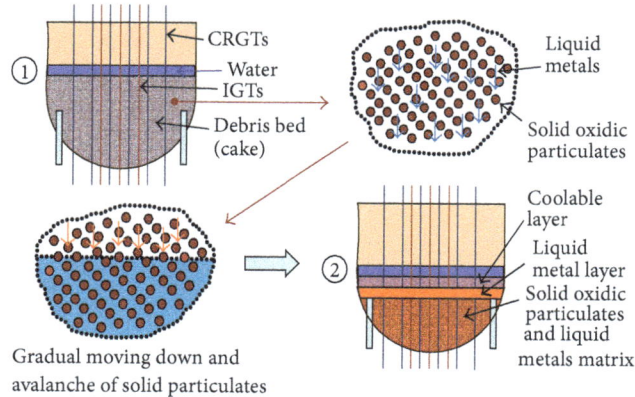

FIGURE 16: A severe accident scenario in the BWR lower head.

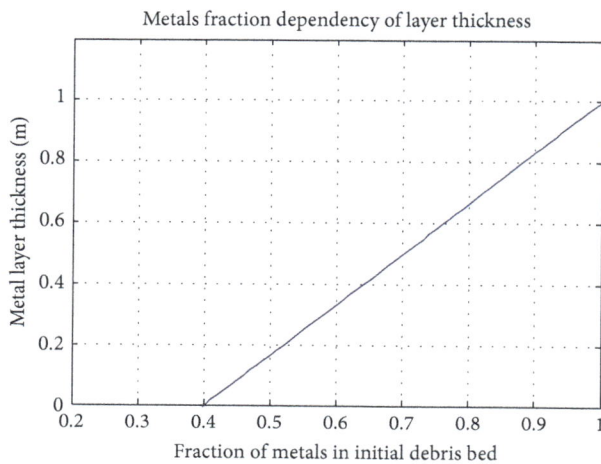

FIGURE 17: Dependency of metal layer thickness on fraction of metallic components in the initial debris bed (~180 tons).

amount of heat generation and pool depth (1.0 m): the first is the considered stratified heterogeneous debris pool ($Q_v = 1\,MW/m^3$ in the lower part and $Q_v = 0\,MW/m^3$ in the metal layer), and the second is the uniform heterogeneous debris pool ($Q_v = 0.72\,MW/m^3$), without a metal layer atop. In the first case, due to effective CRGT cooling, the heat flux to the vessel is low and vessel wall temperature is kept well below 1100°C, while in the second case, even with a smaller volumetric heat generation rate, temperature of the vessel approaches 1200°C. Creep temperature for the reactor vessel steel (e.g., SA533B1) usually assumed at 1100°C [1]. Thus vessel failure is not predicted upon the stratified heterogeneous debris pool scenario. However, for the uniform heterogeneous debris pool, the vessel wall may fail in the place connected to the uppermost region of the debris pool.

Note that the CRGT cooling is provided by the inside water flow; thus it is efficient only when the heat flux to CRGTs is lower than the Critical Heat Flux (CHF) which in turn depends also on water flow rate. It is therefore important to identify whether the heat flux along a CRGT is higher than the CHF.

The PECM simulation shows that, for the stratified heterogeneous debris pool scenario, the steady state heat flux to the CRGTs (234 kW/m²) is lower than the CHF at nominal CRGT purging flow rate of 15 kg/(m²·sec) (around 400 kW/m²). Due to the CRGT cooling efficacy and due to the low heat flux to the metal layer from the below debris pool, the focusing effect is not observed.

Heat flux evolution (Figure 21) shows that the transient heat flux to the CRGT is highest during the first 20 min. The source of the high heat flux is the latent heat released during fast solidification of metals (Figure 21). However, this transient heat flux to the CRGTs is well below the CHF for the CRGTs at water flow rate of 30 kg/(m²·sec) (about 900 kW/m²) [36]. This means the effect of latent heat is small as long as we can provide a sufficient water flow rate to avoid CHF (about ~2–4-folds of nominal water flow rate).

It is worth to mention that at the nominal purging flow rate of ~15 kg/(m²·sec) when the CHF can be considerably smaller [36], failure of the cooled CRGT upon the transient heat flux remains questionable. For more rigorous analysis we envision a need to examine the effect of variable boundary conditions on the transient heat flux to CRGT walls. The study approach is coupling simulation [37] using the metal layer PECM and RELAP which is applied for CRGTs with varying water follow rates to 2–4-folds of the nominal one.

6. Concluding Remarks

The metal layer Effective Convectivity Model (ECM) and Phase-change ECM (PECM) presented in this paper are built on a concept of the previously developed Effective Convectivity Conductivity Model (ECCM) [20] and follows the technical approach of recently developed ECM and PECM for heat transfer simulations of a decay-heated melt pool [6, 7].

Insights gained from the Computational Fluid Dynamics (CFD) study helped to improve the metal layer ECM and PECM. The metal layer PECM is augmented by the mushy zone heat transfer characteristic velocity and compositional convection models for simulation of non-eutectic binary mixtures.

Validation of the developed models is performed against both experimental and CFD-generated data. It is demonstrated that the modified metal layer ECM and PECM are

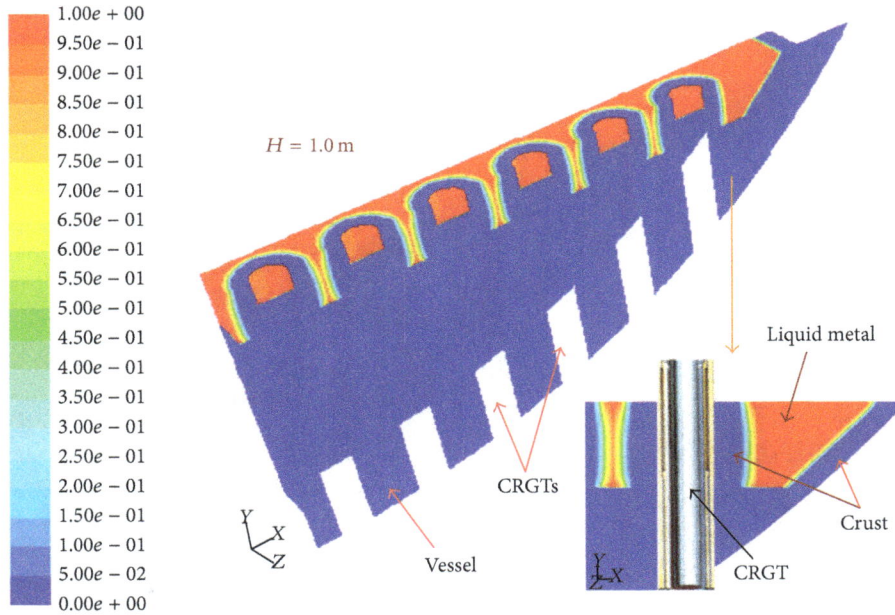

FIGURE 18: Liquid fraction contour of the stratified heterogeneous debris pool (transient).

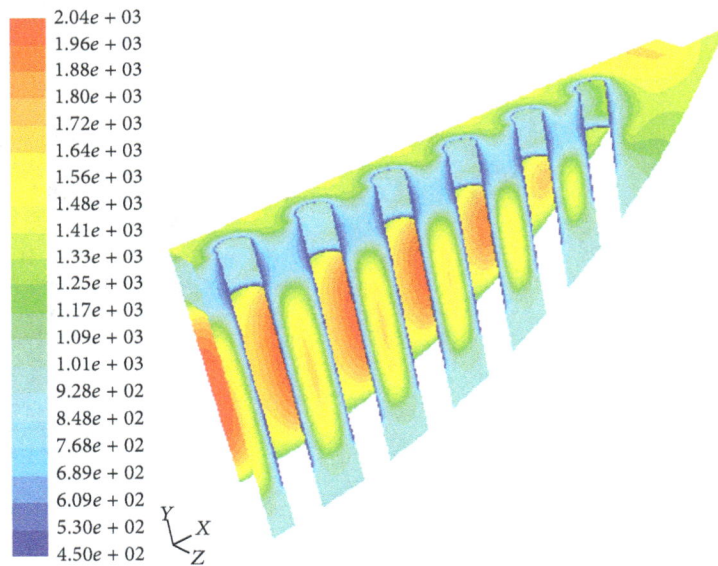

FIGURE 19: Temperature contour of the debris pool and vessel wall (steady-state condition).

capable of predicting energy splitting in metal layers for various heat transfer problems with reasonable accuracy.

The PECM application to heat transfer simulation in a BWR severe accident scenario shows the high potential of CRGT flow cooling as an efficient SAM measure to remove the decay heat from a stratified heterogeneous debris pool, keeping lower head vessel wall temperature well below the thermal creep limit. However, failure of the vessel should be considered using coupled thermo-mechanical creep analysis [38–40]. Further analysis will be performed for assessment of necessary CRGT flow rate which will ensure with sufficient margin no failure of CRGTs.

Nomenclature

Arabic

C: Composition, %
C_p: Specific heat capacity, J/(kg·K)
F_L: Liquid fraction
g: Gravitational acceleration, m/s^2
H: Depth (or thickness) of a fluid layer, m
h: Sensible enthalpy, J/kg
k: Conductivity, W/(m·K)
Nu: Nusselt number, $\text{Nu} = qH_{\text{pool}}/k\Delta T$

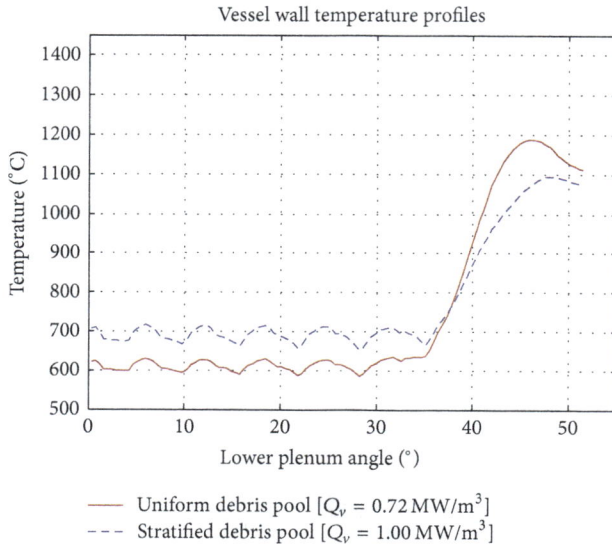

FIGURE 20: Temperature profiles of the vessel external surface.

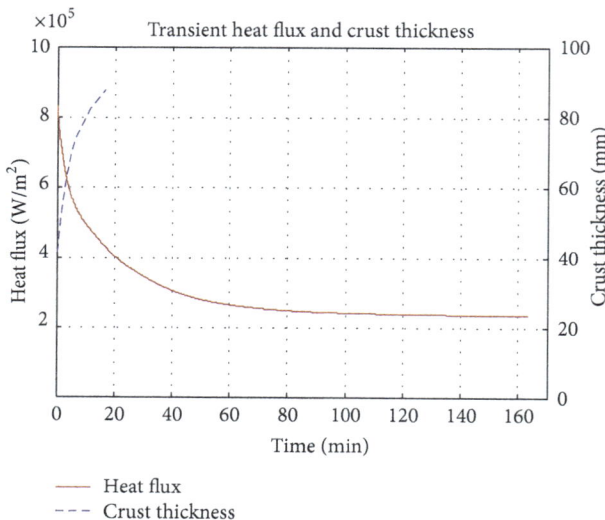

FIGURE 21: Transient crust thickness and heat flux to the CRGT section submerged in the metal layer.

Pr: Prandtl number, $Pr = \nu/\alpha$
Q: Heat, J
Q_v: Volumetric heat generation rate, W/m^3
Ra: External Rayleigh number,
        $Ra = g\beta\Delta T H^3/\nu\alpha$
Ra_y: Local Rayleigh number, $Ra_y = g\beta\Delta T y^3/\nu\alpha$
r_C: Rejectability coefficient
S: Area, m^2
T: Temperature, K
u: Fluid velocity, m/s
U: Characteristic velocity, m/s
U_M: Mushy characteristic velocity, m/s
W: Width of a fluid layer, m
y: Local vertical coordinate, m.

Greek

ΔH: Latent heat, J/kg
ΔT: Temperature difference, K
α: Thermal diffusivity, m^2/s, $\alpha = k/\rho \cdot C_p$
β: Thermal expansion coefficient, 1/K
ν: Kinematics viscosity, m^2/s
ρ: Density, kg/m^3
∞:   Bulk.

Subscripts and Superscripts

C: Compositional
cond: Conduction
conv: Convection
E: Eutectic
eff: Effective
i: Componentindex
LIQ: Liquidus
RB: Rayleigh-Benard
side: Sideward
SOL: Solidus
T: Thermal
up: Upward
x, y, z: Axis direction.

References

[1] J. L. Rempe, S. A. Chavez, G. L. Thinnes et al., "Light water reactor lower head failure analysis," Tech. Rep. NUREG/CR-5642, EGG-2618, Idaho National Engineering Laboratory, Washington, DC, USA, 1993.

[2] P. Kudinov, A. Karbojian, W. M. Ma, and T. N. Dinh, "An experimental study on debris formation with corium stimulant materials," in *Proceedings of the International Congress on Advances in Nuclear Power Plants (ICAPP '08)*, Anaheim, Calif, USA, June 2008.

[3] C. T. Tran and T. N. Dinh, "Application of the phase-change effective convectivity model to analysis of core melt pool formation and heat transfer in a BWR lower head," in *Proceedings of the Annual Meeting of the American Nuclear Society*, pp. 617–618, Anaheim, Calif, USA, June 2008.

[4] V. G. Asmolov, S. V. Bechta, V. B. Khabensky et al., "Partitioning of U, Zr and Fe between molten oxidic and metallic corium," in *Proceedings of the MASCA Seminar*, Aix-en-Provence, France, June 2004.

[5] C. T. Tran, P. Kudinov, and T. N. Dinh, "An approach to numerical simulation and analysis of molten corium coolability in a boiling water reactor lower head," *Nuclear Engineering and Design*, vol. 240, no. 9, pp. 2148–2159, 2010.

[6] C. T. Tran and T. N. Dinh, "An effective convectivity model for simulation of in-vessel core melt progression in boiling water reactor," in *Proceedings of the International Congress on Advances in Nuclear Power Plants (ICAPP '07)*, Nice Acropolis, France, May 2007.

[7] C. T. Tran and T. N. Dinh, "Simulation of core melt pool formation in a Reactor pressure vessel lower head using an Effective Convectivity Model," *Nuclear Engineering and Technology*, vol. 41, no. 7, pp. 929–944, 2009.

[8] T. G. Theofanous, C. Liu, S. Additon, S. Angelini, O. Kymalainen, and T. Salmassi, "In-vessel coolability and retention of a core melt," DOE/ID-1046, 1994.

[9] SCDAP/RELAP5-3D Code Development Team, "SCDAP/RELAP5-3D Code Manual," Report INEEL/EXT-02-00589, Revision 2.2, Idaho National Engineering and Environmental Laboratory, 2003.

[10] R. O. Gauntt, R. K. Cole, C. M. Erickson et al., *MELCOR Computer Code Manual, Core (COR) Package Reference Manuals, NUREG/CR-6119*, vol. 2, Rev. 2, Version 1.8.6, 2005.

[11] S. Globe and D. Dropkin, "Natural-convection heat transfer in liquid confined by two horizontal plates and heated from below," *Journal of Heat Transfer*, vol. 81, pp. 24–28, 1959.

[12] T. G. Theofanous, M. Maguire, S. Angelini, and T. Salmassi, "The first results from the ACOPO experiment," *Nuclear Engineering and Design*, vol. 169, no. 1–3, pp. 49–57, 1997.

[13] *MAAP4 Users Manual*, vol. 2, Fauske Associated, 1999.

[14] S. W. Churchill and H. H. S. Chu, "Correlating equations for laminar and turbulent free convection from a vertical plate," *International Journal of Heat and Mass Transfer*, vol. 18, no. 11, pp. 1323–1329, 1975.

[15] M. Eddi, "Study on heat transfer in lower head of nuclear power plant vessel during a severe accident," in *Proceedings of 9th International Topical Meeting on Nuclear Reactor Thermal Hydraulics (NURETH '99)*, San Francisco, Calif, USA, October 1999.

[16] G. Grotzbach, "Direct numerical simulation of laminar and turbulent benard convection," *Journal of Fluid Mechanics*, vol. 119, pp. 27–53, 1982.

[17] R. Verzicco and R. Camussi, "Prandtl number effects in convective turbulence," *Journal of Fluid Mechanics*, vol. 383, pp. 55–73, 1999.

[18] R. M. Kerr and J. R. Herring, "Prandtl number dependence of Nusselt number in direct numerical simulations," *Journal of Fluid Mechanics*, vol. 419, pp. 325–344, 2000.

[19] I. Otic, G. Grotzbach, and M. Worner, "Analysis and modelling of the temperature variance equation in turbulent natural convection for low-prandtl fluids," *Journal of Fluid Mechanics*, vol. 525, pp. 237–261, 2005.

[20] V. A. Bui and T. N. Dinh, "Modeling of heat transfer in heated-generating liquid pools by an effective diffusivity-convectivity approach," in *Proceedings of 2nd European Thermal-Sciences Conference*, pp. 1365–1372, Rome, Italy, 1996.

[21] B. R. Sehgal, V. A. Bui, T. N. Dinh, and R. R. Nourgaliev, "Heat transfer process in reactor vessel lower plenum during a late phase of in-vessel core melt progression," *Advances in Nuclear Science and Technology*, vol. 26, pp. 103–135, 1998.

[22] E. R. G. Eckert and T. W. Jackson, "Analysis of turbulent free convection boundary layer on flat plate," NACA Technical Note 2207, 1950.

[23] V. R. Voller and C. R. Swaminathan, "General source-based method for solidification phase change," *Numerical Heat Transfer B*, vol. 19, no. 2, pp. 175–189, 1991.

[24] M. G. Worster, "Natural convection in a mushy layer," *Journal of Fluid Mechanics*, vol. 224, pp. 335–359, 1991.

[25] J. S. Wettlaufer, M. G. Worster, and H. E. Huppert, "Natural convection during solidification of an alloy from above with application to the evolution of sea ice," *Journal of Fluid Mechanics*, vol. 344, pp. 291–316, 1997.

[26] R. Trivedi, H. Miyahara, P. Mazumder, E. Simsek, and S. N. Tewari, "Directional solidification microstructures in diffusive and convective regimes," *Journal of Crystal Growth*, vol. 222, no. 1-2, pp. 365–379, 2001.

[27] L. G. Margolin, W. J. Rider, and F. F. Grinstein, "Modeling turbulent flow with implicit LES," *Journal of Turbulence*, vol. 7, pp. 1–27, 2006.

[28] S. Grossmann and D. Lohse, "Scaling in thermal convection: a unifying theory," *Journal of Fluid Mechanics*, vol. 407, pp. 27–56, 2000.

[29] R. J. Goldstein, H. D. Chiang, and D. L. See, "High-Rayleigh-number convection in a horizontal enclosure," *Journal of Fluid Mechanics*, vol. 213, pp. 111–126, 1990.

[30] P. E. Roche, B. Castaing, B. Chabaud, and B. Hebral, "Prandtl and rayleigh numbers dependences in rayleigh-benard convection," *Europhysics Letters*, vol. 58, no. 5, pp. 693–698, 2002.

[31] J. L. O'Toole and P. L. Silveston, "Correlations of convective heat transfer in confined horizontal layers," *AIChE Chemical Engineering Progress Symposium Series*, vol. 57, no. 32, pp. 81–86, 1961.

[32] S. Cioni, S. Ciliberto, and J. Sommeria, "Strongly turbulent Rayleigh-Bénard convection in mercury: comparison with results at moderate Prandtl number," *Journal of Fluid Mechanics*, vol. 335, pp. 111–140, 1997.

[33] C. Gau and R. Viskanta, "Effect of natural convection on solidification from above and melting from below of a pure metal," *International Journal of Heat and Mass Transfer*, vol. 28, no. 3, pp. 573–587, 1985.

[34] A. A. Mohamad and R. Viskanta, "Modeling of turbulent buoyant flow and heat transfer in liquid metals," *International Journal of Heat and Mass Transfer*, vol. 36, no. 11, pp. 2815–2826, 1993.

[35] W. Z. Cao and D. Poulikakos, "Solidification of an alloy in a cavity cooled through its top surface," *International Journal of Heat and Mass Transfer*, vol. 33, no. 3, pp. 427–434, 1990.

[36] C. T. Tran and T. N. Dinh, "Analysis of melt pool heat transfer in a BWR lower head," in *Transactions of ANS Winter Meeting*, pp. 629–631, Albuquerque, NM, USA, November 2006.

[37] F. Cadinu, C. T. Tran, and P. Kudinov, "Analysis of in-vessel coolability and retention with control rod guide tube cooling in boiling water reactors," in *Proceedings of the NEA/SARNET In-Vessel Coolability (IVC) Workshop*, Issy-les-Moulineaux, France, October 2009.

[38] W. Villanueva, C.-T. Tran, and P. Kudinov, "Coupled thermo-mechanical creep analysis for boiling water reactor pressure vessel lower head," *Nuclear Engineering and Design*, vol. 249, pp. 146–153, 2012.

[39] C. Torregrosa, W. Villanueva, C.-T. Tran, and P. Kudinov, "Coupled 3D thermo-mechanical analysis of a nordic BWR vessel failure and timing," in *Proceedings of the 15th International Topical Meeting on Nuclear Reactor Thermal Hydraulics (NURETH '13)*, Pisa, Italy, May 2013, Paper 495.

[40] A. Goronovski, W. Villanueva, C.-T. Tran, and P. Kudinov, "The Effect of internal pressure and debris bed thermal properties on BWR vessel lower head failure and timing," in *Proceedings of the 15th International Topical Meeting on Nuclear Reactor Thermal Hydraulics (NURETH '13)*, Pisa, Italy, May 2013, Paper 500.

20

Spreading of Excellence in SARNET Network on Severe Accidents: The Education and Training Programme

Sandro Paci[1] and Jean-Pierre Van Dorsselaere[2]

[1] *Dipartimento di Ingegneria Civile e Industriale (DICI), Università di Pisa, Via Diotisalvi 2, 56126 Pisa, Italy*
[2] *Nuclear Safety Division, Institut de Radioprotection et de Sûreté Nucléaire (IRSN), Cadarache, BP3, 13115 Saint-Paul-Lez-Durance Cedex, France*

Correspondence should be addressed to Sandro Paci, sandro.paci@ing.unipi.it

Academic Editor: Leon Cizelj

The SARNET2 (severe accidents Research NETwork of Excellence) project started in April 2009 for 4 years in the 7th Framework Programme (FP7) of the European Commission (EC), following a similar first project in FP6. Forty-seven organisations from 24 countries network their capacities of research in the severe accident (SA) field inside SARNET to resolve the most important remaining uncertainties and safety issues on SA in water-cooled nuclear power plants (NPPs). The network includes a large majority of the European actors involved in SA research plus a few non-European relevant ones. The "Education and Training" programme in SARNET is a series of actions foreseen in this network for the "spreading of excellence." It is focused on raising the competence level of Master and Ph.D. students and young researchers engaged in SA research and on organizing information/training courses for NPP staff or regulatory authorities (but also for researchers) interested in SA management procedures.

1. Introduction

Despite the accident prevention measures adopted in modern NPPs, some accident scenarios, in very low probability circumstances, may result in a SA, as shown in the accident that occurred at the Fukushima Daiichi NPP in March 2011, with nuclear fuel melting and dispersal of radioactive materials into the external environment, thus constituting a hazard for the public health and for the environment. Significant progress has been achieved since the 1980s in the understanding of a SA progress, thanks in particular to the several research actions carried out inside the different EC FPs, but several "open issues" still need further research activities to reduce uncertainties and consolidate SA management plans [1, 2].

Facing the increasing reduction of the national budgets on SA researches, in 2003 EC judged necessary a better coordination of the single national efforts in this field to preserve and optimise the use of the available expertise and of the experimental facilities, in order to resolve the remaining "open issues" for enhancing the safety of existing and future NPPs. In April 2004, 51 worldwide organisations involved in R&D on SA, including research organisations, safety authorities, technical safety organisations (TSO), industries, utilities, and universities, decided to network together in SARNET, in the framework of the EC FP6, linking their capacities of research in the SA area in a consolidated manner [3] in a network of excellence, that is, an instrument for strengthening excellence by tackling the fragmentation of the European research, where the main deliverable is a durable structuring and shaping of the way that research is carried out on the topic of the network. A second phase of this excellence network (SARNET2) has started in April 2009, again supported by EC during the present FP7, for a duration of 4 years and again under the coordination of the French "Institut de Radioprotection et de Sûreté Nucléaire" (IRSN). The SARNET2 partners contribute to a joint programme of activities (JPA) including, in the framework of education and training, the following main actions:

(i) development of education courses on SA for students and researchers, and training courses for specialists,

(ii) promotion of personnel mobility amongst various European organisations,

(iii) organisation of large international conferences on SA researches.

This education and training programme has been set up to disseminate the excellence and knowledge in the SA area, including out of the SARNET2 network perimeter. It is intended to be an in-depth treatment so that the university students and researchers will be able to (a) understand and (b) develop the methodology in the topics further and (c) use analysis tools more effectively. In particular, the mobility programme under which students (including students from European universities out of the SARNET2 perimeter) and researchers will be able to go into different laboratories of SARNET2 partners for training will complete this education and training programme to develop a common safety culture throughout Europe. Links with the ENEN (European Nuclear Education Network) association have been also strengthened, ENEN being a nonprofit international organisation established on September 2003 under the French Law of 1901 with the mission of the preservation and further development of expertise in the nuclear fields by higher education and training.

2. The SARNET Excellence Network

Forty-seven partners from Europe, Canada, India, South Korea, United States, and Japan participate currently in SARNET with an overall manpower that represents about 40 persons per year (230 researchers and more than 20 Ph.D. students are involved in the network). This excellence network has been defined in order to optimize the use of the available research budget and to constitute a sustainable consortium in which common research programmes in the SA area and a common SA code (ASTEC) are developed. The ASTEC integral code [4] plays a key role in the network, by capitalizing SA knowledge through the implementation inside the code of the new physical models that are produced by the network itself.

In the SARNET network structure (Figure 1), a steering committee is in charge of strategy and decisions, advised by an advisory committee, composed of external end-user organisations (utilities or safety authorities). A general assembly, composed of one representative of each SARNET partner, plus the EC representative, is called periodically for information and consultation on the progress of the activities, the work orientations and the steering committee decisions. A management team, composed of the network coordinator and of the work package (WP) leaders, is entrusted with the day-to-day management of the network. The performed JPA is broken down into 8 WPs, pertaining to the three following types of activities:

(a) integrating activities, aiming at the management of the network and at strengthening links among the partner organisations (WP1 on management, WP3 on information systems and WP4 on ASTEC code),

(b) joint research activities (from WP5 to WP8),

(c) spreading of excellence in WP2 (analysed in detail in the following).

About 320 papers related to SARNET activities in the last 7 years have been presented in conferences or published in scientific journals. The dissemination of internal and public information is also done through periodic open newsletters, two websites (the public one and the restricted one) and the participation of the members to public events. After the mid-term of the project, the paper in [5] presents a more detailed SARNET2 description and a few recent outcomes of joint research done by the network members.

2.1. Integrating Activities

2.1.1. ASTEC Code Assessment and Improvements (WP4).
IRSN and German Gesellschaft für Anlagen und Reaktorsicherheit mbH (GRS) jointly develop the ASTEC code to describe the complete evolution of a SA in a water-cooled reactor, including the behaviour of engineered safety systems and procedures used in SA management (SAM). The series of code versions V2 aims at covering all European NPPs and the new Gen. III light water reactor (LWR) designs; it can simulate the EPR (European pressurised reactor), especially its external core-catcher, and it includes the advanced core degradation models of the IRSN ICARE2 mechanistic code.

Both organisations also assure the code maintenance and the support to the code users, notably through annual ASTEC Users' Club meetings. In fact, twenty-eight organisations actively collaborate on the development and in the assessment of the successive ASTEC versions. The code assessment activity mainly consists in covering a broad matrix of reactor applications, aiming at the most important SA scenarios for 4 types of water reactors (PWR, BWR, VVER, and CANDU) present in Europe. In complement to these reactor applications, ASTEC assessment continues through calculations of a wide range of experiments in different SA areas, such as the Phébus FP experiments [6] and of real plant accidents such as TMI2 and Fukushima-Daiichi sequences.

2.1.2. Information Systems (WP3).
A public website (http://www.sar-net.eu/) is open since 2009 in order to provide information on SARNET2 activities or events and, more generally, on the SA research field to the general public. For the internal communication and file sharing among all the network members, the e-collaborative internet advanced communication tool (ACT) is used.

One of the main SARNET goals is also to collect and preserve, for future utilizations, the large quantity of today's available SA experimental data, in order to ensure their preservation, exchange, and processing, including all the related documentation. The data collected inside DATANET are both previous experimental data that partners are voluntary willing to share and all new experimental data produced within SARNET since 2004. DATANET is actually

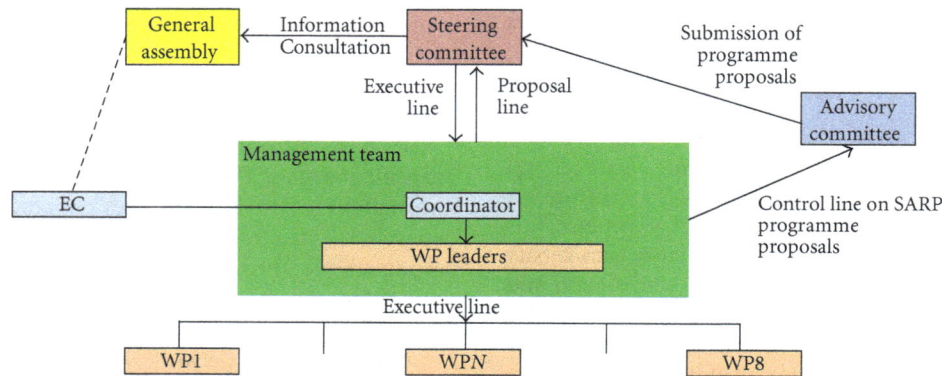

Figure 1: Structure of the SARNET network.

based on the STRESA tool [7], managed by the JRC Institute for Energy in Petten (Netherlands).

2.2. Joint Research Activities. One of the achievements of SARNET consisted in obtaining a European consensus on six high priority issues on which research was still considered as necessary. These six open safety issues, selected at the beginning of the SARNET2 project, are analysed inside the WPs 5 to 8:

(1) core coolability during reflooding and debris cooling—WP5,

(2) ex-vessel melt pool configuration during molten corium concrete interaction (MCCI), ex-vessel corium coolability by top flooding—WP6,

(3) melt relocation into water, ex-vessel fuel coolant interaction (FCI)—WP7,

(4) hydrogen mixing and combustion in containment—WP7,

(5) oxidising impact on source term—WP8,

(6) iodine chemistry in reactor coolant system (RCS) and in containment—WP8.

The experimental efforts carried out in SARNET2 are mainly devoted to two of these open issues, for which a real progress toward the "closure" of the issue itself is expected: "corium/debris coolability" and MCCI. For all these six issues, the same method is adopted: review and selection of available relevant experiments, contribution to the definition of new test matrices, synthesis of the interpretation of experimental data, benchmark exercises among available codes, review of models, and synthesis and proposals of new or improved models to be implemented inside ASTEC.

The ranking of R&D priorities has been recently reviewed and updated to take into account the Fukushima-Daiichi accident that took place in Japan in March 2011. No strong evolution of priorities was concluded, but a few new subjects were identified where the efforts must be increased in the next years.

Indeed a key integration aspect of the SARNET network is the setup of the "technical circles," each covering a specific detailed topic. These technical circles, experimented with success in the recent past for the PHEBUS FP experimental programme [8, 9], bring experimenters and modellers closer together, concerning test definition, interpretation and model development.

2.2.1. Corium and Debris Coolability (WP5). The major motivation is to solve the remaining uncertainties on the possibility of cooling structures and materials during a SA sequence, either in the core or the vessel bottom head or in the reactor cavity, in order to limit the accident progression. This could be achieved by water injection either by ensuring corium retention within the vessel or at least slowing down the corium progression and limiting the flow rates of corium release into the reactor cavity. These issues are covered within SAM for current reactors and also within the scope of the design and safety evaluation of future reactors. The following 3 key situations and processes are analysed inside this WP5:

(a) reflooding and coolability of a degraded core,

(b) remelting of debris, melt pool formation, and coolability,

(c) ex-vessel debris formation and coolability.

2.2.2. Molten Corium Concrete Interaction (WP6). The addressed main situation is the reactor pit initially dry but with the possibility of water injection later during MCCI. The work programme is complementary with the OECD/NEA (Nuclear Energy Agency) MCCI project that finished in 2011 [10]. Recent experiments [11] have provided new results that questioned the reliability of the available models and their extrapolation to reactor conditions. As an example, it becomes clear that new effects have to be taken into account to be able to describe the ablation anisotropy observed in case of silica-rich concrete and the different behaviour of limestone common sand concrete. This ablation anisotropy was also present in the ablation of the Chernobyl silica-rich concrete. The intention is thus to gain sufficient experimental data in order to determine which phenomena are responsible for the observed isotropy/anisotropy of the concrete ablation.

2.2.3. Containment Issues (WP7). The considered issue is the threat to the containment integrity due to two types of highly energetic phenomena: steam explosion and hydrogen combustion. Steam explosion may be caused by ex-vessel FCI due to a vessel failure and pouring of the corium in the flooded reactor cavity. Hydrogen combustion may be caused by ignition of a gas mixture with high local hydrogen concentrations, which may be due to the imperfect mixing of the containment atmosphere. Phenomena, as the containment atmosphere stratification, influencing these threats are considered as well.

Several benchmarks were launched inside this WP7, involving simulation codes on different physical phenomena or safety systems influencing the behaviour of the containment atmosphere, such as containment sprays [12], hydrogen combustion, steam condensation [13], interaction between passive autocatalytic recombiners and containment atmosphere, and a theoretical benchmark on the thermohydraulics behaviour of a generic NPP containment [14] for in-deep comparison of several codes and code users.

2.2.4. Source Term (WP8). The overall objective is to reduce the uncertainties associated with calculating the potential releases of radiotoxic fission products to the external environment. WP8 activities are concentrated on iodine and ruthenium, given their high radio-toxicity, noting that the ruthenium release is enhanced in oxidising atmospheres, such as those that may follow air ingress into the RCS. The research treats the transport of these elements through the RCS and their behaviour in the containment. Of particular importance is the prediction of volatile iodine and ruthenium species in the containment, forms that are hard to remove by containment sprays or by filtration while the venting of the containment system is actuated.

Full advantage is taken of cooperation with international programmes such as Phébus FP, International Source Term Programme (ISTP) [6], and programmes of OECD/NEA/CSNI, to avoid duplication of experiments, to help the consistency of the programmes and to identify the remaining needs in this field.

3. The SARNET Education and Training Programme

The "Education and Training" Programme, included in the WP2 "spreading of excellence," is focusing on the following.

(a) Raising the competence level of the European university students (Master and Ph.D.) and young researchers engaged in SA research. Towards this purpose, in the streamline of what was done in the SARNET FP6 project, new *educational* courses have been developed on the SA phenomenology. Four one-week educational courses, the last one held in Pisa (Italy) in January 2011, were organised since 2005, gathering from 40 to 100 young participants.

(b) Organisation of *information/training* short courses for staff of NPPs or regulatory authorities (but also

for interested researchers) mainly involved in SA management (SAM) procedures. In this kind of course, the emphasis is not only on SA phenomenology, as in educational ones, but also on identifying what these SAM procedures are based on and why they are effective for the plant protection. The last 2-day information/training course was organized in Karlsruhe (Germany) by the Karlsruhe Institute of Technology (KIT) in July 2012, gathering 32 participants from 11 different countries, including 3 non EU countries (Canada, Mexico, and Russia).

(c) Publication at the beginning of 2012 of the SARNET text book on SA phenomenology [15] of 700 pages long written internally to the SARNET framework that covers in a complete manner the historical aspects of water-cooled reactors safety principles and phenomena concerning in-vessel accident progression, early and late containment failure, fission product release, and transport in RCS and containment. It contains also a description of reference analysis tools or computer codes, of management and termination of a SA sequence, as well as of the environmental management. This unique reference book emphasizes the prevention and management of a SA, in order to teach nuclear professionals how to mitigate potential risks to the public and the external environment to the maximum extent possible.

Other "spreading of excellence" activities foreseen inside this WP2 are the organisation of periodic European Review Meetings on Severe Accident Research (ERMSAR Conferences) and the mobility programme (MOB), under which university students and young researchers can go to internship programmes. These particular activities will be illustrated in the following part of the present paper.

3.1. The SA Courses. The SARNET *one-week educational courses* are mainly courses that focus on disseminating the knowledge gained on SA phenomenology in the last two decades principally to Master and Ph.D. students, young engineers and researchers while the *two-day information and training courses* are mainly foreseen for industry managers and senior scientists. So, there is a clear indication and separation for the audience of the specific course, to better satisfy the different requests and needs from the participants.

During the previous SARNET FP6 project, three short courses (1 week) on the different aspects of SA research were organized, by different partners, with a mixed content (not only phenomenology but also SAM and SA codes):

(i) SA phenomenology, including some description of the SAM,

(ii) SA progression (analysis, data and uncertainties) to give the order of magnitude of physical phenomena occurring during a SA progression to researchers and engineers working in industry and regulatory organisations,

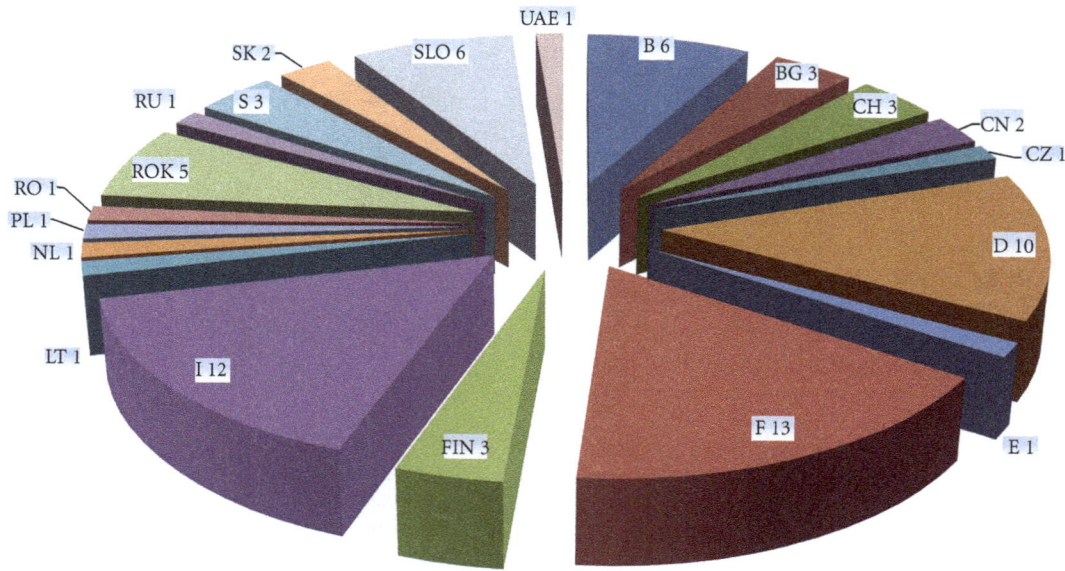

FIGURE 2: Nationalities of the participants in the Pisa 2011 course.

(iii) nuclear reactor SA analysis: application and management guidelines, focused more on SA methodology (models, codes), analyses, and SAM.

The first educational SARNET2 *"Severe Accident Phenomenology Short Course"* was organized in January 2011 by CEA and Pisa University and was hosted by the Engineering Faculty of Pisa University (Italy). The participation was quite high, with about 100 students from 20 worldwide countries (Figure 2) and a practical balance of gender for the participants. This Pisa course was again a one-week course for education on phenomenology because the goal was also to refresh participant memories after 5 years and SARNET new outcomes, with a program covering not only SA phenomenology and progression in current water-cooled Gen. II NPPs, but also the different design solutions in Gen. III ones. The purpose of this course [16] was to describe Gen. III designs addressing SA (i.e., the "in-vessel" melt retention concept or the "ex-vessel" core catcher concept). SA phenomenology has been described through its progression in the core and in the vessel lower head up to vessel failure, followed by the ex-vessel accident progression, with the loadings which can cause an early containment failure (i.e., direct containment heating, hydrogen combustion in containment, steam explosion), or the late containment failure (i.e., MCCI, corium coolability, etc.) The source term with fission products release from the core and transport in the RCS and containment has been specially emphasized. Lecturers were experts from 8 different countries, with large skills and knowledge on Gen. III NPPs and on the progression of a SA. The presence

of lecturers from nuclear industry was utilized to describe how the different NPPs would react during a SA, keeping in mind that an introductory course would not allow lengthy discussions or computer simulations. The Pisa course was open to university students with a very strong discount fee and contributed for 3 ECTS (with a mandatory written work) as an advanced course for Master students, with a strong link among SARNET, ENEN, and European Master of Science in Nuclear Engineering (EMSNE).

An evaluation form was distributed during the last days of the course to the participants to analyse the course and its impact. The evaluation was generally made on a scale from 1 (very bad) to 5 (very good). The evaluation of the general organisation of the course was very good for the overall mark: 4.32/5. Participants have not expressed a main field of interest for reactor applications, with a quite equilibrated distribution of this interest between Gen. II and Gen. III reactor design and the lectures had an average mark of 3.86/5, with no BAD or VERY BAD marks. Finally, it can be stated that the course met very well the participant expectations in terms of personal objectives in the field of SA as can be highlighted from the distribution of answers at the main question *"You have participated to the course in order to reach your own objectives in the field of SA, after the course, do you estimate that these objectives have been reached?"*: the MAINLY answer had 70.5% while the TOTALLY 17.9%, without NOT AT ALL mark. The suggestions reported in the forms also provide an indication for the planning of the future courses, again on SA phenomenology, reducing the focus on codes (more models, less codes).

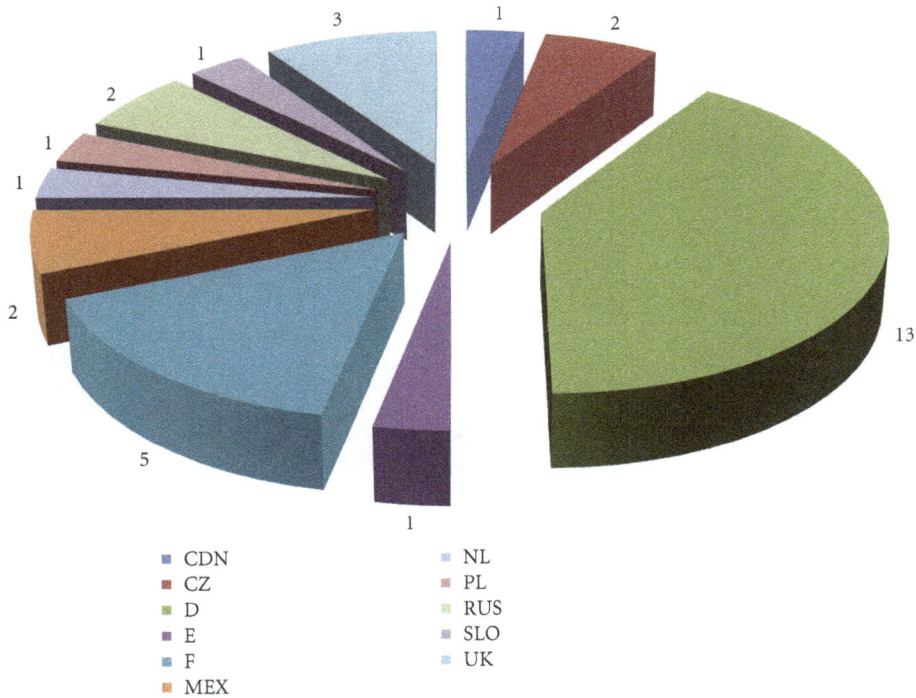

FIGURE 3: Nationalities of the participants in the KIT 2012 course.

The second SARNET two-day course was an information and training one. It was entitled *"Severe Accident Phenomenology and Management"* and was hosted and organized in Karlsruhe (Germany) by KIT, in cooperation with various SARNET partners, in July 2012. The goal of this last course was quite different respect to the previous one because it was mainly focused on disseminating the knowledge gained on severe accidents in the last two decades primarily to managers and senior scientists. The KIT course program covered SA phenomenology and progression in current water-cooled Gen. II and III NPPs with the following main topics:

(i) historical overview including a review of TMI, Chernobyl, and—above all—Fukushima accidents,

(ii) short introduction to the SA phenomenology,

(iii) overview of the main SA codes (in priority ASTEC and MELCOR),

(iv) severe Accident Management Guidelines,

(v) backfitting of Gen. II NPP and SA mitigation for Gen. III plants,

(vi) radiological consequences to the environment and to the public resulting from a SA.

Lectures have been given by 10 international experts from major nuclear institutes, industries, and universities working on the SA topic. As expected, the participation at this course was good but lower respect to the Pisa course, considering the particular high level target of the audience, with 32 participants but from 11 worldwide countries (see Figure 3 for the distribution).

The next SARNET course will be again an educational one-week course, based on the skeleton of the Pisa one, following the strong requests to repeat this kind of phenomenological course received from different worldwide organisations after the Fukushima accident.

3.2. The Mobility Programme. The Mobility Programme (MOB) aims at training young researchers and students through a delegation towards SARNET research teams, in order to enhance the exchanges and the dissemination of knowledge on SA area. In this MOB programme, the long-term goal is to build and strengthen teams which would engage together in a certain activity of the excellence network. A total of 33 mobility actions (with an average duration of 3 months) were completed at the end of the SARNET FP6 project, where most of the delegates were from Eastern Europe countries (as defined in the EU EuroVoc multilingual thesaurus) going to laboratories and universities in Western Europe countries and a large fraction of the delegates were women. The dominant area of training was on the use of the ASTEC code.

In SARNET2 FP7 an increased financial support covering partially the delegation costs (maximum of 2000€/month) is provided. To apply for mobility support, the following conditions shall be fulfilled: organisations shall be SARNET members (or only the hosting one for the delegation of an university master/Ph.D. students), the work to be

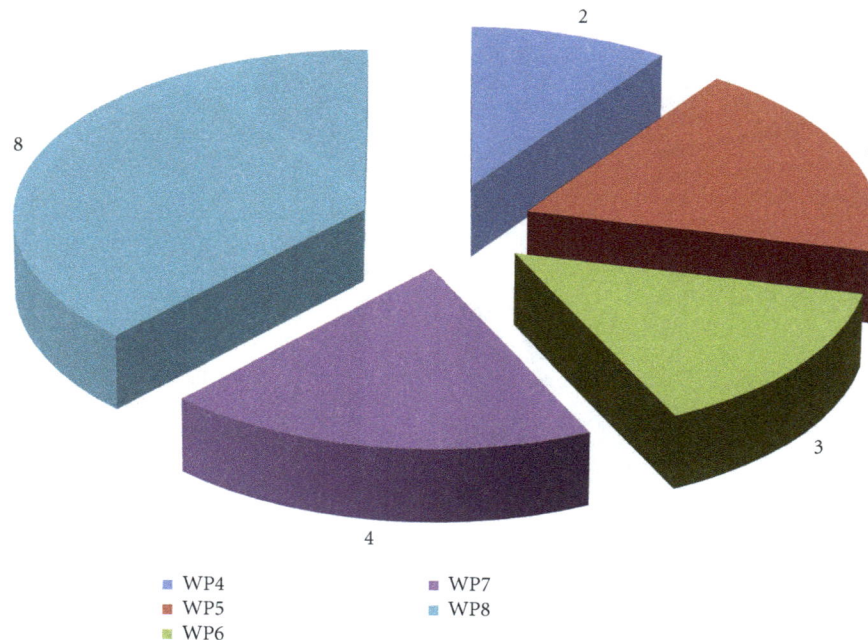

FIGURE 4: Distribution of the SARNET2 MOB actions.

performed during the delegation period shall be relevant to SARNET objectives and be part of the JPA, the delegation shall contribute to the transfer of knowledge/know-how. The allocation of funds is subject to the final delivery of an open delegation report describing the main technical achievements. The number of these mobility actions is one of the several integration indicators defined in order to assess the progress of the SARNET2 FP7 project and the success of the network integration: in total 21 mobility actions have been authorized, with the distribution among the different WPs reported in Figure 4, without the request for a dominant topic as in SARNET FP6 (although a predominance of WP8 is present), a good balance of gender (Figure 5), and an average time duration of about 4 months.

For the success of this SARNET MOB programme, strong partners' efforts were needed to highlight research projects which could be of interest to young researchers or students and to disseminate internally and externally the understanding of the benefits of this MOB program. In this context an attractive idea has been to arrange, for the university delegates, to perform a master thesis and/or to take courses in SA technology in the EMSNE framework, funding the master thesis stages in the SA field to be performed to obtain the 20 ECTS necessary for the EMSNE achievement.

3.3. ERMSAR Conferences. Five ERMSAR (European Review Meetings on Severe Accident Research) conferences have been organized successively in France, Germany (twice), Bulgaria, and Italy as an exchange forum for the whole

international SA community. The proceedings of these ERMSAR Conferences are available for download on the SARNET2 public website.

The 4th ERMSAR conference was hosted by ENEA in Bologna (Italy) on May 11-12, 2010. It gathered 98 participants from 24 different nationalities (8 non-EC countries and 15 EC ones). The SARNET partners presented papers on their joint work, including some first conclusions of the work done in the first period.

The latest 5th ERMSAR conference was the first one fully open to the international community. It has been hosted by GRS in Cologne (Germany) on 21–23 March 2012, gathering 157 participants with 52 presented papers, not only by SARNET partners. The participants came from 60 organisations and 27 countries (42 of them were coming from non-SARNET partners). This seminar aimed at presenting the current status of R&D on SAs both in Europe and elsewhere, as an "open conference." Therefore, beside the presentations of the current work by SARNET members in the network JPAs, non-SARNET organisations were invited to contribute through papers and presentations to the seminar or the SARNET members were also able to present SA individual work conducted outside the JPAs. The four previous ERMSAR conferences provided insights on the SARNET network activities and in particular on the SA research priorities established in 2008. On the contrary, the global objective of this 5th seminar has been to present the recent progress of international knowledge on SAs, of course including the work done within the SARNET network in the last two years. It was also an opportunity to discuss future

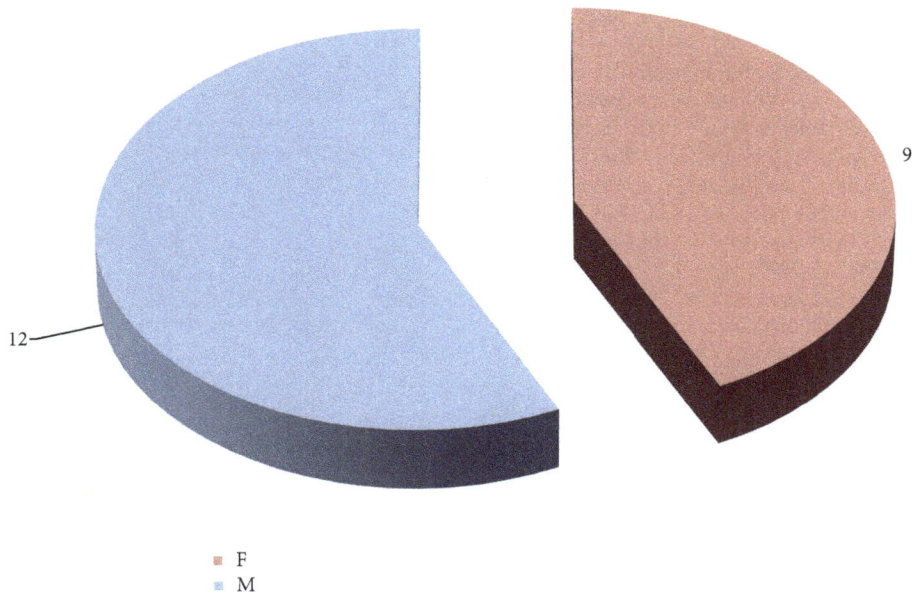

FIGURE 5: Distribution of gender for MOB actions.

R&D priorities on SAs and, in particular, how the feedback of the Fukushima accident can be taken into account in planning the new research needs.

The final ERMSAR 6th Conference will present all the results of the whole SARNET project.

4. Conclusions

After a first phase of four and a half years, the SARNET network of excellence continues from April 2009 for four years more on the different SA research lines. SARNET aims both at solving the latest "open" issues in this field for current NPPs safety and at consolidating the sustainable integration of the European SA research capacities. Efforts have also continued on the preservation and the transfer of knowledge to younger generations through the different "spreading of excellence" activities, including the publication of the SA text book, the organisation of the education and training courses, and other different actions.

The work done on the ASTEC code assessment and improvements will reinforce ASTEC position of European reference SA code. It will allow preservation of knowledge produced by thousands of person-years of R&D and dissemination to end-users.

Links with OECD/NEA and other programmes co-funded by the EC will be maintained and reinforced, in particular with ENEN association for education and training courses for students and young researchers, and with NUGENIA (a non-profit-making association dedicated to the research and development of nuclear fission technologies, with a focus on Gen. II and III NPPs, gathering stakeholders from industry, research, safety organisations, and academia and committed to develop joint R&D projects in this field)

and SNETP (the European stakeholder forum for nuclear technology). Finally, in order to keep the competence in SA and SAM alive after the SARNET2's end in 2013, a self-sustaining entity in the field of SA research is under construction in close links with NUGENIA.

Acknowledgment

The authors thank the European Commission for funding the SARNET network in FP7 (Project SARNET2 no. 231747 in the area "Nuclear Fission and Radiation Protection").

References

[1] D. Magallon, A. Mailliat, J. M. Seiler et al., "European expert network for the reduction of uncertainties in severe accident safety issues (EURSAFE)," *Nuclear Engineering and Design*, vol. 235, no. 2-4, pp. 309–346, 2005.

[2] B. Schwinges, "Ranking of SA Research Priorities in the Frame of SARNET," in *Proceedings of the European Review Meeting on Severe Accident Research (ERMSAR '08)*, Nesseber, Bulgaria, 2008.

[3] J. C. Micaelli, J. P. Van Dorsselaere, B. Chaumont et al., "SARNET: a European Cooperative effort on LWR severe accident research," in *Proceedings of the European Nuclear Conference*, Versailles, France, 2005.

[4] J. P. Van Dorsselaere, C. Seropian, P. Chatelard et al., "The ASTEC integral code for severe accident simulation," *Nuclear Technology*, vol. 165, no. 3, pp. 293–307, 2009.

[5] J. P. Van Dorsselaere, A. Auvinen, D. Beraha et al., "The European Research on Severe Accidents in Generation-II and -III Nuclear Power Plants," *Science and Technology of Nuclear Installations*, vol. 2012, Article ID 686945, 12 pages, 2012.

[6] B. Clément and R. Zeyen, "The Phébus FP and international source term programmes," in *Proceedings of the International Conference on Nuclear Energy for New Europe*, Bled, Slovenia, 2005.

[7] R. Zeyen, "European approach for a perennial storage of Severe Accident Research experimental data, as resulting from EU projects like SARNET, Phébus FP and ISTP," in *Proceedings of the ANS Winter Meeting*, Washington, DC, USA, 2009.

[8] K. Mueller, S. Dickinson, C. de Pascale et al., "Validation of severe accident codes on the phebus fission product tests in the framework of the PHEBEN-2 project," *Nuclear Technology*, vol. 163, no. 2, pp. 209–227, 2008.

[9] B. Clément, T. Haste, E. Krausmann et al., "Thematic network for a Phebus FPT1 international standard problem (THEN-PHEBISP)," *Nuclear Engineering and Design*, vol. 235, no. 2-4, pp. 347–357, 2005.

[10] M. T. Farmer, "A summary of findings from Melt Coolability and Concrete Interaction (MCCI) Program," in *Proceedings of the International Congress on Advances in Nuclear Power Plants (ICAPP '07)*, Nice, France, 2007.

[11] C. Journeau, J. F. Haquet, P. Piluso, and J. M. Bonnet, "Differences between silica and limestone concretes that may affect their interaction with corium," in *Proceedings of the International Conference on Advances in Nuclear Power Plants (ICAPP '08)*, pp. 1233–1240, Anaheim, Calif, USA, June 2008.

[12] J. Malet, L. Blumenfeld, S. Arndt et al., "Sprays in containment: final results of the SARNET spray benchmark," *Nuclear Engineering and Design*, vol. 241, no. 6, pp. 2162–2171, 2011.

[13] W. Ambrosini, M. Bucci, N. Forgione, F. Oriolo, and S. Paci, "Quick look report on SARNET2 condensation benchmark-2 results," Report DIMNP RL 1252, University of Pisa, Pisa, Italy, 2010.

[14] S. Kelm, P. Broxtermann, S. Krajewski, and H. J. Allelein, *Report on the Generic Containment Code-To-Code Comparison-Run0*, Forschungszentrum, Jülich, Germany, 2010.

[15] B. Sehgal, Ed., *Nuclear Safety in Light Water Reactors—Severe Accident Phenomenology*, Academic Press, 2012.

[16] S. Paci and P. Piluso, *SARNET2 Short Course on Severe Accident Phenomenology*, Dedizioni, Pisa, Italy, 2012.

Permissions

The contributors of this book come from diverse backgrounds, making this book a truly international effort. This book will bring forth new frontiers with its revolutionizing research information and detailed analysis of the nascent developments around the world.

We would like to thank all the contributing authors for lending their expertise to make the book truly unique. They have played a crucial role in the development of this book. Without their invaluable contributions this book wouldn't have been possible. They have made vital efforts to compile up to date information on the varied aspects of this subject to make this book a valuable addition to the collection of many professionals and students.

This book was conceptualized with the vision of imparting up-to-date information and advanced data in this field. To ensure the same, a matchless editorial board was set up. Every individual on the board went through rigorous rounds of assessment to prove their worth. After which they invested a large part of their time researching and compiling the most relevant data for our readers. Conferences and sessions were held from time to time between the editorial board and the contributing authors to present the data in the most comprehensible form. The editorial team has worked tirelessly to provide valuable and valid information to help people across the globe.

Every chapter published in this book has been scrutinized by our experts. Their significance has been extensively debated. The topics covered herein carry significant findings which will fuel the growth of the discipline. They may even be implemented as practical applications or may be referred to as a beginning point for another development. Chapters in this book were first published by Hindawi Publishing Corporation; hereby published with permission under the Creative Commons Attribution License or equivalent.

The editorial board has been involved in producing this book since its inception. They have spent rigorous hours researching and exploring the diverse topics which have resulted in the successful publishing of this book. They have passed on their knowledge of decades through this book. To expedite this challenging task, the publisher supported the team at every step. A small team of assistant editors was also appointed to further simplify the editing procedure and attain best results for the readers.

Our editorial team has been hand-picked from every corner of the world. Their multi-ethnicity adds dynamic inputs to the discussions which result in innovative outcomes. These outcomes are then further discussed with the researchers and contributors who give their valuable feedback and opinion regarding the same. The feedback is then collaborated with the researches and they are edited in a comprehensive manner to aid the understanding of the subject.

Apart from the editorial board, the designing team has also invested a significant amount of their time in understanding the subject and creating the most relevant covers. They scrutinized every image to scout for the most suitable representation of the subject and create an appropriate cover for the book.

The publishing team has been involved in this book since its early stages. They were actively engaged in every process, be it collecting the data, connecting with the contributors or procuring relevant information. The team has been an ardent support to the editorial, designing and production team. Their endless efforts to recruit the best for this project, has resulted in the accomplishment of this book. They are a veteran in the field of academics and their pool of knowledge is as vast as their experience in printing. Their expertise and guidance has proved useful at every step. Their uncompromising quality standards have made this book an exceptional effort. Their encouragement from time to time has been an inspiration for everyone.

The publisher and the editorial board hope that this book will prove to be a valuable piece of knowledge for researchers, students, practitioners and scholars across the globe.

List of Contributors

Pavan K. Sharma, B. Gera, R. K. Singh and K. K. Vaze
Reactor Safety Division, Bhabha Atomic Research Centre, Engg. Hall-7, Trombay, Mumbai 400085, India

Bolade A. Adetula and Pavel M. Bokov
Research and Development Division, The South African Nuclear Energy Corporation (Necsa), Building 1900, P.O. Box 582, Pretoria 0001, South Africa

M. Valette
CEA Grenoble, Commissariat `a l'´Energie Atomique et aux ´Energies Alternatives, DEN, DM2S/SMTH, 38054 Grenoble, France

Christian P. Deck, H. E. Khalifa, B. Sammuli and C. A. Back
General Atomics, P.O. Box 85608, San Diego, CA 92186-5608, USA

B.Merk and V. Glivici-Cotruta
Helmholtz-Zentrum Dresden-Rossendorf, Institut fur Sicherheitsforschung, Postfach 51 01 19, 01314 Dresden, Germany

F. Alvarez-Velarde and E. M. Gonzalez-Romero
Centro de Investigaciones Energ´eticas, Medioambientales y Tecnol´ogicas (CIEMAT), Avenida. Complutense 40. Ed. 17, 28040 Madrid, Spain

Shengyi Si
Shanghai Nuclear Engineering Research & Design Institute, 29 Hongcao Road, Shanghai 200233, China

Maria Pusa
VTT Technical Research Centre of Finland, P.O. Box 1000, VTT 02044, Finland

J. Freixa and A.Manera
Laboratory for Reactor Physics and Systems Behavior (LRS), Paul Scherrer Institut (PSI), 5232 Villigen, Switzerland

Bruno Merk
Department of Reactor Safety, Institute of Resource Ecology, Helmholtz-Zentrum Dresden-Rossendorf, Postfach 51 01 19, 01314 Dresden, Germany

Andrea Achilli, Cinzia Congiu and Roberta Ferri
SIET S.p.A., UdP, Via Nino Bixio 27/c, 29121 Piacenza, Italy

Fosco Bianchi and Paride Meloni
ENEA, UTFISSM, Via Martiri di Monte Sole 4, 40129 Bologna, Italy

Davor Grgic
FER, University of Zagreb, Unska 3, 10000 Zagreb, Croatia

Milorad Dzodzo
Research and Technology Unit, Westinghouse Electric Company LLC, Cranberry Township, PA 16066, USA

Takashi Futatsugi, Shigeo Hosokawa and Akio Tomiyama
Department of Mechanical Engineering, Faculty of Engineering, Kobe University, 1-1 Rokkodai, Nada, Hyogo Kobe, 657-8501, Japan

Chihiro Yanagi and Michio Murase
Institute of Nuclear Safety System Inc. (INSS), 64 Sata, Mihama-cho, Mikata-gun, Fukui 919-1205, Japan

Jin Huang
Chongqing Institute of Automobile, Chongqing University of Technology, Chongqing 400054, China
The Key Laboratory of Manufacture and Test Techniques for Automobile Parts, Chongqing University of Technology, Chongqing 400054, China

Ping Wang and Guochao Wang
Chongqing Institute of Automobile, Chongqing University of Technology, Chongqing 400054, China

Jianzuo Ma and Hongyu Shu
College of Mechanical Engineering, Chongqing University, Chongqing 400044, China

Jin Huang
Chongqing Automobile College, Chongqing University of Technology, Chongqing 400054, China
The Key Laboratory of Manufacture and Test Techniques for Automobile Parts, Chongqing University of Technology, Chongqing 400054, China

Gilberto Espinosa-Paredes and Alejandro Nunez-Carrera
Area de Ingenieria en Recursos Energeticos, Universidad Autonoma Metropolitana- Iztapalapa, Avenida San Rafael Atlixco 186 Col. Vicentina, 09340 Mexico City, DF, Mexico

Raul Camargo-Camargo
Nuclear Safety Division, Comision Nacional de Seguridad Nuclear y Salvaguardias, Doctor Barragan 779, Col. Narvarte, 03020 Mexico City, DF, Mexico

Gustavo A.Montoya
Chemical Engineering Department, Simon Bolivar University, Valle de Sartenejas, Baruta, 1080A Caracas, Venezuela

Deendarlianto
Department of Mechanical and Industrial Engineering, Faculty of Engineering, Gadjah Mada University, Jalan Grafika No. 2, Yogyakarta 55281, Indonesia

Dirk Lucas, Thomas Hohne and Christophe Vallee
Helmholtz-Zentrum Dresden-Rossendorf e.V., Institute of Safety Research, P.O. Box 510 119, 01314 Dresden, Germany

Seyed AlirezaMousavi Shirazi
Department of Physics, Sciences College, Islamic Azad University, South Tehran Branch, Tehran 1581819411, Iran

Dariush Sardari
Department of Nuclear Engineering, Technical College, Islamic Azad University, Science and Research Branch, Tehran, Iran

Pavel N. Alekseev, Yury M. Semchenkov and Alexander L. Shimkevich
NRC "Kurchatov Institute", 1 Kurchatov Square, Moscow 123182, Russia

Chi-Thanh Tran and Pavel Kudinov
Division of Nuclear Power Safety, Royal Institute of Technology, Roslagstullsbacken 21, D5, 10691 Stockholm, Sweden

Sandro Paci
Dipartimento di Ingegneria Civile e Industriale (DICI), Universit`a di Pisa, Via Diotisalvi 2, 56126 Pisa, Italy

Jean-Pierre Van Dorsselaere
Nuclear Safety Division, Institut de Radioprotection et de Sˆuretˆe Nuclˆeaire (IRSN), Cadarache, BP3, 13115 Saint-Paul-Lez-Durance Cedex, France